Interferogram Analysis for Optical Testing

Second Edition

OPTICAL ENGINEERING

Founding Editor
Brian J. Thompson
University of Rochester
Rochester, New York

Interferogram Analysis for Optical Testing

Second Edition

Daniel Malacara
Centro de Investigaciones de Optica
Leon, Mexico

Manuel Servín
Centro de Investigaciones de Optica
Leon, Mexico

Zacarias Malacara
Centro de Investigaciones de Optica
Leon, Mexico

CRC Press
Taylor & Francis Group
Boca Raton London New York

CRC Press is an imprint of the
Taylor & Francis Group, an **informa** business

A TAYLOR & FRANCIS BOOK

CRC Press
Taylor & Francis Group
6000 Broken Sound Parkway NW, Suite 300
Boca Raton, FL 33487-2742

First issued in paperback 2019

© 2005 by Taylor & Francis Group, LLC
CRC Press is an imprint of Taylor & Francis Group, an Informa business

No claim to original U.S. Government works

ISBN-13: 978-1-57444-682-1 (hbk)
ISBN-13: 978-0-367-39319-9 (pbk)

Library of Congress Card Number 2004056966

Library of Congress Cataloging-in-Publication Data

Malacara, Daniel, 1937–
    Interferogram analysis for optical testing / Daniel Malacara, Manuel Servín, Zacarias Malacara.
        p. cm. -- (Optical engineering ; 84)
    Includes bibliographical references and index.
    ISBN 1-57444-682-7
1. Optical measurements. 2. Interferometry. 3. Interferometers. 4. Diffraction patterns—Data processing. I. Servín, Manuel. II. Malacara, Zacarias, 1948–. III. Title. IV. Optical engineering (Marcel Dekker, Inc.) ; v. 84.

QC367.M25 2005
681.'25--dc22   2004056966

Visit the Taylor & Francis Web site at
http://www.taylorandfrancis.com

and the CRC Press Web site at
http://www.crcpress.com

Contents

Contents

1

Review and Comparison of the Main Interferometric Systems

1.1 TWO-WAVE INTERFEROMETERS AND CONFIGURATIONS USED IN OPTICAL TESTING

Two-wave interferometers produce an interferogram by superimposing two wavefronts, one of which is typically a flat reference wavefront and the other a distorted wavefront whose shape is to be measured. The literature (e.g., Malacara, 1992; Creath, 1987) provides many descriptions of interferometers; here, we will just describe some of the more important aspects.

An interferometer can measure small wavefront deformations with a high accuracy, of the order of a fraction of the wavelength. The accuracy in a given interferometer depends on many factors, such as the optical quality of the components, the measuring methods, the light source properties, and disturbing external factors, such as atmospheric turbulence and mechanical vibrations. It has been shown by Kafri (1989), however, that the accuracy of any interferometer is limited. He proved that, if everything else is perfect, a short coherence length and a long sampling time can improve the accuracy. Unfortunately, a short coherence length and long measuring

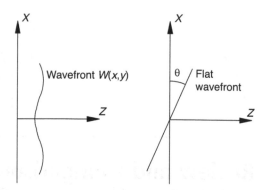

Figure 1.1   Two interfering wavefronts.

time combined make the instrument more sensitive to mechanical vibrations. In conclusion, the uncertainty principle imposes a fundamental limit to the accuracy that depends on several parameters but is of the order of 1/1000 of the wavelength of the light.

To study the main principles of interferometers, let us consider a two-wave interferogram with a flat wavefront that has a positive tilt about the y-axis and a wavefront under analysis, for which the deformations with respect to a flat wavefront without tilt are given by $W(x,y)$. This tilt is said to be positive when the wavefront is as shown in Figure 1.1. The complex amplitude in the observation plane, where the two wavefronts interfere, is the sum of the complex amplitudes of the two waves as follows:

$$E_1(x,y) = A_1(x,y)\exp ikW(x,y) + A_2(x,y)\exp i(kx\sin\theta) \quad (1.1)$$

where A_1 is the amplitude of the light beam at the wavefront under analysis, A_2 is the amplitude of the light beam with the reference wavefront, and $k = 2\pi/\lambda$. Hence, the irradiance is:

$$E_1(x,y)\cdot E_1^*(x,y) = A_1^2(x,y) + A_2^2(x,y) +$$
$$+2A_1(x,y)A_2(x,y)\cos k[x\sin\theta - W(x,y)] \quad (1.2)$$

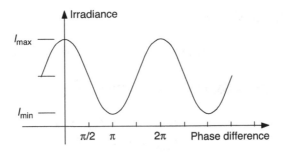

Figure 1.2 Irradiance as a function of phase difference between the two waves along the light path.

where the symbol * denotes the complex conjugate of the electric field. Here, we have introduced optional tilt θ about the y-axis between the two wavefronts. The irradiance function, $I(x,y)$, may then be written as:

$$I(x,y) = I_1(x,y) + I_2(x,y) +$$
$$+2\sqrt{I_1(x,y)I_2(x,y)}\,\cos k[x\sin\theta - W(x,y)] \tag{1.3}$$

where $I_1(x,y)$ and $I_2(x,y)$ are the irradiances of the two beams, and the phase difference between them is given by $\phi = k(x\sin\theta - W(x,y))$. This function is shown graphically in Figure 1.2.

For convenience, Equation 1.3 is frequently written as:

$$I(x,y) = a(x,y) + b(x,y)\cos k[x\sin\theta - W(x,y)] \tag{1.4}$$

Assuming that the variations in the values of $a(x,y)$ and $b(x,y)$ inside the interferogram aperture are smoother than the variations of the cosine term, the maximum irradiance in the vicinity of the point (x,y) in this interferogram is given by:

$$I_{max}(x,y) = \left(A_1(x,y) + A_2(x,y)\right)^2$$
$$= I_1(x,y) + I_2(x,y) + 2\sqrt{I_1(x,y)I_2(x,y)} \tag{1.5}$$

and the minimum irradiance in the same vicinity is given by:

$$I_{\min}(x,y) = \left(A_1(x,y) - A_2(x,y)\right)^2$$
$$= I_1(x,y) + I_2(x,y) - 2\sqrt{I_1(x,y)I_2(x,y)}$$

(1.6)

The fringe visibility, $v(x,y)$, is defined by:

$$v(x,y) = \frac{I_{\max}(x,y) - I_{\min}(x,y)}{I_{\max}(x,y) + I_{\min}(x,y)}$$

(1.7)

Hence, we may find:

$$v(x,y) = \frac{2\sqrt{I_1(x,y)I_2(x,y)}}{I_1(x,y) + I_2(x,y)} = \frac{b(x,y)}{a(x,y)}$$

(1.8)

Using the fringe visibility, Equation 1.3 is sometimes also written as:

$$I(x,y) = I_0(x,y)\left(1 + v(x,y)\cos k[x\sin\theta - W(x,y)]\right)$$

(1.9)

where $I_0(x,y) = a(x,y)$ is the irradiance for a fringe-free field, when the two beams are incoherent to each other. This irradiance, as a function of the phase difference between the two interfering waves, is shown in Figure 1.2.

Several basic interferometric configurations are used in optical testing procedures, but almost all of them are two-wavefront systems. Both wavefronts come from a single light source, separated by amplitude. Furthermore, most modern interferometers use a helium–neon laser as the light source. The main advantage of using a laser as the source of light is that fringe patterns may be easily obtained because of the great coherence of the laser. In fact, this advantage can also be a serious disadvantage, as spurious diffraction patterns and secondary fringe patterns are easily obtained. Special precautions must be taken into account to achieve a clean interference pattern. In this chapter, we review some of these interferometers, but greater detail about these systems may be found in many books (e.g., Malacara, 1992).

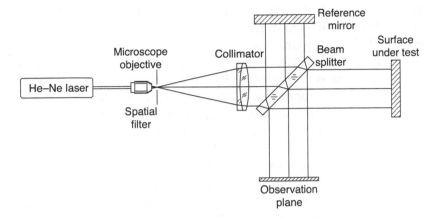

Figure 1.3 Basic configuration in a Twyman–Green interferometer.

1.2 TWYMAN–GREEN INTERFEROMETER

The basic configuration of the Twyman–Green interferometer, invented by F. Twyman and A. Green (Twyman, 1918), is illustrated in Figure 1.3. The fringes in a Twyman–Green interferometer are of equal thickness. The light from the laser is expanded and collimated by means of a telescopic system that usually includes a microscope objective and collimator. To obtain a clean wavefront, without diffraction rings on the field, the optical components must be as clean as possible. For an even cleaner beam, a spatial filter (pinhole) may be used at the focal plane of the microscope objective. The quality of the wavefront produced by this telescope does not need to be extremely high, because its deformations will appear on both interfering wavefronts and not produce any fringe deviations. If the optical path difference between both interfering beams is large, the tolerance on the wavefront deformations in the illuminating telescope may be drastically reduced; in this case, the illuminating wavefront must be quite flat, within a fraction of the wavelength.

If the beam splitter is nonabsorbing, the main interference pattern is complementary to the one returning to the source, due to the conservation of energy principle, even though the optical path difference is the same for both patterns. Phase shifts upon reflection on dielectric interfaces may explain this complementarity.

The beam splitter must be of high quality with regard not only to its surfaces but also to the material, which must be extremely homogeneous. The reflecting surface must be of the highest quality — flat, with an accuracy of about twice the required interferometer accuracy. The quality of a nonreflecting surface may be relaxed by a factor of four with respect to a reflecting face. To prevent spurious interference fringes, the nonreflecting surface must not reflect any light. One way to accomplish this is by coating the surface with an anti-reflection multilayer coating. Another possible method is for the beam splitter to have an incidence angle equal to the Brewster angle and which properly polarizes the incident light beam; however, this solution substantially increases the size of the beam splitter, making it more difficult to construct and hence more expensive.

Many different optical elements may be tested using a Twyman–Green interferometer, as described by Malacara (1992). For example, a plane-parallel plate of glass may be tested as shown in Figure 1.4a. The *optical path difference* (*OPD*) introduced by this glass plate is:

$$OPD = 2(n-1)t \qquad (1.10)$$

where n is the refractive index and t is the plate thickness. The interferometer is first adjusted so no fringes are observed before introducing the plate into the light beam, thus ensuring that all fringes that appear are due to the plate. If the field remains free of fringes after introducing the plate, we can say that the quantity $(n-1)t$ is constant over the entire plate aperture. If the fringes are straight, parallel, and equidistant and we may assume that the glass is perfectly homogeneous so that n is constant, then the fringes are produced by a small angle between the two flat faces of the plate. If the fringes are not straight but are distorted, we may conclude that either the refractive index is not constant or the surfaces are not flat, or both. We can only be sure that $(n-1)t$ is not constant. To

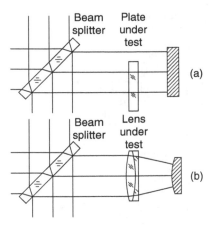

Figure 1.4 Testing a glass plate and a lens in a Twyman–Green interferometer.

measure the n and t separately, we must augment the results from this test with another measurement made in a Fizeau interferometer, which measures the values of nt.

The optical arrangements in Figure 1.4b can be used to test a convergent lens. A convex spherical mirror with its center of curvature at the focus of the lens is used for lenses with long focal lengths, and a concave spherical mirror is used for lenses with short focal lengths. A small, flat mirror located at the focus of the lens can also be employed. The portion of the flat mirror being used is so small that its surface does not need to be very accurate; however, the wavefront is rotated 180°, thus the spatial coherence requirements are stronger and odd aberrations are canceled out.

Concave or convex optical surfaces may also be tested using a Twyman–Green interferometer with the configurations shown in Figure 1.5. Even large astronomical mirrors can be tested. For this purpose, an unequal-path interferometer for optical shop testing was designed by Houston et al. (1967). When the beam-splitter plate is at the Brewster angle, it has a wedge angle of 2 to 3 arc min between the surfaces. The reflecting surface of this plate is located to receive the rays returning from the test specimen in such a way as to

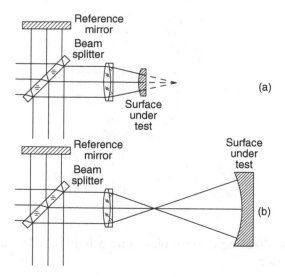

Figure 1.5 Twyman–Green interferometer configurations to test a convex or concave optical surface.

preclude astigmatism and other undesirable effects. A two-lens beam diverger can be placed in one arm of the interferometer. It is made of high-index glass with all the surfaces being spherical and has the capability for testing a surface as fast as $f/1.7$.

1.3 FIZEAU INTERFEROMETERS

Like the Twyman–Green interferometer, the Fizeau interferometer is a two-beam interferometer with fringes of equal thickness (see Figure 1.6). The optical path difference (OPD) introduced when testing a plane-parallel glass plate placed in the light beam is:

$$\text{OPD} = 2nt \tag{1.11}$$

which, as we may notice, is different from the corresponding expression for the Twyman–Green interferometer. In this sense, the two interferometers are complementary, so that the

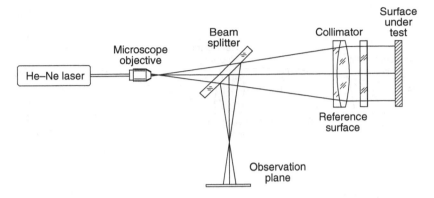

Figure 1.6 Basic Fizeau interferometer configuration.

constancy of thickness t and refractive index n may be tested only when both interferometers are used.

A large concave optical surface may also be tested with a Fizeau interferometer, as shown in Figure 1.7. If the concave surface is aspherical, the spherical aberration may be compensated if the converging lens has the opposite aberration. The reference surface is placed between the collimator and the converging lens.

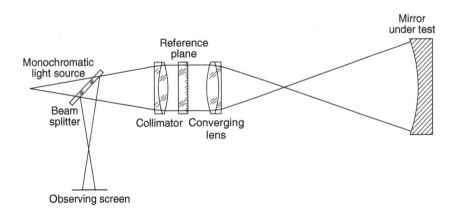

Figure 1.7 Fizeau interferometer to test a concave surface using a flat reference surface.

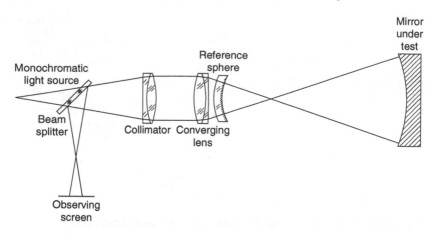

Figure 1.8 Fizeau interferometer to test a concave surface using a concave reference surface.

When the reference surface is flat, as in Figure 1.7, no off-axis configuration appears when the concave mirror under analysis is tilted to introduce many tilt fringes (linear carrier). A perfect focusing lens is required, however, because the lens is located inside the cavity; thus, the wavefront under analysis passes through this lens but not the reference wavefront. Any error in the focusing lens will be apparent in the interferogram. A second possible source of errors appears when a flat reference is used. In this case, the reference wavefront returns to the collimator lens at an angle with respect to the optical axis, and the collimator has to be corrected for some field angle.

As shown in Figure 1.8, a spherical reference surface is sometimes used. In this case, the linear carrier can be introduced by tilting the concave sphere under analysis or the reference sphere. This arrangement prevents the presence of any optical elements inside the interferometer cavity, between the reference surface and the surface being analyzed, thus relaxing the requirements for good focusing and collimating optics. These lenses still have to be corrected for some small field angle, but their degree of correction does not need to be very high. Even better, if the whole optical system formed by the focusing lens and the collimator is made symmetrical, correction of the coma aberration is automatic. In such a configuration, some wavefront aberrations may appear when the linear carrier is introduced, due to the large tilt in the spherical mirror, in addition to the well-known primary astigmatism.

With this arrangement, an off-axis configuration results when a large tilt is applied to an interferometer to introduce a linear carrier with more than 200 fringes in the interferogram (Kuchel, 1990). The linear carrier is obtained by tilting the reference. The surface being tilted may be the concave mirror under analysis or the spherical reference. We have seen that, in addition to introduction of the primary astigmatic aberration (due to off-axis testing), spherical and high-order (ashtray) astigmatism is also generated; however, we may see that even for a large number of fringes the wavefront aberration remains small for all practical purposes so we may introduce as many fringes as desired.

Another source of wavefront errors in the spherical cavity configuration, when testing a high-aperture optical element, may be introduced by large axial displacements of the concave surface under analysis with respect to the spherical reference sphere. In addition to the expected defocusing, a spherical aberration is introduced in the wavefront. A common variation of the Fizeau interferometer is the Shack–Fizeau interferometer (Figure 1.9), which is used to test a large concave surface with a spherical reference surface.

1.4 TYPICAL INTERFEROGRAMS IN TWYMAN–GREEN AND FIZEAU INTERFEROMETERS

Interferograms produced by the primary aberrations have been described by Kingslake (1925–1926). A wavefront with primary aberrations, as measured with respect to a sphere with its center of curvature at the Gaussian image point, is given by:

$$W(x,y) = A\left(x^2 + y^2\right)^2 + By\left(x^2 + y^2\right) + C\left(x^2 - y^2\right) +$$
$$+ D\left(x^2 + y^2\right) + Ex + Fy + G \tag{1.12}$$

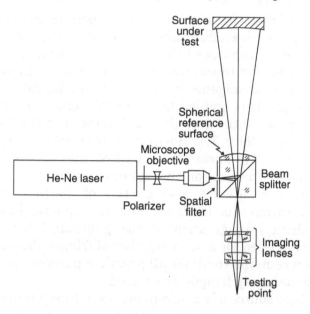

Figure 1.9   Shack–Fizeau interferometer.

where:

   A = spherical aberration coefficient.
   B = coma coefficient.
   C = astigmatism coefficient.
   D = defocusing coefficient.
   E = tilt about the y-axis coefficient (image displacement along the x-axis).
   F = tilt about the x-axis coefficient (image displacement along the y-axis).
   G = piston or constant term.

This expression may also be written in polar coordinates (θ, ρ). For simplicity, when computing typical interferograms of primary aberrations, a normalized entrance pupil with unit semidiameter can be taken. Some typical interference pattern are shown in Figure 1.10; a more complete set of illustrations may be found in Malacara (1992).

   Diagrams of typical interferograms can be simulated in a computer using beams of fringes of equal inclination on a

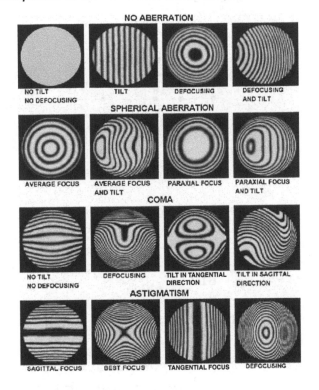

Figure 1.10 Some Twyman–Green interferograms.

Michelson interferometer (Murty, 1964) using the OPDs introduced by a plane-parallel plate and cube-corner prisms instead
of mirrors, or by electronic circuits on a cathode ray tube (CRT)
(Geary et al., 1978; Geary, 1979).

Twyman–Green interferograms were analyzed by Kingslake
(1925–1926) by measuring the optical path difference at several points using fringe sampling. Then, solving a system of
linear equations, he computed the OPD coefficients A, B, C,
D, E, and F. Another similar method for analyzing a Twyman–
Green interferogram was proposed by Saunders (1961). He
found that the measurement of nine appropriately chosen
points is sufficient to determine any of the three primary
aberrations. The points were selected as shown in Figure 1.11,
and the aberration coefficients were calculated with:

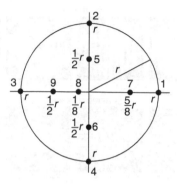

Figure 1.11   Selected points for evaluation of primary aberrations.

$$A = \frac{128}{81r^2}\left[W_1 - W_9 + 2(W_8 - W_7)\right] \qquad (1.13)$$

$$B = \frac{128}{3r^2}\left[W_2 - W_4 + 2(W_6 - W_5)\right] \qquad (1.14)$$

and

$$C = \frac{1}{4r^2}\left[W_2 + W_4 - W_1 - W_3\right] \qquad (1.15)$$

where W_i is the estimated wavefront deviation at the point I.

The aberration coefficients can be determined by direct reading on the interferogram setting, looking for interference patterns with different defocusing settings and tilts. Vazquez-Montiel et al. (2002) have developed a method to determine the wavefront deformation for these primary aberrations from the interferogram using an iterative trial-and-error method which they refer to as an evolution strategy.

1.5 LATERAL SHEAR INTERFEROMETERS

A lateral shear interferogram does not require any reference wavefront; instead, the interference takes place between two identical aberrated wavefronts, laterally sheared with respect to each other as shown in Figure 1.12. The optical path difference is:

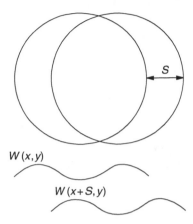

Figure 1.12 Two laterally sheared wavefronts.

$$OPD = W(x,y) - W(x \times S, y) \qquad (1.16)$$

where S is the lateral shear in the sagittal (x) direction.

Let us now assume that lateral shear S is sufficiently small such that the wavefront slopes in the x direction may be considered almost constant in an interval S. This is equivalent to the condition when the fringe spatial frequency in the x direction is almost constant in an interval S. Then, we may expand in a Taylor series to obtain:

$$OPD = W(x+S,y) - W(x,y) = \frac{\partial W(x,y)}{\partial x} S \qquad (1.17)$$

A bright fringe occurs when:

$$OPD = \frac{\partial W(x,y)}{\partial x} S = \frac{TA_x(x,y)}{r} = m\lambda \qquad (1.18)$$

where $TA_x(x,y)$ is the transverse aberration of the ray perpendicular to the wavefront, measured at a plane containing the center of curvature of the wavefront, and m is an integer number.

Thus, we can conclude that a lateral shearing interferometer does not measure the wavefront deformation, $W(x,y)$, in a

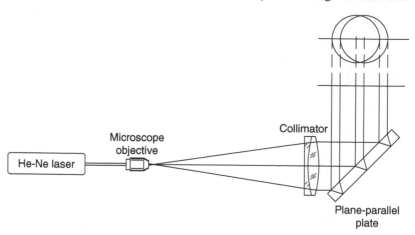

Figure 1.13   Murty's lateral shear interferometer.

direct manner but rather its slope, or transverse aberration, in the direction of the lateral shear. To measure the two components of the transverse aberrations we must utilize two laterally sheared interferograms in perpendicular directions.

The derivative of a function reduces the power of the function by one; thus, the slopes of the function are also reduced, and we can see that, if a wavefront is highly aspheric (with large slopes) in the lateral shearing interferometer, then these slopes are greatly reduced, producing greater fringe separations. This is an important advantage when testing highly aspheric surfaces with a lateral shearing interferometer. Of course, an important consequence of such an approach is that the sensitivity is also reduced.

Many practical configurations are available for laterally sheared interferometers. The most popular, due to its simplicity, is the Murty interferometer (Murty, 1964), which is illustrated in Figure 1.13.

1.5.1   Primary Aberrations

Lateral shear interferograms for the primary aberrations can be obtained by using the expression for the primary aberrations, Equation 1.12, which is now discussed in greater detail.

1.5.1.1 Defocus

The interferogram with a defocused wavefront is given by:

$$2DxS = m\lambda \qquad (1.19)$$

This is a system of straight, parallel, and equidistant fringes that are perpendicular to the lateral shear direction. When the defocusing is large, the spacing between the fringes is small. On the other hand, in the absence of defocus, no fringes occur in the field.

1.5.1.2 Spherical Aberration

In this case the interferogram is given by:

$$4A\left(x^2 + y^2\right)xS = m\lambda \qquad (1.20)$$

If this aberration is combined with defocus, we may write instead:

$$\left[4A\left(x^2 + y^2\right)x + 2Dx\right]S = m\lambda \qquad (1.21)$$

Then, the interference fringes are cubic curves.

1.5.1.3 Coma

In the case of the coma aberration, the interferogram is given by:

$$2BxyS = m\lambda \qquad (1.22)$$

when the lateral shear is S in the sagittal (x) direction. If the lateral shear is T in the tangential (y) direction, the fringes are given by:

$$B\left(x^2 + 3y^2\right)T = m\lambda \qquad (1.23)$$

1.5.1.4 Primary Astigmatism

In the case of astigmatism, when the lateral shear is S in the sagittal (x) direction, the fringes are given by:

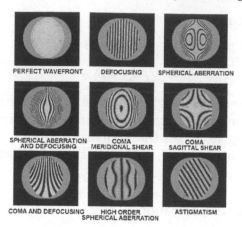

Figure 1.14 Some lateral shear interferograms.

$$(2Dx + 2Cx)S = m\lambda \qquad (1.24)$$

and for the lateral shear T in the tangential (y) direction we have:

$$(2Dy - 2Cy)T = m\lambda \qquad (1.25)$$

The fringes are straight and parallel, as in the case of defocus, but the interferograms have different separations.

Some lateral shear interferograms for primary aberrations are shown in Figure 1.14. Yang and Oh (2001) have proposed a method to identify these primary aberrations in a lateral shear interferogram using a neural network to obtain a mapping function. The neural network is a network of non-linear functions between the input, formed by line images, and the output, or the primary aberrations.

1.5.2 Rimmer–Wyant Method To Evaluate Wavefronts

The Rimmer–Wyant method (Rimmer, 1974; Rimmer and Wyant, 1975) performs a polynomial interpolation while determining the wavefront shape from a set of lateral-shear interferogram sampled points. The wavefront is represented by $W(x,y)$ and may be expressed by the xy polynomial with degree k:

$$W(x,y) = \sum_{n=0}^{k}\sum_{m=0}^{n} B_{nm}x^{m}y^{n-m} \qquad (1.26)$$

with $N = (k + 2)(k + 1)/2$ coefficients B_{nm}. The expression for the laterally sheared wavefront by distance S in the x direction is:

$$W(x+S,y) = \sum_{n=0}^{k}\sum_{m=0}^{n} B_{nm}(x+S)^{m}y^{n-m} \qquad (1.27)$$

and, similarly, the sheared wavefront by distance T in the y direction is:

$$W(x,y+T) = \sum_{n=0}^{k}\sum_{m=0}^{n} B_{nm}x^{m}(y+T)^{n-m} \qquad (1.28)$$

On the other hand, the Newton binomial theorem is:

$$(x+S)^{m} = \sum_{j=0}^{m}\binom{m}{j}x^{m-j}S^{j} \qquad (1.29)$$

where:

$$\binom{m}{j} = \frac{m!}{(m-j)!\,j!} \qquad (1.30)$$

Thus, Equations 1.27 and 1.28 may be written:

$$W(x+S,y) = \sum_{n=0}^{k}\sum_{m=0}^{n}\sum_{j=0}^{m} B_{nm}\binom{m}{j}x^{m-j}y^{n-m}S^{j} \qquad (1.31)$$

and

$$W(x,y+T) = \sum_{n=0}^{k}\sum_{m=0}^{n}\sum_{j=0}^{n-m} B_{nm}\binom{n-m}{j}x^{m}y^{n-m-j}T^{j} \qquad (1.32)$$

Hence, by subtracting Equation 1.26 from Equation 1.31 we obtain:

$$\Delta W_S = W(x + S, y) - W(x, y) = \sum_{n=0}^{k-1} \sum_{m=0}^{n} C_{nm} x^m y^{n-m} \quad (1.33)$$

and by subtracting Equation 1.26 from Equation 1.32 we obtain:

$$\Delta W_T = W(x, y + T) - W(x, y) = \sum_{n=0}^{k-1} \sum_{m=0}^{n} D_{nm} x^m y^{n-m} \quad (1.34)$$

with $k(k + 1)/2$ coefficients C_{nm} and the same number of coefficients D_{nm} given by:

$$C_{nm} = \sum_{j=1}^{k-n} \binom{j+m}{j} S^j B_{j+nj+m} \quad (1.35)$$

and

$$D_{nm} = \sum_{j=1}^{k-n} \binom{j+n-m}{j} T^j B_{j+n,m} \quad (1.36)$$

The values of C_{nm} and D_{nm} are obtained from the two laterally sheared interferograms in orthogonal directions by means of a two-dimensional, least-squares fit to the measured values of ΔW_S and ΔW_T. Then, the values of all coefficients B_{nm} are calculated by solving the system of linear equations defined by Equations 1.35 and 1.36, each with a matrix of dimensions $N \times M$. The Rimmer–Wyant method to find the wavefront using Zernike polynomials has been further developed by Okuda et al. (2000) to improve its accuracy.

1.5.3 Saunders Method To Evaluate Interferograms

When evaluating an unknown wavefront it is possible to determine its shape from a lateral shearing interferogram. To illustrate the method proposed by Saunders (1961), let us consider Figure 1.15, assuming that $W_1 = 0$. Then, we can write:

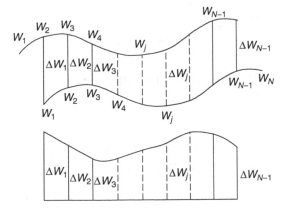

Figure 1.15 Saunders method to obtain the wavefront in a lateral shearing interferogram.

$$W_1 = 0$$
$$W_2 = \Delta W_1 + W_1$$
$$W_3 = \Delta W_2 + W_2 \qquad (1.37)$$
$$\cdots$$
$$W_n = \Delta W_{n-1} - W_{n-1}$$

The primary problem with this method is that the wavefront is evaluated only at points separated by a distance S. Intermediate values are not measured and must be interpolated. Orthogonal polynomials, as described in Chapter 4 in this book, may be used to some advantage to represent the wavefront in a lateral shearing interferometer. The accuracy of this mathematical representation has been studied by Wang and Ling (1989).

1.5.4 Spatial Frequency Response of Lateral Shear Interferometers

Unlike Twyman–Green interferometers, lateral shearing interferometers have a nonuniform response to spatial frequencies (Fourier components) in the wavefront deformations function. This response may be analyzed as illustrated in Figure 1.16.

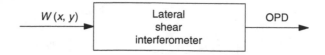

Figure 1.16 The lateral shearing interferometer, considered to be an electronic system.

The spatial frequency content of the lateral shearing optical path difference function, which is the interferometer output OPD, is given by:

$$F\{OPD\} = F\{W(x,y) - W(x - S, y)\} \qquad (1.38)$$

or

$$F\{OPD\} = F\{W(x,y)\} - F\{W(x - S, y)\} \qquad (1.39)$$

where $F\{g\}$ is the Fourier transform of g. Using the lateral displacement theorem of Fourier theory, this expression is transformed into:

$$F\{OPD\} = F\{W(x,y)\} - F\{W(x,y)\}\exp(-i2\pi f S) \qquad (1.40)$$

where f is the spatial frequency of a Fourier component, or

$$F\{OPD\} = F\{W(x,y)\}\left[1 - \exp(-i2\pi f S)\right] \qquad (1.41)$$

from which we may obtain:

$$F\{OPD\} = 2i\sin(\pi f S)F\{W(x,y)\}\exp(-i\pi f S) \qquad (1.42)$$

The spatial frequency sensitivity of the interferometer $R(f)$ may now be defined as:

$$R(f) = \frac{F\{OPD\}}{F\{W(x,y)\}} = 2i\sin(\pi f S)\exp(-i\pi f S) \qquad (1.43)$$

which may also be written as:

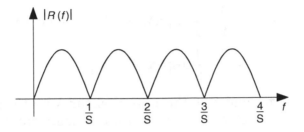

Figure 1.17   Lateral shear interferometer sensitivity as a function of the spatial frequency.

$$R(f) = 2\sin(\pi f S)\exp\left[-i\pi\left(fS - \frac{1}{2}\right)\right]$$ (1.44)

This function has zeros at $\pi f S = m\pi$. Thus, the lateral displacement interferometer is not sensitive to spatial frequencies given by:

$$f = \frac{m}{S}$$ (1.45)

where m is an integer, as shown in Figure 1.17. This result implies that the wavefront deformations, $W(x,y)$, are not obtained with the same precision for all spatial frequencies. A larger uncertainty in the calculation will be encountered for recovery of spatial frequency components close to the zeros in Equation 1.44. Elster and Weingärtner (1999a,b) have proposed a method to obtain the wavefront from two lateral shear interferograms taken with two different shears that avoids the loss of some spatial frequencies.

1.5.5   Regularization Method To Obtain Wavefronts

In lateral shearing interferometry, the interference pattern is formed with two mutually laterally displaced copies of the wavefront under analysis. The mathematical form of the irradiance of a lateral shear fringe pattern may be written as:

$$I_x(x,y) = \frac{1}{2} + \frac{1}{2}\cos\left[k\big(W(x-S,y) - W(x,y)\big)\right]$$
$$= \frac{1}{2} + \frac{1}{2}\cos\left[k\Delta_x W(x,y)\right] \tag{1.46}$$

where $k = 2\pi/\lambda$ and S is the lateral shear. We also need the orthogonally displaced interferogram to completely describe the wavefront under analysis. The orthogonal interferogram may be written as:

$$I_y(x,y) = \frac{1}{2} + \frac{1}{2}\cos\left[k\big(W(x,y-T) - W(x,y)\big)\right]$$
$$= \frac{1}{2} + \frac{1}{2}\cos\left[k\Delta_y W(x,y)\right] \tag{1.47}$$

where T is the lateral shear, orthogonal to S. The fringe patterns in Equations 1.46 and 1.47 may be transformed into carrier-frequency interferograms by introducing a large and known amount of defocusing to the testing wavefront (Mantravadi, 1992). Having obtained linear carrier fringe patterns, we can proceed to their demodulation using standard techniques of fringe carrier analysis as provided in this book.

The demodulated and unwrapped difference wavefront may be integrated using the path-independent integration procedure presented here. Assume that we have already estimated and unwrapped the interesting phase of the two orthogonally sheared interferograms. Using this information, the least-squares wavefront reconstruction may be stated to minimize the following merit function:

$$U(\hat{W}) = \sum_{(x,y)\in L_x} \left[\hat{W}(x-S,y) - \hat{W}(x,y) - \Delta_x W(x,y)\right]^2 +$$
$$+ \sum_{(x,y)\in L_y} \left[\hat{W}(x,y-T) - \hat{W}(x,y) - \Delta_y W(x,y)\right]^2 \tag{1.48}$$
$$= \sum_{(x,y)\in L_x} U_x^2(x,y) + \sum_{(x,y)\in L_y} U_y^2(x,y)$$

where the "hat" function represents the estimated wavefront, and L_x and L_y are two-dimensional lattices containing valid phase data in the x and y shearing directions. However, the minimization problem stated in Equation 1.48 is not well posed, because the matrix that results from setting the gradient of U equal to zero is not invertible. Fortunately, we may apply classical regularization to this inverse problem to find the expected smooth solution of the problem (Thikonov, 1963). In classical regularization theory, the regularizer consists of a linear combination of the squared magnitudes of derivatives of the estimated wavefront inside the domain of interest. In particular, we may use a discrete approximation to the Laplacian to obtain the second-order potentials:

$$R_x(x_i, y_j) = \hat{W}(x_{i-1}, y_j) - 2\hat{W}(x_i, y_j) + \hat{W}(x_{i+1}, y_j)$$
$$R_y(x_i, y_i) = \hat{W}(x_i, y_{j-1}) - 2\hat{W}(x_i, y_j) + \hat{W}(x_i, y_{j+1})$$

$$(1.49)$$

Therefore, the regularized merit function becomes:

$$U(\hat{W}) = \sum_{(x,y)\in L_x} U_x^2(x, y) + \sum_{(x,y)\in L_y} U_y^2(x, y) +$$
$$+ \lambda \sum_{(x,y)\in Pupil} \left[R_x^2(x, y) + R_y^2(x, y) \right]$$

$$(1.50)$$

where *Pupil* refers to the two-dimensional lattice inside the pupil of the wavefront being tested. The estimated wavefront obtained using these second-order potentials as regularizers makes the solution behave like a thin metallic plate attached to the observations by linear springs. The regularizing potentials discourage large changes in the estimated wavefront among neighboring pixels. As a consequence, the searched solution will be relatively smooth. The λ parameter controls the amount of smoothness of the estimated wavefront. If the observations have a negligible amount of noise, then λ may be set to a small value (~0.1); if the observations are noisy, then λ may be set to a higher value (in the range of 0.5 to 11.0) to filter out some noise. It should be noted that the use of regularizing potentials in this case is a must, even for noise-free observations, to yield a stable solution of the least-squares integration for lateral displacements greater than two pixels. As analyzed by Servín et al. (1996), this is because the inverse operator that performs the least-squares integration has poles in the frequency domain.

The estimated wavefront may be calculated using a simple gradient descent:

$$\hat{W}^{k+1}(x,y) = \hat{W}^{k}(x,y) - \tau \left(\frac{\partial U(\hat{W})}{\partial \hat{W}(x,y)} \right) \qquad (1.51)$$

applied to all pixels, where τ is the convergence rate. This optimizing method is not very fast, so we normally use faster algorithms, such as the conjugate gradient.

1.6 RONCHI TEST

In the Ronchi test (Cornejo, 1992), the screen is a ruling placed near the point of convergence of the returning aberrated wavefront, as shown in Figure 1.18. An imaging optical system is used to observe the projected shadows of the ruling lines over the surface being analyzed. This imaging system may be the eye in qualitative tests but may be a lens in quantitative tests. By measuring the fringe deformations in the projected shadows, the transverse aberration in the direction perpendicular to the ruling lines is easily computed. If the ruling lines are along the y-axis, the transverse aberration TA_x is measured. If the ruling lines are along the x-axis, the transverse aberration TA_y is measured. In other words, two different measurements with two orthogonal ruling orientations are necessary to measure the two components of the transverse aberration.

Another system that measures the wavefront slopes is the lateral shearing interferometer (Mantravadi, 1992) described earlier, where the lateral shear is small compared with the period of the maximum spatial frequency to be detected in the wavefront deformations. Under these conditions the lateral shearing interferometer is identical to the Ronchi test.

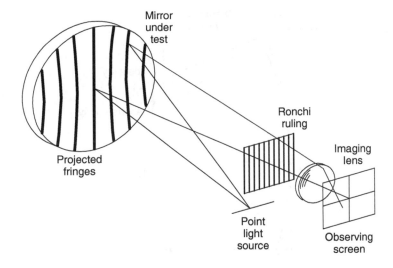

Figure 1.18 Optical arrangement in the Ronchi test.

Thus, in these tests, we measure the transverse aberrations at an observing plane located at a distance L from the wavefront being measured, as shown in Figure 1.19. These transverse aberrations are related to the wavefront slopes in the x and y directions by:

$$\frac{\partial W(x,y)}{\partial x} = -\frac{TA_x}{L - W(x,y)} \approx -\frac{TA_x}{L} \qquad (1.52)$$

and

$$\frac{\partial W(x,y)}{\partial y} = -\frac{TA_y}{L - W(x,y)} \approx -\frac{TA_y}{L} \qquad (1.53)$$

As mentioned before, a linear grating fringe pattern is easier to analyze using standard carrier fringe detecting procedures, such as the Fourier method, the synchronous method, or the spatial phase-shifting method. These techniques are described later in this book.

We may start with a simplified mathematical model for the transmittance of a linear grating:

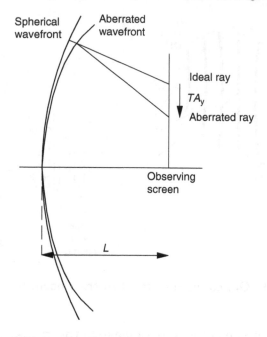

Figure 1.19 Measuring the transverse aberration in an aberrated wavefront.

$$T_x(x,y) = \frac{\left(1 + \cos(\omega_0 x)\right)}{2} \tag{1.54}$$

(Ronchi rulings are normally made of binary transmittance, not sinusoidal, but for mathematical simplicity we have considered here a sinusoidal ruling.) The linear ruling is placed at the plane where the aberrated wavefront is to be measured. If we place a light detector at a distance L from the plate, due to the wavefront aberrations we will obtain a distorted irradiance pattern that will be approximately given by:

$$I_x(x,y) = \frac{1}{2} + \frac{1}{2}\cos\left(\omega_0 x + \omega_0 L \frac{\partial W(x,y)}{\partial y}\right) \tag{1.55}$$

The irradiance, $I_x(x,y)$, will be a distorted version of the transmittance, $T_x(x,y)$. The shadow of the ruling, when illuminated

Figure 1.20   Typical Ronchi pattern with spherical aberration.

with a wavefront with spherical aberration, produces a shadow over a charge-coupled device (CCD) video array, as shown in Figure 1.20.

As pointed out before, in the absence of rotational symmetry, it is necessary to detect two orthogonal shadow patterns to completely describe the gradient field of the wavefront being analyzed. The second linear ruling is located at the same testing plane, but with its strip lines oriented orthogonally to that of the first ruling. That is,

$$T_y(x,y) = \frac{(1 + \cos(\omega_0 y))}{2} \qquad (1.56)$$

The lines in this transparency are perpendicular to the first one.

Thus, the distorted image of the Ronchi ruling at the collecting data plane will be given by:

$$I_y(x,y) = \frac{1}{2} + \frac{1}{2}\cos\left(\omega_0 y + \omega_0 L \frac{\partial W(x,y)}{\partial x}\right) \qquad (1.57)$$

We may use any of the carrier fringe methods described in this book to demodulate these two Ronchigrams.

Once the detected and unwrapped phase of the ruling's shadows has been obtained, we need to integrate the resulting gradient field. To integrate this phase gradient we may use path-independent integration, such as least squares. Least-squares integration of the gradient field may be considered to be the function that minimizes the following quadratic merit function:

$$U(\hat{W}) = \sum_{(x,y)\in L}\left[\hat{W}(x_{i+1},y_j)-\hat{W}(x_i,y_j)-\left(\frac{\partial W(x,y)}{\partial x}\right)_{x=x_i,y=y_j}\right]^2$$

$$+ \sum_{(x,y)\in L}\left[\hat{W}(x_i,y_{j+1})-\hat{W}(x_i,y_j)-\left(\frac{\partial W(x,y)}{\partial y}\right)_{x=x_i,y=y_j}\right]^2 \tag{1.58}$$

where the "hat" function $\hat{W}$ is the estimated wavefront, and we have approximated the derivative of the searched phase along the x- and y-axes as first-order differences of the estimated wavefront. The least-squares estimator may be obtained from U by a simple gradient descent applied to all pixels:

$$\hat{W}^{k+1}(x,y) = \hat{W}^k(x,y) - \tau\frac{\partial U(\hat{W})}{\partial \hat{W}(x,y)} \tag{1.59}$$

or by using a faster algorithm such as conjugate gradient or transform methods (Fried, 1977; Hunt, 1979).

1.7 HARTMANN TEST

The Hartmann test is a well-known technique for testing large optical components (Ghozeil, 1992). It uses a screen with holes or strips lying perpendicular to the propagation direction of the wavefront being analyzed, as shown in Figure 1.21. A screen with an array of circular holes is placed over the concave reflecting surface being analyzed. Each of the narrow beams of light reflected on each hole returns back to an observing screen called the *Hartmann plate*. Here, we measure the deviation of the reflected light beams on the Hartmann plate with respect to the ideal positions. These deviations are the transverse aberrations TA_x and TA_y measured along the x- and y-axes, respectively. Thus, to obtain the shape of the testing wavefront we must use one of the many possible integration procedures. One method is use of the trapezoidal rule, which can be mathematically expressed by:

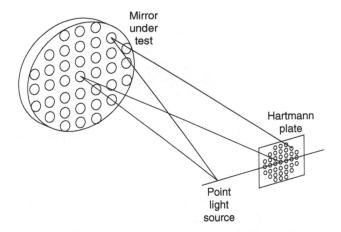

Figure 1.21 Optical arrangement in the Hartmann test.

$$W(x, y) = -\frac{1}{L} \int_0^x TA_x dx$$

$$= -\frac{1}{L} \sum_2^N \left(\frac{TA_{x(n)} + TA_{x(n-1)}}{2} \right)(x_n - x_{n-1})$$

(1.60)

Another method is to first interpolate the transverse discrete measurements of the aberration by means of a two-dimensional polynomial fitting and then performing the integration analytically, as described by Cornejo (1992). Still another approach is applying a least-squares solution to the integration problem. This integration procedure has the advantage of being path independent and robust to noise.

The Hartmann technique samples the wavefront being analyzed using a screen of uniformly spaced holes situated at the pupil plane:

$$HS(x, y) = \sum_{n=N/2}^{N/2} \sum_{m=-N/2}^{N/2} h(x - nd, y - md)$$

(1.61)

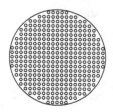

Figure 1.22 Typical Hartmann screen used in the Hartmann screen test.

where $HS(x,y)$ is the Hartmann screen, and $h(x,y)$ represents the small holes that are uniformly spaced in the Hartmann screen. Finally, d is the space among the holes of the screen. A typical Hartmann screen is shown in Figure 1.22.

The collimated rays of light that pass through the screen holes (Equation 1.61) are then captured by a photographic plate at some distance L from it. The uniformly spaced array of holes at the pupil of the instrument is then distorted at the photographic plate by the spherical aberration of the wavᵉfront under analysis. The screen deformations are then proportional to the slope of the aspherical wavefront; that is, we have:

$$H(x,y) = \left[\sum_{(n,m)=-N/2}^{N/2} h'\left(\begin{array}{c} x - nd - L\dfrac{\partial W(x,y)}{\partial x}, \\ y - md - L\dfrac{\partial W(x,y)}{\partial x} \end{array} \right) \right] P(x,y) \quad (1.62)$$

where $H(x,y)$ is the Hartmanngram obtained at distance L from the Hartmann screen. The function $h'(x,y)$ is an image of the screen holes, $h(x,y)$, as projected at the Hartmanngram plane. Finally, $P(x,y)$ is the pupil of the wavefront being tested. As Equation 1.62 shows, only one Hartmanngram is necessary to fully estimate the gradient of the wavefront. The frequency content of the estimated wavefront will be limited by the sampling theorem to the inverse of the period d of the screen holes. Figure 1.23 shows the Hartmanngram of a 62-cm paraboloidal mirror.

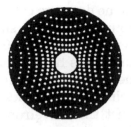

Figure 1.23   Hartmanngram of 62-cm paraboloidal primary mirror.

Traditionally, these Hartmanngrams (distorted images of the screen at the plane of the photographic plate) are analyzed by measuring the centroid of the spot images $h'(x,y)$ generated by the screen holes, $h(x,y)$. Deviations of these centroids from their uniformly spaced positions (unaberrated positions) are recorded. As Equation 1.62 shows, these deviations are proportional to the slope of the aspherical aberration. The coordinates of the centroid give a two-dimensional discrete field of the wavefront gradient which requires integration and interpolation over regions without data. Integration of the gradient field of the wavefront is normally done by applying the trapezoidal rule — that is, by following several independent integration paths and averaging their outcomes. In this way, we may approach a path-independent integration. Using this integration procedure, the wavefront is known only at the position of the hole. Although this integration technique may provide a good wavefront estimation, a determination of the positions of the Harmann spots could be a time-consuming process. Finally, a polynomial or spline wavefront fitting is necessary to estimate values of the wavefront at places other than the discrete points where the gradient data are collected. A two-dimensional polynomial for the wavefront gradient is then fitted by least-squares to the slope data. This polynomial must contain every possible type of wavefront aberration; otherwise, some unexpected features (especially at the edges) of the wavefront may be filtered out. On the other hand, if one uses a high-degree polynomial (to avoid filtering out any wavefront aberration), the estimated continuous wavefront may oscillate wildly in regions where no data are collected. The performance of the Hartmann test and the lateral shearing interferometer has been compared by Welsh et al. (1995).

Many similar procedures have been developed to obtain the wavefront from measurements of transverse aberrations. For example, Rubinstein and Wolansky (2001) have proposed a method to reconstruct the wavefront shape from a set of first-order, partial-differential equations.

1.8 FRINGE PROJECTION

For a fringe projection, a periodic ruling is projected onto a solid body, then the image of this body with the fringes over its surface is imaged over another periodic ruling to form moiré fringes. The shape of a solid body can be measured by projecting a periodic structure or ruling over the body (Idesawa et al., 1977; Takeda, 1982; Doty, 1983; Gåsvik, 1983; Creath and Wyant, 1988). The fringes may be projected onto the body by a lens or slide projector (Takasaki, 1970, 1973; Parker, 1978; Pirodda, 1982; Gåsvik, 1983; Cline et al., 1984; Reid, 1984; Suganuma and Yoshisawa, 1991). In another method, the interference fringes produced by two tilted, flat wavefronts are projected over the body (Brooks and Heflinger, 1969). A slightly different method, shadow moiré, produces the moiré fringes between a Ronchi ruling and the shadow of the ruling projected over a solid body located just behind the ruling. This method makes it possible to find the shape of nearly flat surfaces (Jaerisch and Makosch, 1973; Pirodda, 1982).

Let us now consider a straight fringe that is projected from point A with height z_a to point C on the plane $z = 0$, as shown in Figure 1.24. This fringe is observed from point B with height z_b over the plane $z = 0$. If the surface to be measured is located over the plane $z = 0$, this surface will intersect the fringe at point D. As observed from point B, the fringe appears to be at point E on the plane $z = 0$. The separation between points E and C allows us to calculate the object height over the plane $z = 0$. Obviously, the lines AC and BE are on a common plane, as they intersect at D. Nevertheless, this plane is not necessarily perpendicular to

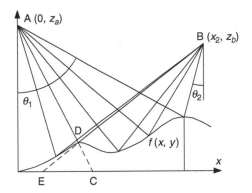

Figure 1.24 Projecting a periodic structure over a solid body to measure its shape.

the plane $z = 0$. This geometry is completely general. The shape of the body is determined if the three-dimensional coordinates of point D are calculated from measurements of the coordinates of point E on the plane $x = 0$ for many positions on the projected fringes.

This is the general configuration for fringe projection, but a simpler analysis can be made if both the lens projector and the observer are optically placed at infinite distances from the body to be measured, as shown in Figure 1.25. The observer is located in a direction parallel to the z-axis. In this case, the object heights are given by:

$$f(x,y) + \frac{x}{\tan\theta} - \frac{s}{\sin\theta} = \frac{m\,d}{\sin\theta} \qquad (1.63)$$

where angle θ is the inclination of the illuminator; m is the fringe number, with the fringe $m = 0$ being located at the origin ($x = 0$); and distance d is the fringe period in a plane perpendicular to the illuminating light beam.

The equivalent two-beam interferometric expression for the wavefront deformation, $W(x)$, is:

$$W(x,y) + \frac{\lambda}{p}x = m\lambda \qquad (1.64)$$

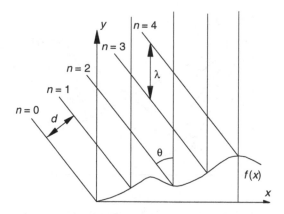

Figure 1.25 Projecting a periodic structure over a solid body to measure its shape, with both the projector and observer at infinity.

Hence, the surface deformation $f(x,y) = 2W(x,y)$ when tested in a Fizeau interferometer is:

$$f(x,y) + \frac{2\lambda}{p} x + a = 2m\lambda \qquad (1.65)$$

where m is the order of interference, p is the fringe period introduced by tilting the reference wavefront, and a is a constant. By comparing these two expressions, we see that we may consider fringe projection with this geometry as Fizeau interferometry with wavelength λ given by:

$$\lambda = \frac{d}{2\sin\theta} \qquad (1.66)$$

These projected fringes may then be considered Fizeau fringes with a large linear carrier (tilt) introduced. This body, with the fringes or interferogram, is imaged on the observing plane by means of an optical system, photographic camera, or television camera. This interferogram with tilt may be analyzed by any of the traditional methods, but one common method applies the moiré techniques, as described later in Chapter 9. The image is then superimposed on a linear ruling with approximately the same frequency as the fringes on the interferogram. This linear ruling may be real or computer generated.

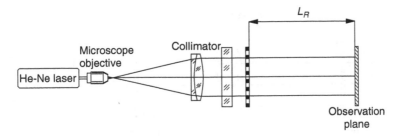

Figure 1.26 Autoimage formation of a ruling, illuminated with a collimated beam of light.

Moiré methods are not really interferometric; nevertheless, their fringe analyses are so similar that a description of these methods is convenient. Whenever two slightly different periodic structures are superimposed, a "beating" between the two structures is observed in the form of another periodic structure with a lower spatial frequency. These fringes are moiré fringes.

Moiré techniques have been used in metrology for a long time, with many different configurations and purposes (see reviews by Sciammarella, 1982; Reid, 1984; Patorski, 1988). They are discussed in more detail in Chapter 9, primarily as tools for the analysis of interferograms. Here, we briefly consider the basic moiré configurations.

1.9 TALBOT INTERFEROMETRY AND MOIRÉ DEFLECTOMETRY

Another method commonly used to measure wavefront deformations uses the Talbot autoimaging procedure, illustrated in Figure 1.26. A ruling is illuminated with a collimated, convergent, or divergent beam of light. The shadow of the ruling is projected upon a screen placed at some distance from the ruling, where another ruling is placed to form the moiré. Talbot (1836) discovered that when a linear ruling is illuminated with a collimated beam of light, perfect images of this ruling are formed without any lenses, at distances that are integer multiples of a distance called the Rayleigh (1881) distance (L_R), as shown in Figure 1.26.

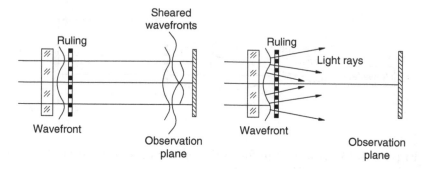

Figure 1.27   Formation of autoimages with distorted or spherical wavefronts.

If the illuminating wavefront is not flat but spherical or distorted, the fringes in the autoimage are distorted, not straight. The interferometric explanation assumes that the diffracted wavefronts produce a lateral shearing interferogram, as shown in Figure 1.27a. On the other hand, the geometric interpretation considers the fringes to be shadows of the ruling lines, projected in a direction perpendicular to the wavefront (Figure 1.27b). Both models are equivalent.

When the moiré pattern between the fringe image represented by the autoimage and a superposed linear ruling is formed, we speak of a Talbot interferometer. Talbot interferometry has been described by many researchers, such as Yokoseki and Susuki (1971a,b), Takeda and Kobayashi (1984), and Rodríguez-Vera et al. (1991). These authors interpreted the fringe using interferometric models such as multiple-beam lateral shearing interferometry. Kafri (1980, 1981) applied this method from a geometrical point of view and referred to it as *moiré deflectometry*. Glatt and Kafri (1988), Stricker (1985), and Vlad et al. (1991) have described this method and some applications. Interferometric and geometric interpretations may be proved to be equivalent, as pointed out by Patorski (1988). This procedure is closely analogous to the Ronchi test (Cornejo, 1992).

In moiré deflectometry, or Talbot interferometry, as previously described, the observing plane is located at the first

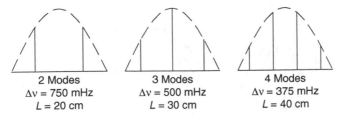

2 Modes	3 Modes	4 Modes
$\Delta v = 750$ mHz	$\Delta v = 500$ mHz	$\Delta v = 375$ mHz
$L = 20$ cm	$L = 30$ cm	$L = 40$ cm

Figure 1.28 Spectrum of light (longitudinal modes) from a gas laser.

Talbot autoimage of the ruling; thus, distance d_T is equal to the Rayleigh distance L_R, as given by:

$$L_R = \frac{2d^2}{\lambda} \tag{1.67}$$

The resulting deflectograms, or Talbot interferograms, may be analyzed in the same way as the Ronchigrams.

1.10 COMMON LIGHT SOURCES USED IN INTERFEROMETRY

By far the most common light source in interferometry is the helium–neon laser. The great advantage of this light source is its large coherence length and monochromaticity; however, these characteristics can sometimes be a significant problem when many spurious fringes are also formed, unless great precautions are taken. When a laser light source is used, extremely large OPDs can be introduced (Morokuma et al., 1963). As shown in Figure 1.28, the light emitted by a gas laser usually consists of several equally spaced spectral lines (longitudinal modes) with a frequency separation equal to:

$$\Delta v = \frac{c}{2L} \tag{1.68}$$

where L is the laser cavity length. If cavity length L of a laser changes because of thermal expansion or contraction or mechanical vibrations, the lines move along the frequency

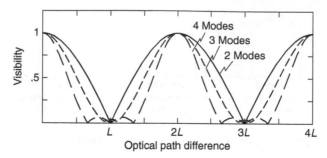

Figure 1.29 Visibility in a Twyman–Green interferometer using a helium–neon laser, as a function of the optical path difference, for three different lengths of the laser cavity.

scale to preserve their relative separations, but the intensities remain under the power-gain curve, as shown in Figure 1.29.

Single-mode or single-frequency lasers produce a perfectly monochromatic wavetrain, but because of instabilities in the cavity length the frequency may be unstable. Servomechanisms have allowed the commercial production of single-frequency lasers that have extremely stable frequencies. These lasers are the ideal source for interferometry because an OPD as long as desired can be introduced without any loss in contrast.

The fringe visibility in an interferometer using a laser source with several longitudinal modes is a function of the optical path difference. For good contrast, the OPD has to be an integral multiple of 2*L*. A laser with two longitudinal modes is sometimes stabilized to avoid contrast changes by a method recommended by Bennett et al. (1973), Gordon and Jacobs (1974), and Balhorn et al. (1972).

Another laser frequently used in interferometers is the laser diode. Creath and Wyant (1985), Ning et al. (1989), and Onodera and Ishii (1996) have studied the most important characteristics of these lasers for use in interferometers. Their low coherence length (of the order of 1 millimeter) is a great advantage in many applications, and other advantages include their low price and small size.

1.11 ASPHERICAL COMPENSATORS
AND ASPHERIC WAVEFRONTS

The most common types of interferometer, with the exception of lateral or rotational shearing interferometers, produce interference patterns in which the fringes are straight, equidistant, and parallel when the wavefront under analysis is perfect and spherical, with the same radius of curvature as the reference wavefront. If the surface being analyzed does not have a perfect shape, the fringes will not be straight and their separations will be variable. Deformations of the wavefront may be determined by a mathematical examination of the shapes of the fringes. Because the fringe separations are not constant, in some places the fringes will be widely spaced but in some others the fringes will be too close together. It is desirable to compensate in some way for the spherical aberrations of wavefronts being analyzed so that the fringes appear straight, parallel, and equidistant for perfect wavefronts. The necessary *null test* may be accomplished utilizing some special configurations that may be used to test a conical surface. Almost all of these surfaces have rotational symmetry. An aspherical or null compensator is an optical element with spherical aberrations designed to compensate for spherical aberrations in an aspherical wavefront. It is beyond the scope of this book to discuss them further here, but they have been described in detail in the literature (e.g., Offner and Malacara, 1992). A typical example of such compensators, the well-known Offner compensator, is illustrated in Figure 1.30.

1.12 IMAGING OF THE PUPIL ON
THE OBSERVATION PLANE

An aberrated wavefront continuously changes its shape as it travels; thus, if the optical system is not perfect, then the interference pattern will also continuously change as the beam advances, as shown in Figure 1.31. The change in shape of a traveling wavefront has been studied and calculated by Józwicki (1990), who has taken into account the effects of diffraction. The errors of an instrument are represented by

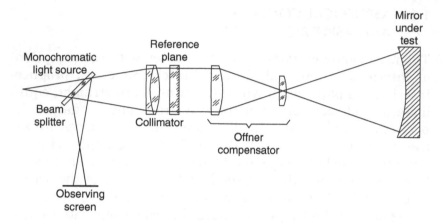

Figure 1.30 Offner compensator.

wavefront distortions on the pupil; hence, the interferogram should be taken at that place.

1.12.1 Imaging the Pupil Back on Itself

When testing a lens with any of the configurations described earlier, the wavefront travels twice through the lens, the second time after being reflected at the small mirror in front of the lens. If the aberration is small, the total wavefront deformation is twice the deformation introduced in a single pass through the lens; however, if the aberration is large, this is not so because the wavefront changes while traveling from the lens to the mirror and back to the lens. If the spot on the surface where the defect is located is not imaged back onto itself by the concave or convex mirror, the ray will not pass through this defect a second time. Great confusion then results with regard to interpretation of the interferogram, as the defect is not precisely duplicated by the double pass through the lens (Dyson, 1959).

It may be shown that the image of the lens is formed at a distance S from the lens given by:

$$S = \frac{2(F - r)^2}{2F - r} \tag{1.69}$$

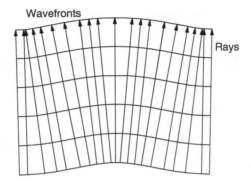

Figure 1.31 Change in the shape of a wavefront as it travels.

where F is the focal length, and r is the radius of curvature of the surface ($r > 0$ for a convex mirror, $r < 0$ for a concave mirror). We can see that the ideal mirror is convex and very close to the lens ($r \sim F$).

An appropriate optical configuration has to be used if the lens being analyzed has a large aberration in order to image its pupil back on itself. Any auxiliary lenses or mirrors must be used to preserve the wavefront shape. Some examples of these arrangements are provided in Figure 1.32 (Malacara and Menchaca, 1985). For microscope objectives, however, these solutions are not satisfactory because the ideal place to observe the fringes is at the back focal plane. In this case, the Dyson system illustrated in Figure 1.33 is an ideal solution. It is interesting to point out that Dyson's system can be used to place the self-conjugate plane at a concave or convex surface while maintaining the concentricity of the surfaces.

1.12.2 Imaging the Pupil on the Observing Screen

The second problem is to image the interference pattern on the observing detector, screen, or photographic plate. The imaging lens does not need to preserve the wavefront shape, as it is generally placed after the beam splitter so both inter- fering wavefronts pass through this lens; however, this lens has to be designed in such a way that the interference pattern

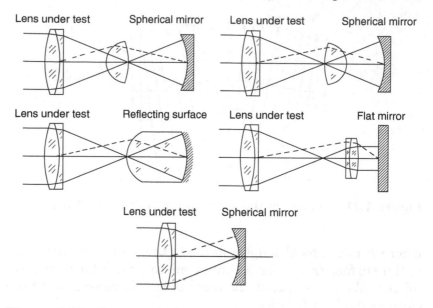

Figure 1.32 Some optical arrangements to test a lens, imaging its pupil back on itself.

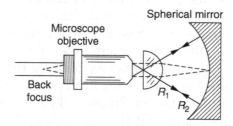

Figure 1.33 Dyson's system to test microscope objectives.

is imaged without any distortion, assuming that the pupil of the system is at the closest image of the light source, as shown in Figure 1.34a. A rotating ground glass in the plane of the interferogram might be useful sometimes in order to reduce the noise due to speckle and dust in the optical components. Ideally, this rotating glass should not be completely ground in order to reduce the loss of brightness and to maintain the stop of the imaging lens at the original position, as shown in

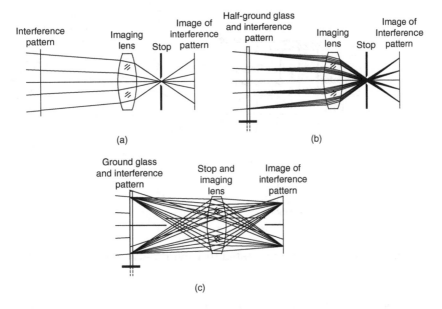

Figure 1.34 Imaging the interferogram on the observation plane: (a) without any rotating ground glass, (b) with a rotating half-ground glass, and (c) with a rotating ground glass.

Figure 1.34b. If the rotating glass is completely ground, the stop of the imaging lens should be shifted to the lens in order to use all available light, but then the lens must be designed to take into consideration this new stop position, as shown in Figure 1.34c.

When a distorted wavefront propagates in space its shape is not preserved but changes continuously along its trajectory. From a geometrical point of view (that is, neglecting diffraction), only a spherical or flat wavefront keeps its shape, with only the radius of curvature changing. This is a well-known fact that should be taken into account in the interferometry of wavefronts. As an example, let us consider the Twyman–Green interferometer shown in Figure 1.35. A conic or spherical mirror is tested by means of this interferometer. If the mirror has a conical shape, the spherical aberration is compensated with a lens having the proper amount of spherical aberration with the opposite sign.

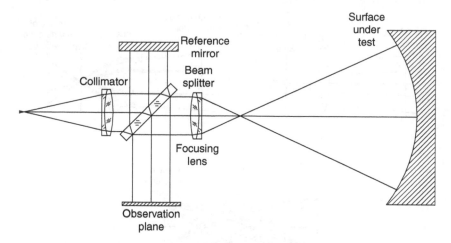

Figure 1.35   Conic mirror tested in Twyman–Green interferometer.

The wavefront reflected on the surface is combined at the beam splitter with a perfectly flat reference wavefront. The focusing lens has to be designed so that the returning wavefront is perfectly flat if the surface has no defects. If the surface has a distorted shape, the reflected wavefront is also distorted; thus, the wavefront going out of the focusing lens and returning to the beam splitter will not be flat but distorted. The deformations in the wavefront going out of the focusing lens, however, are not the same as the deformations at the surface.

1.12.3   Requirements on the Imaging Lens

To obtain an interference pattern that is directly related to the wavefront deformations on the surface, the pattern must be observed at a plane that is conjugate to this surface, as has been described in the literature (e.g., Slomba and Figoski, 1978; Malacara and Menchaca, 1985; Selberg, 1987; Józwicki, 1989, 1990; Malacara, 1992). This is the purpose of the projection lens, which has to form an image of the surface being analyzed on the observing screen. The following two requirements must be satisfied by this lens (see Figure 1.36):

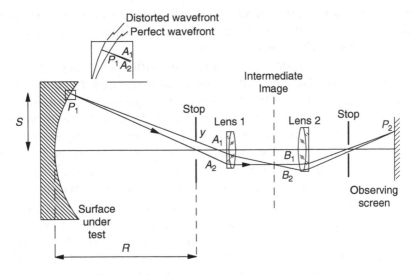

Figure 1.36 Optical system to image the pupil of a system on the observing plane.

1. The height of point P_2 over the optical axis should be strictly linear with the height of the point P_1 over the optical axis; in other words, there should be no distortion. This assures us that a straight fringe on the surface being analyzed is also a straight fringe on the observing screen. This condition is not absolutely necessary if the fringe distortion is taken into account during computer analysis of the fringes.

2. Point object P_1 must correspond to point image P_2. By Fermat's principle, then, the optical path through A_1B_2 is equal to the optical path through A_2B_1. Let us assume that a perfect surface sends the reflected ray from P_1 through A_1. A distorted wavefront sends a ray that passes through P_1 toward A_2. Both rays then arrive together at point P_2. Because the optical paths are equal, any phase difference between the two rays at point P_1 is the same when they arrive at point P_2.

If these conditions are satisfied the interferograms are identical. It must be noted that it is not necessary for lens 2 to produce a perfect wavefront, as both wavefronts are refracted on this lens, and any deformations are introduced in both wavefronts in the same amount.

The imaging lens design must include a complete system, with all lenses between the surface and the observing screen. The points where the light beams converge may be considered the stops of the lens system, so the system may have two or more virtual stops. An intermediate image occurs, as shown in Figure 1.36; however, the observing plane cannot be located at this position for two reasons: (1) it is very unlikely that it has the required dimensions, and (2) the system would be so asymmetric that the distortion would be extremely large.

A complete system, with lenses 1 and 2, is more symmetric, making it easier to correct the distortion. The stop diameter is given by the maximum transverse aberration at the stop. This maximum transverse aberration is a function of three factors: (1) the degree of asphericity of the surface under analysis, (2) the deformation error in this surface, and (3) the tilt between the wavefront under analysis and the reference wavefront. In general, this aperture is extremely small, even with large transverse aberrations.

Let us now analyze the degree of correction required for each of the five Seidel aberrations.

- *Spherical aberration.* This aberration increases with the fourth power of the aperture; thus, it does not have to be highly corrected as the aperture is very small. A large amount of spherical aberration may be tolerated.
- *Coma.* This aberration increases with the cube of the aperture in the tangential plane and with the square of the aperture in the sagittal plane; thus, correction of this aberration is more necessary than that of the spherical aberration, the most important being the sagittal coma. If a large tilt is introduced in the interferogram, resulting in straight fringes perpendicular to the tangential plane, the fringes in the vicinity of this plane are affected by coma to a lesser degree than the fringes on the sagittal plane.

- *Petzval curvature.* Ideally, the curvature of the surface under analysis must be taken into account by curving the object plane by the same amount. The wavefront aberration due to this aberration increases with the square of the aperture; however, this aberration is not so important as long as the ray transverse aberration in the observing plane remains small, as we will see later.
- *Astigmatism.* The wavefront aberration produced by astigmatism, as for the Petzval curvature, increases with the square of the aperture. So, the important criterion here should also be the magnitude of the ray transverse aberration.
- *Distortion.* This aberration, as we explained before, may be ignored if the compensation is made in the computer analysis of the fringes; however, it is always easier to correct it on the lens. Again, the important criterion is the magnitude of the ray transverse aberration.

The slope of the aberrated wavefront with respect to the ideal wavefront (reference wavefront) is:

$$\left(\frac{\partial W}{\partial S}\right) = \frac{\Delta W}{\Delta S} \qquad (1.70)$$

where ΔW is the change in the wavefront deformation if the height of point P_1 changes by an amount ΔS. Let us assume that the magnification of the entire lens system is m. Then, the magnitude of the transverse ray aberration (TA) on the observing plane corresponds to the object height shift, ΔS, given by:

$$m = \frac{TA}{\Delta S} \qquad (1.71)$$

Thus, we may see that

$$TA = \frac{m\Delta W}{\left(\dfrac{\partial W}{\partial S}\right)} \qquad (1.72)$$

To find the maximum allowable ray transverse aberration (TA_{max}) we see that if ΔW_{max} is the maximum permissible error in the wavefront measurement, the corresponding maximum value of this ray transverse aberration is:

$$TA_{max} = \frac{m\Delta W_{max}}{\left(\dfrac{\partial W}{\partial y}\right)} \qquad (1.73)$$

If the minimum separation between two consecutive fringes on the surface is σ_1 and ΔW_{max} is a fraction ($1/n$) of the wavelength ($\Delta W_{max} = \lambda/n$), we may write:

$$TA_{max} = \frac{m\sigma_1}{n} \qquad (1.74)$$

Hence, if the minimum separation between two consecutive fringes in the observation plane is σ_2 (given by $\sigma_2 = m\sigma_1$), we see that

$$TA_{max} = \frac{\sigma_2}{n} \qquad (1.75)$$

which means that the maximum permissible transverse aberration in the projecting optical system is equal to a predetermined fraction of the minimum separation between the fringes in the observation plane.

When the interferogram is observed with a two-dimensional detector, a wavefront tilt or aberration may be introduced to the limit imposed by the detector. Then, the maximum transverse aberration is approximately equal to the resolution power of the detector, given by the separation between two consecutive pixels, or detector elements.

The stop semiaperture y may be obtained by using the minimum fringe separation as follows:

$$y = \frac{\lambda R}{\sigma_1} = \frac{m\lambda R}{\sigma_2} \qquad (1.76)$$

where R is the radius of curvature of the mirror, as shown in Figure 1.36.

If the distortion aberration is not compensated for during computer analysis, then the transverse aberration must be measured from the Gaussian image position; otherwise, it is measured from the center of gravity of the image. If the magnification of the system is much less than 1, the interferogram in the observation plane is very small and the requirement for a small transverse aberration may be quite strong.

The principles to be used in the design of projecting lenses for interferometry have been described using the Twyman–Green interferometer as an example, but they may be applied to Fizeau interferometers as well.

1.13 MULTIPLE-WAVELENGTH INTERFEROMETRY

In phase-shifting interferometry, the phase is calculated modulo 2π, so a phase wrapping occurs during the calculation. To unwrap the phase, the phase between two adjacent measured points in the interferogram must be smaller than 2π which limits the maximum wavefront slope and hence the maximum asphericity being measured. Wyant (1971), Polhemus (1973), Cheng and Wyant (1984), Wyant et al. (1984), Creath et al. (1985), Creath and Wyant (1986), Gushov and Solodkin (1991), and Onodera and Ishii (1999) have studied the problem of phase determination when two or more different wavelengths are used. If two different wavelengths (λ_a and λ_b) are simultaneously used, the wavetrain is modulated as shown in Figure 1.37, with the group length (λ_{eq}) given by:

$$\lambda_{eq} = \frac{\lambda_a \lambda_b}{|\lambda_b - \lambda_a|} \qquad (1.77)$$

Wyant (1971) described two methods that utilize two wavelengths. In the first method, a photographic recording of an interferogram is taken with one wavelength, then another interferogram is formed with the second wavelength and the photograph of the first interferogram is placed over the second one. In this manner, a moiré between the photograph of one interferogram and the real-time image of the second is

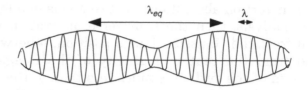

Figure 1.37 Wavetrain formed by two wavelengths.

obtained. High frequencies of this moiré are then filtered out with a pinhole. In the second method, images of the two interferograms are taken simultaneously, one on top of the other, by illuminating with the two wavelengths. The high spatial frequencies of the resulting moiré are also filtered with a pinhole.

Polhemus (1973) described a real-time, two-wavelength interferometer using a television camera to detect the moiré pattern. Figure 1.38 shows the interferograms obtained with two wavelengths, the resulting moiré pattern, and its filtered pattern. The resulting pattern is the image of an interferogram taken with the equivalent wavelength.

Cheng and Wyant (1984), Creath et al. (1985), and Creath and Wyant (1986) implemented phase-shifting interferometers using two wavelengths. Two separate wrapped-phase maps are obtained by taking two independents sets of measurements, using each of the two wavelengths. We assume that the Nyquist limit has been exceeded, due to the high wavefront asphericity. With one wavelength the phase unwrapping would be impossible, but it can be achieved with two wavelengths. The two wavefront deformations are different if the scale is the phase, because the wavelengths are different; however, they must be equal if the optical path difference is used instead of the phase. Thus, we have:

$$\text{OPD}_a(x, y) = \text{OPD}_b(x, y) \qquad (1.78)$$

We may also write:

$$\text{OPD}_a(x, y) = \left(\frac{\phi_a(x, y)}{2\pi} + m_a \right) \lambda_a \qquad (1.79)$$

Figure 1.38 Moiré of interferograms taken with two wavelengths: first wavelength, $\lambda_a = 0.633$; second wavelength, $\lambda_b = 0.594$; equivalent wavelength, $\lambda_{eq} = 9.714$.

and

$$\text{OPD}_b(x,y) = \left(\frac{\phi_b(x,y)}{2\pi} + m_b \right)\lambda_b \qquad (1.80)$$

where m_a and m_b are integers. Thus, using Equation 1.78 we have:

$$\left(\frac{\phi_a(x,y)}{2\pi} + m_a \right)\lambda_a = \left(\frac{\phi_b(x,y)}{2\pi} + m_b \right)\lambda_b \qquad (1.81)$$

We have one equation with two unknowns (m_a and m_b). The system may be solved if we assume that the difference of order numbers between two adjacent pixels is the same for both wavelengths. This hypothesis is valid if the asphericity is not extremely high. Thus, we may obtain:

$$\Delta\text{OPD}_{n+1} = \frac{1}{2\pi}\left(\Delta\phi_{(n+1)a} - \Delta\phi_{(n+1)b} \right)\lambda_{eq} \quad \text{if } \lambda_b > \lambda_a \quad (1.82)$$

The OPD values for all pixels in a row may be obtained if we take $\text{OPD}_1 = 0$. Figure 1.39 illustrates the phase unwrapping procedure using two different wavelengths with a ratio of 6 to 5. The only possible valid points when unwrapping the wavefront are the thick circles, where the two wavelengths coincide. The result is that, even with subsampling, the unwrapping presents no ambiguities.

Cheng and Wyant (1985) enhanced the capability of two-wavelength interferometry by introducing a third wavelength so even steeper wavefront slopes can be measured. Löfdahl and Eriksson (2001) developed a mathematical algorithm for resolving with a good certainty the 2π ambiguities when using any number of wavelengths.

Hariharan and Roy (1994) proposed using white light and measuring the contrast function in the frequency domain. The interferometer has to be designed using an achromatic phase shifter in order to avoid a change in the contrast function when changing the phase. This achromatic phase shifter allows a change in the phase between the two beams for different wavelengths, without a change in the optical path difference. The mathematical procedure involves two Fourier transforms, forward and inverse, along the direction of change of the phase for each pixel in the interferogram. White-light interferometry has developed impressively to the point that many opaque materials such as ceramics, plastics, and even paper can be measured like specular materials (Wyant, 1993; Harasaki and Wyant, 2000; Harasaki et al., 2000; de Groot et al., 2002).

REFERENCES

Balhorn, R., Kunzmann, H., and Lebowsky, F., Frequency stabilization of internal mirror helium–neon lasers, *Appl. Opt.*, 11, 742–744, 1972.

Bennett, S.J., Ward, R.E., and Wilson, D.C., Comments on frequency stabilization of internal mirror helium–neon lasers, *Appl. Opt.*, 12, 1406–1406, 1973.

Brooks, R.E. and Heflinger, L.O., Moiré gauging using optical interference fringes, *Appl. Opt.*, 8, 935–939, 1969.

Burge, J., Fizeau interferometry for large convex surfaces, *Proc. SPIE*, 2536, 127–137, 1995.

Cheng, Y.-Y. and Wyant, J.C., Two-wavelength phase shifting interferometer, *Appl. Opt.*, 23, 4539–4543, 1984.

Cline, H.E., Lorensen, W.E., and Holik, A.S., Automatic moiré contouring, *Appl. Opt.*, 23, 1454–1459, 1984.

Cornejo, A., Ronchi test, in *Optical Shop Testing*, Malacara, D., Ed., John Wiley & Sons, New York, 1992.

Creath, K., Interferometric investigation of a laser diode, *Appl. Opt.*, 24, 1291–1293, 1985.

Creath, K., Wyko systems for optical metrology, *Proc. SPIE*, 816, 111–126, 1987.

Creath, K. and Wyant, J.C., Direct phase measurement of aspheric surface contours, *Proc SPIE*, 645, 101–106, 1986.

Creath, K. and Wyant, J.C., Aspheric measurement using phase shifting interferometry, *Proc SPIE*, 813, 553–554, 1987.

Creath, K. and Wyant, J.C., Comparison of interferometric contouring techniques, *Proc. SPIE*, 954, 174–182, 1988.

Creath, K., Cheng, Y.-Y., and Wyant, J.C., Contouring aspheric surfaces using two-wavelength phase shifting interferometry, *Opt. Acta*, 32, 1455–1464, 1985.

de Groot, P., Colona de Lega, J., Kramer, J., and Turzhitsky, M., Determination of fringe order in white-light interference microscopy, *Appl. Opt.*, 41, 4571–4578, 2002.

Dörband, B. and Tiziani, H.J., Testing aspheric surfaces with computer generated holograms: analysis of adjustment and shape errors, *Appl. Opt.*, 24, 2604–2611, 1985.

Doty, J.L., Projection moiré for remote contour analysis, *J. Opt. Soc. Am.*, 73, 366–372, 1983.

Dyson, J., Unit magnification optical system without Seidel aberrations, *J. Opt. Soc. Am.*, 49, 713–716, 1959.

Elster, C. and Weingärtner, I., Solution to the shearing problem, *Appl. Opt.*, 38, 5024–5031, 1999a.

Elster, C. and Weingärtner, I., Exact wave-front reconstruction from two lateral shearing interferograms, *J. Opt. Soc. Am. A*, 16, 2281–2285, 1999b.

Fienup, J.R. and Wackermann, C.C., Phase-retrieval stagnation problems and solutions, *J. Opt. Soc. Am. A*, 3, 1897–1907, 1986.

Fischer, D.J., Vector formulation for Ronchi shear surface fitting, *Proc. SPIE*, 1755, 228–238, 1992.

Freischlad, K., Wavefront integration from difference data, *Proc. SPIE*, 1755, 212–218, 1992.

Freischlad, K. and Koliopoulos, C.L., Wavefront reconstruction from noisy slope or difference data using the discrete Fourier transform, *Proc. SPIE*, 551, 74–80, 1985.

Fried, D.L., Least-squares fitting of a wave-front distortion estimate to an array of phase-difference measurements, *J. Opt. Soc. Am.*, 67, 370–375, 1977.

García-Márquez, J., Malacara, D., and Servín, M., Limit to the degree of asphericity when testing wavefronts using digital interferometry *Proc. SPIE*, 2263, 274–281, 1995.

Gåsvik, K.J., Moiré technique by means of digital image processing, *Appl. Opt.*, 22, 3543–3548, 1983.

Geary, J.M., Real-time interferogram simulation, *Opt. Eng.*, 18, 39–45, 1979.

Geary, J.M., Holmes, D.H., and Zeringue, Z., Real-time interferogram simulation, in *Optical Interferograms: Reduction and Interpretation*, American Society for Testing and Materials, West Conshohocken, PA, 1978.

Ghozeil, I., Hartmann and other screen tests, in *Optical Shop Testing*, Malacara, D., Ed., John Wiley & Sons, New York, 1992.

Glatt, I. and Kafri, O., Moiré deflectometry: ray tracing interferometry, *Opt. Lasers Eng.*, 8, 227–320, 1988.

Gordon, S.K. and Jacobs, S.F., Modification of inexpensive multimode lasers to produce a stabilized single-frequency beam, *Appl. Opt.*, 13, 231–231, 1974.

Gushov, V.I. and Solodkin, Y.N., Automatic processing of fringe patterns in integer interferometers, *Opt. Lasers Eng.*, 14, 311–324, 1991.

Harasaki, A. and Wyant, J.C., Fringe modulation skewing effect in white-light vertical scanning interferometry, *Appl. Opt.*, 39, 2101–2106, 2000.

Harasaki, A., Schmit, J., and Wyant, J.C., Improved vertical scanning interferometry, *Appl. Opt.*, 39, 2107–2115, 2000.

Hardy, J.W. and MacGovern, A.J., Shearing interferometry: a flexible technique for wavefront measuring, *Proc. SPIE*, 816, 180–195, 1987.

Hariharan, P. and Roy, M., White-light phase-stepping interferometry for surface profiling, *J. Mod. Optics*, 41, 2197–2201, 1994.

Horman, M.H., An application of wavefront reconstruction to interferometry, *Appl. Opt.*, 4, 333–336, 1965.

Houston, J.B., Jr., Buccini, C.J., and O'Neill, P.K., A laser unequal path interferometer for the optical shop, *Appl. Opt.*, 6, 1237, 1967.

Hudgin, R.H., Wave-front reconstruction for compensated imaging, *J. Opt. Soc. Am,.* 67, 375–378, 1977.

Hung, Y.Y., Shearography: a new optical method for strain measurement and nondestructive testing, *Opt. Eng.*, 21, 391–395, 1982.

Hunt, B.R., Matrix formulation of the reconstruction of phase values from phase differences, *J. Opt. Soc. Am.*, 69, 393–399, 1979.

Idesawa, M., Yatagai, T., and Soma, T., Scanning moiré method and automatic measurement of 3D shapes, *Appl. Opt.*, 16, 2152–2162, 1977.

Jaerisch, W. and Makosch, G., Optical contour mapping of surfaces, *Appl. Opt.*, 12, 1552–1557, 1973.

Józwicki, R., Telecentricity of the interferometric imaging system and its importance in the measuring accuracy, *Optica Applicata*, 19, 469–475, 1989.

Józwicki, R., Propagation of an aberrated wave with nonuniform amplitude distribution and its influence upon the interferometric measurement accuracy, *Optica Applicata*, 20, 229–252, 1990.

Kafri, O., Noncoherent method for mapping phase objects, *Opt. Lett.*, 5, 555–557, 1980.

Kafri, O., High sensitivity moiré deflectometry using a telescope, *Appl. Opt.*, 20, 3098–3100, 1981.

Kafri, O., Fundamental limit on the accuracy in interferometers, *Opt. Lett.*, 14, 657–658, 1989.

Kingslake, R., The interferometer patterns due to the primary aberrations, *Trans. Opt. Soc.*, 27, 94, 1925–1926.

Kuchel, M., The new Zeiss interferometer, *Proc. SPIE*, 1332, 655–663, 1990.

Löfdahl, M.T. and Eriksson, H., Algorithm for resolving 2π ambiguities in interferometric measurements by use of multiple wavelengths, *Opt. Eng.*, 40, 984–990, 2001.

Malacara, D., Ed., *Optical Shop Testing*, 2nd ed., John Wiley & Sons, New York, 1992.

Malacara, D. and Menchaca, C., Imaging of the wavefront under test in interferometry, *Proc. SPIE*, 540, 34–40, 1985.

Malacara-Hernández, D., Malacara-Hernández, Z., and Servín, M., Digitization of interferograms of aspheric wavefronts, *Opt. Eng.*, 35, 2102–2105, 1996.

Mantravadi, M.V., Lateral shearing interferometers, in *Optical Shop Testing*, Malacara D., Ed., John Wiley & Sons, Inc., New York, 1992.

Murty, M.V.R.K., The use of a single plane parallel plate as a lateral shearing interferometer with a visible gas laser source, *Appl. Opt.*, 3, 531–351, 1964.

Morokuma, T., Neflen, K.F., Lawrence, T.R., and Klucher, T.M., Interference fringes with a long path difference using He–Ne laser, *J. Opt. Soc. Am.*, 53, 394, 1963.

Ning, Y., Grattan, K.T.V., Meggitt, B.T., and Palmer, A.W., Characteristics of laser diodes for interferometric use, *Appl. Opt.*, 28, 3657–3661, 1989.

Noll, R.J., Phase estimates from slope-type wavefront sensors, *J. Opt. Soc. Am.*, 68, 139–140, 1978.

Offner, A. and Malacara, D., Null tests using compensators, in *Optical Shop Testing*, Malacara, D., Ed., John Wiley & Sons, New York, 1992.

Okuda, S., Nomura, T., Kamiya, K., Miyashiro, H., Yoshikawa, K., and Tashiro, H., High-precision analysis of a lateral shearing interferogram by use of the integration method and polynomials, *Appl. Opt.*, 39, 5179–5186, 2000.

Omura, K. and Yatagai, T., Phase measuring Ronchi test, *Appl. Opt.*, 27, 523–528, 1988.

Ono, A., Aspherical mirror testing with an area detector array, *Appl. Opt.*, 26, 1998–2004, 1987.

Onodera, R. and Ishii, Y., Phase-extraction analysis of laser-diode phase-shifting interferometry that is insensitive to changes in laser power, *J. Opt. Soc. Am. A*, 13, 139–146, 1996.

Onodera, R. and Ishii, Y., Two-wavelength interferometry based on a Fourier-transform technique, *Proc. SPIE*, 3749, 430–431, 1999.

Parker, R.J., Surface topography of nonoptical surfaces by oblique projection of fringes from diffraction gratings, *Opt. Acta*, 25, 793–799, 1978.

Patorski, K., Moiré methods in interferometry, *Opt. Lasers Eng.*, 8, 147–170, 1988.

Pirodda, L., Shadow and projection moiré techniques for absolute and relative mapping of surface shapes, *Opt. Eng.*, 21, 640–649, 1982.

Polhemus, C., Two-wavelength interferometry, *Appl. Opt.*, 12, 2071–2078, 1973.

Reid, G.T., Moiré fringes in metrology, *Opt. Lasers Eng.*, 5, 63–93, 1984.

Rayleigh, Lord, *Philos. Mag.*, 11, 196, 1881.

Rodriguez-Vera, R., Kerr, D., and Mendoza-Santoyo, F., Three-dimensional contouring of diffuse objects by Talbot projected fringes, *J. Mod. Opt.*, 38, 1935–1945, 1991.

Reid, G.T., Moiré fringes in metrology, *Opt. Lasers Eng.*, 5, 63–93, 1984.

Rimmer, M.P., Method for evaluating lateral shearing interferometer, *Appl. Opt.*, 13, 623–629, 1974.

Rimmer, M.P. and Wyant, J.C., Evaluation of large aberrations using a lateral shear interferometer having variable shear, *Appl. Opt.*, 14, 142–150, 1975.

Rubinstein, J. and Wolansky, G., Reconstruction of surfaces from ray data, *Opt. Rev.*, 8, 281–283, 2001.

Saunders, J.B., Measurement of wavefronts without a reference standard: the wavefront shearing interferometer, *J. Res. Natl. Bur. Stand.*, 65B, 239, 1961.

Sciammarella, C.A., The moiré method: a review, *Exp. Mech.*, 22, 418–433, 1982.

Selberg, L.A., Interferometer accuracy and precision, *Proc. SPIE*, 749, 8–18, 1987.

Seligson, J.L., Callari, C.A., Greivenkamp, J.E., and Ward, J.W., Stability of lateral-shearing heterodyne Twyman–Green interferometer, *Opt. Eng.*, 23, 353–356, 1984.

Servín, M., Malacara, D., and Marroquín, J.L., Wave-front recovery from two orthogonal sheared interferograms, *Appl. Opt.*, 35, 4343–4348, 1996.

Slomba, A.F. and Figoski, J.W., A coaxial interferometer with low mapping distortion, *Proc. SPIE*, 153, 156–161, 1978.

Stricker, J., Electronic heterodyne readout of fringes in moiré deflectometry, *Opt. Lett.*, 10, 247–249, 1985.

Suganuma, M. and Yoshisawa, T., Three-dimensional shape analysis by use of a projected grating image, *Opt. Eng.*, 30, 1529–1533, 1991.

Takasaki, H., Moiré topography, *Appl. Opt.*, 9, 1467–1472, 1970.

Takasaki, H., Moiré topography, *Appl. Opt.*, 12, 845–850, 1973.

Takeda, M., Fringe formula for projection-type moiré topography, *Opt. Lasers Eng.*, 3, 45–52, 1982.

Takeda, M. and Kobayashi, S., Lateral aberration measurements with a digital Talbot interferometer, *Appl. Opt.*, 23, 1760–1764, 1984.

Talbot, W.H.F., Facts relating to optical science, *Phil. Mag.*, 9, 401, 1836.

Thikonov, A.N., Solution of incorrectly formulated problems and the regularization method, *Sov. Math. Dokl.*, 4, 1035–1038, 1963.

Twyman, F., Correction of optical surfaces, *Astrophys. J.*, 48, 256, 1918.

Vazquez-Montiel, S., Sánchez-Escobar, J.J., and Fuentes, O., Obtaining the phase of an interferogram by use of an evolution strategy, part I, *Appl. Opt.*, 41, 3448–3452, 2002.

Vlad, V., Popa, D., and Apostol, I., Computer moiré deflectometry using the Talbot effect, *Opt. Eng.*, 30, 300–306, 1991.

Wan, D.-S. and Lin, D.-T., Ronchi test and a new phase reduction algorithm, *Appl. Opt.*, 29, 3255–3265, 1990.

Wang, G.-Y. and Ling, X.-P., Accuracy of fringe pattern analysis, *Proc. SPIE*, 1163, 251–257, 1989.

Welsh, B.M., Ellerbroek, B.L., Roggemann, M.C., and Pennington, T.L., Fundamental performance comparison of a Hartmann and a shearing interferometer wave-front sensor, *Appl. Opt.*, 34, 4186–4195, 1995.

Wyant, J.C., Testing aspherics using two-wavelength holography, *Appl. Opt.*, 10, 2113–2118, 1971.

Wyant, J.C., How to extend interferometry for rough-surface tests, *Laser Focus World.*, September, 131–135, 1993.

Wyant, J.C., Oreb, B.F., and Hariharan, P., Testing aspherics using two wavelength holography: use of digital electronic techniques, *Appl. Opt.*, 23, 4020–4023, 1984.

Yang, T.-S. and Oh, J.H., Identification of primary aberrations on a lateral shearing interferogram of optical components using neural network, *Opt. Eng.*, 40, 2771–2779, 2001.

Yatagai, T., Fringe scanning Ronchi test for aspherical surfaces, *Appl. Opt.*, 23, 3676–3679, 1984.

Yatagai, T. and Kanou, T., Aspherical surface testing with shearing interferometer using fringe scanning detection method, *Opt. Eng.*, 23, 357–360, 1984.

Yokoseki, S. and Susuki, T., Shearing interferometer using the grating as the beam splitter, part 1, *Appl. Opt.*, 10, 1575–1580, 1971a.

Yokoseki, S. and Susuki, T., Shearing interferometer using the grating as the beam splitter, part 2, *Appl. Opt.*, 10, 1690–1693, 1971b.

2

Fourier Theory Review

2.1 INTRODUCTION

Fourier theory is an important mathematical tool for the digital processing of interferograms; hence, it is logical to begin this chapter with a review of this theory. Extensive treatments of this theory may be found in many textbooks, such as those by Bracewell (1986) and by Gaskill (1978). The topic of digital processing of images has been also treated in several textbooks — for example, Gonzales and Wintz (1987), Jain (1989), and Pratt (1978).

2.1.1 Complex Functions

Complex functions are very important tools in Fourier theory. Before beginning the study of Fourier theory let us review a brief summary of complex functions. A complex function may be plotted in a *complex plane* by means of a so-called *phasor diagram*, where the real part of the function is plotted on the horizontal axis and the imaginary part on the vertical axis. A complex function may be written as:

$$g(x) = \text{Re}\{g(x)\} + i\,\text{Im}\{g(x)\} \qquad (2.1)$$

where Re(*g*) stands for the real part of *g* and Im(*g*) stands for the imaginary part of *g*.

The phase of this complex number is the angle with respect to the horizontal axis of the line from the origin to the complex function value being plotted. Thus, the phase of any complex function *g*(*x*) may be obtained with:

$$\phi = \tan^{-1}\left[\frac{\text{Im}\{g(x)\}}{\text{Re}\{g(x)\}}\right] \tag{2.2}$$

This phase has a wrapping effect, however, because if both the real and the imaginary parts are negative, the ratio is the same as if both quantities are positive. Thus, this phase is within the limits $0 \le \phi \le \pi$. The *magnitude* of this complex number is defined by:

$$|g(x)| = \left[\left(\text{Re}\{g(x)\}\right)^2 + \left(\text{Im}\{g(x)\}\right)^2\right]^{1/2} \tag{2.3}$$

which is always positive. This complex function may also be written as:

$$g(x) = \text{Am}(g(x))\exp(i\phi) \tag{2.4}$$

where Am(*g*(*x*)) is the *amplitude* of the complex function or, in terms of the magnitude |*g*(*x*)|:

$$g(x) = |g(x)|\exp(i\phi) \tag{2.5}$$

The phase ϕ has a value between 0 and 2π.

To understand the difference between these two representations of the complex function, let us consider the complex function represented in Figure 2.1. In the complex plane in Figure 2.1a, the complex function passes through the origin. Figure 2.1b shows the amplitude and phase vs. position *s* along the function, and Figure 2.1c provides a plot of the magnitude and phase vs. the distance *s*. We can see that when the function passes through the origin of the complex plane, the amplitude and its derivative (slope) as well as the phase are continuous.

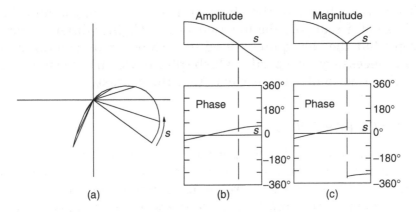

Figure 2.1 (a) Plotting a complex function that passes through the origin in the complex plane, (b) amplitude and phase vs. s, and (c) magnitude and phase vs. s.

On the other hand, we see that neither the derivative of the magnitude nor its corresponding phase is continuous.

Explained another way, let us consider, for example, the real function $g(x) = x$, which is a horizontal line along the axis on the complex plane. Using this expression, it has to be written as $g(x) = |x|$ for $x \geq 0$ and as $g(x) = |x| \exp(\pi)$ for $x \leq 0$. To avoid this discontinuity, both on the derivative of the function and on the phase, we use the amplitude instead of the magnitude, in which case the derivative of the function $g(x)$ and the phase will be continuous for all values of x. This amplitude is the equivalent of the radial coordinate in polar coordinates. A change in the sign of the amplitude is equivalent to a change of π in the phase.

The phase, as plotted in the phasor diagram, of a periodic real function such as the functions $\sin\phi$ and $\cos\phi$, is zero, because the function is real; however, another concept of phase ϕ is associated with real sinusoidal functions. Frequently, we refer to these real functions as *stationary waves*, and their phase in the phasor diagram is zero. On the other hand, on the phase diagram the plot of the function $\exp i\phi = \cos\phi + i \sin\phi$ is a unit circle and its phase may be represented there. For this reason, this function is sometimes called a *traveling wave*.

These two phases — the phase of a complex function and the phase of a real periodic function — are slightly different concepts but they are quite related to each other. In general, it is not necessary to specify which phase we are considering because normally that is clear from the context.

2.2 FOURIER SERIES

A real, infinitely extended periodic function with fundamental frequency f_1 may be decomposed into a sum of real (stationary) sinusoidal functions with frequencies that are multiples of the fundamental, referred to as *harmonics*. Thus, we may write:

$$g(x) = \frac{a_0}{2} + \sum_{n=1}^{\infty} \left[a_n \cos(2\pi n f_1 x) + b_n \sin(2\pi n f_1 x) \right] \quad (2.6)$$

The coefficients a_n and b_n are the amplitudes of each of the sinusoidal components. If the function $g(x)$ is real, these coefficients are also real. Multiplying this expression first, by $\cos(2\pi m f_1 x)$ and then, by $\sin(2\pi m f_1 x)$ and making use of the well-known orthogonality properties for the trigonometric functions we may easily obtain, after integrating for a full period, an analytical expression for the coefficients which may be calculated from $g(x)$ by:

$$a_n = \frac{1}{x_0} \int_{-x_0}^{x_0} g(x) \cos(2\pi n f_1 x) dx \quad (2.7)$$

and

$$b_n = \frac{1}{x_0} \int_{-x_0}^{x_0} g(x) \sin(2\pi n f_1 x) dx \quad (2.8)$$

where the fundamental frequency is equal to twice the inverse of the period length $2x_0$ ($f_1 = 1/2x_0$). We may see that the frequency components have a constant separation equal to the fundamental frequency f_1. If the function is symmetrical (i.e., $g(x) = g(-x)$), then only the coefficients a_n may be different from zero, but, if the function is antisymmetrical

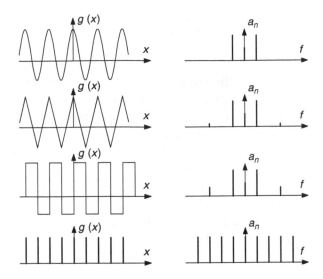

Figure 2.2 Some periodical functions and their spectra.

(i.e., $g(x) = -g(-x)$), then only the coefficients b_n may differ from zero. If the function is asymmetrical, both coefficients a_n and b_n may be different from zero. The coefficients a_n and b_n always correspond to positive frequencies. Figure 2.2 shows some common periodical functions and their Fourier transforms.

Fourier series may also be written in terms of complex functions. The periodic functions just described are represented by a sum of real (stationary) sinusoidal functions. In order to describe complex functions, the coefficients a_n and b_n must be complex. An equivalent expression in terms of complex (traveling) sinusoidal functions $\exp(i2\pi n f_1 x)$ and $\exp(-i2\pi n f_1 x)$ using complex exponential functions instead of real trigonometric functions is:

$$g(x) = \sum_{n=-\infty}^{\infty} c_n e^{i2\pi n f_1 x} \tag{2.9}$$

where the coefficients c_n may be real, imaginary, or complex. These exponential functions are also orthogonal, as are the trigonometric functions. The coefficients can be calculated as:

$$c_n = \int_{-x_0}^{x_0} g(x) e^{i2\pi n f_1 x} dx \tag{2.10}$$

In this case, the coefficients c_n correspond to positive (phase is increasing in the negative direction of x) as well as to negative (phase is increasing in the positive direction of x) frequencies. Thus, the number n may be positive as well as negative. In general, the coefficients c_n are complex. If the function $g(x)$ is symmetrical, the coefficients c_n are real, with $c_n = c_{-n} = 2a_n$. On the other hand, if the function $g(x)$ is antisymmetrical, the coefficients c_n are imaginary, with $c_n = -c_{-n} = -2ib_n$. Table 2.1 shows some periodical functions and their coefficients a_n and b_n.

2.3 FOURIER TRANSFORMS

If the period of the function $g(x)$ is increased, separation of the sinusoidal components decreases. In the limit when the period becomes infinity, the frequency interval among harmonics tends to zero. Any nonperiodical function may be regarded as a periodical function with an infinite period. Thus, a nonperiodical continuous function may be represented by an infinite number of sinusoidal functions, transforming the series in Equation 2.5 into an integral, where the frequency separation f_1 becomes df. This leads us to the concept of the Fourier transform.

Let $g(x)$ be a continuous function of a real variable x. The Fourier transform of $g(x)$ is $G(f)$, defined by:

$$G(f) = \int_{-\infty}^{\infty} g(x) e^{-i2\pi f x} dx \tag{2.11}$$

This Fourier transform function $G(f)$ is also called the *amplitude spectrum* of $g(x)$, and its magnitude is the *Fourier spectrum* of the function $g(x)$. This Fourier transform of $g(x)$ may also be represented by $\mathrm{F}\{g(x)\}$. For example, a perfectly sinusoidal function $g(x)$ without any constant term added has a single frequency component. The spectrum is a pair of Dirac

TABLE 2.1 Some Periodical Functions and Their Coefficients a_n and b_n

Function	Coefficients
Cosinusoidal:	
$g(x) = A + B\cos(2\pi f_1 x)$	$a_0 = 2A$
	$a_1 = B; \quad b_n = 0$
	$a_n = 0; \quad n \geq 2$
Triangular:	
$g(x) = A + B(1 + 4f_1 x); \quad -x_0 \leq x \leq 0$	$a_0 = 2A; \quad b_n = 0$
$g(x) = A + B(1 - 4f_1 x); \quad 0 \leq x \leq x_0$	$a_n = \dfrac{2B}{n^2 \pi}; \quad n \text{ odd}$
	$a_n = 0; \quad n \text{ even}$
Square:	
$g(x) = A - B; \quad -x_0 \leq x \leq 0$	$a_0 = 2A$
$g(x) = A + B; \quad 0 \leq x \leq x_0$	$b_n = \dfrac{2B}{n\pi}; \quad n \text{ odd}$
	$b_n = 0; \quad n \text{ even}$
Comb:	
$g(x) = \displaystyle\sum_{n=-\infty}^{\infty} \delta(x - nx_0)$	$a_0 = \dfrac{\delta(f)}{2}; \quad b_n = 0$
	$a_n = \delta(f - nf_1); \quad n \neq 0$

delta functions located symmetrically with respect to the origin, at its corresponding frequency. Given $G(f)$, the function $g(x)$ may be obtained by its inverse Fourier transform, defined by:

$$g(x) = \int_{-\infty}^{\infty} G(f)e^{i2\pi f x} df \qquad (2.12)$$

We may notice that Equation 2.10 is similar to Equation 2.11 and that Equation 2.12 is similar to Equation 2.9 when

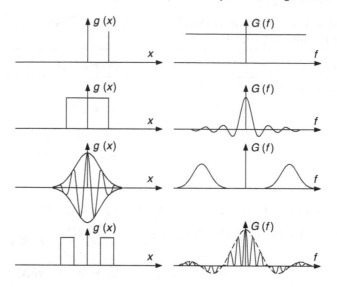

Figure 2.3 Some Fourier transform pairs.

the fundamental frequency tends to zero. Here, x is the space variable, and its domain is referred to as the *space domain*. On the other hand, f is the frequency variable, and its domain is the *frequency* or *Fourier domain*. A *Fourier transform pair* is defined by Equations 2.11 and 2.12. Both functions, $g(x)$ and $G(f)$ may be real or complex. Figure 2.3 and Table 2.2 provide some examples of Fourier transform pairs.

The magnitude $|G(f)|$ as we mentioned before, is called the *Fourier spectrum* of $g(x)$, and the square of this magnitude is the *power spectrum*, sometimes also known as the *spectral density*.

The phase ϕ at the origin $(x = 0)$ of a real cosinusoidal function, $\cos(\omega_S x + \phi)$, is equal to the complex phase at the origin of its spectral component $\exp i(\omega_S x + \phi)$, which in turn is equal to the complex phase of the Fourier transform $[\delta(\omega - \omega_S)\exp i\phi]$ of the cosine function at the frequency $\omega = \omega_S$. An important and useful conclusion is that the phase of the real cosinusoidal Fourier components of a real function is equal to the complex phase of its Fourier transform at the frequency of that component.

TABLE 2.2 Some Fourier Transform Pairs

Space Domain Function	Frequency Domain Function
Dirac delta (impulse) function:	Constant:
$g(x) = \delta(x - x_0)$	$G(f) = Ae^{-i2\pi fx_0}$
Square function:	Sinc function:
$g(x) = A; \quad \|x\| \le a$ $g(x) = 0; \quad \|x\| > a$	$G(f) = 2Aa\dfrac{\sin(2\pi fx_0)}{2\pi fx_0}$
Gaussian modulated wave:	Gaussian function:
$g(x) = A\cos(2\pi f_0 x)e^{-x^2/a^2}$	$G(f) = \dfrac{Aa\sqrt{\pi}}{2}e^{-2\pi a^2(f-f_0)^2/4} +$ $+\dfrac{Aa\sqrt{\pi}}{2}e^{-2\pi a^2(f+f_0)^2/4}$
Pair of square functions:	Sinc modulated wave:
$g(x) = A; \quad b - a \le \|x\| \le b + a$ $g(x) = 0; \quad \|x\| < b - a \wedge \|x\| > b + a$	$G(f)4Aa\cos(2\pi fb)\dfrac{\sin(2\pi fa)}{2\pi fa}$

2.3.1 Parseval Theorem

An important theorem is the Parseval theorem, which may be written as:

$$\int_{-\infty}^{\infty} |g(x)|^2 dx = \int_{-\infty}^{\infty} |G(f)|^2 df \qquad (2.13)$$

This theorem may be described by saying that the total power in the space domain is equal to the total power in the frequency domain.

2.3.2 Central Ordinate Theorem

From Equation 2.11 we can see that

$$\left[\int_{-\infty}^{\infty} g(x)e^{-i2\pi fx}dx\right]_{f=0} = G(0) = \int_{-\infty}^{\infty} g(x)dx \qquad (2.14)$$

Thus, the integral of a function is equal to the central ordinate of the Fourier transform. An immediate consequence is that, because any lateral translation of the function $g(x)$ does not change the area, the central ordinate value also does not change.

2.3.3   Translation Property

Another useful property of the Fourier transform is the translation property, which states that a translation of the input function $g(x)$ changes the phase of the transformed function as follows:

$$\mathrm{F}\{g(x+x_0)\} = G(f)\exp(i2\pi fx_0) \qquad (2.15)$$

or in the frequency domain:

$$G(f+f_0) = \mathrm{F}\{g(x)\exp(i2\pi f_0 x)\} \qquad (2.16)$$

A consequence of this theorem is that the Fourier transform of any function with any kind of symmetry can be made to be real, imaginary, or complex by means of a proper translation of the function $f(x)$.

2.3.4   Derivative Theorem

If $g'(x)$ is the derivative of $g(x)$, then the Fourier transform of this derivative is given by:

$$\int_{-\infty}^{\infty} g'(x)\exp-(2\pi ifx)dx = \int_{-\infty}^{\infty} \lim_{\Delta x \mapsto 0} \frac{g(x+\Delta x) - g(x)}{\Delta x}\exp-(i2\pi fx)dx$$

$$= \lim_{\Delta x \mapsto 0} \frac{\exp(i2\pi f\Delta x)G(f) - G(f)}{\Delta x} \qquad (2.17)$$

$$= i2\pi f\, G(f)$$

or

$$g'(x) = \mathrm{F}^{-1}\{i2\pi fG(f)\} \qquad (2.18)$$

Thus, the Fourier transform of the derivative of function $g(x)$ is equal to the Fourier transform of the function multiplied by $i2\pi f$. Now, using the convolution expression in Equation 2.25, to be described below, we may write:

$$g'(x) = \mathrm{F}^{-1}\{G(f)H(f)\} = g(x) * h(x) \qquad (2.19)$$

with

$$h(x) = \mathrm{F}^{-1}\{i2\pi f\} = \mathrm{F}^{-1}\{H(f)\} \qquad (2.20)$$

This means that the derivative of $g(x)$ may be calculated with the convolution of this function with the function $h(x)$. By taking the inverse Fourier transform, this function $h(x)$ is equal to:

$$h(x) = \frac{2f}{x}\cos(2\pi f_0 x) - \frac{1}{\pi x^2}\sin(2\pi f_0 x)$$

$$= \lim_{f_0 \to \infty} 2f_0 \frac{d}{dx}[\mathrm{sinc}(2\pi f_0 x)] \qquad (2.21)$$

2.3.5 Symmetry Properties of Fourier Transforms

A function $g(x)$ is symmetric or even if $g(x) = g(-x)$, antisymmetric or odd if $g(x) = -g(-x)$, or asymmetric if it is neither symmetric nor antisymmetric. An asymmetric function may always be expressed by the sum of a symmetric function plus an antisymmetric function. A complex function is Hermitian if the real part is symmetrical and the imaginary part is antisymmetrical. For example, the function $\exp(ix)$ is Hermitian. The complex function is anti-Hermitian if the real part is antisymmetrical and the imaginary part symmetrical. These definitions are illustrated in Figure 2.4.

The Fourier transform has many interesting properties, as shown in Table 2.3. The fact that the Fourier transform of a real asymmetrical function is Hermitian is referred to as

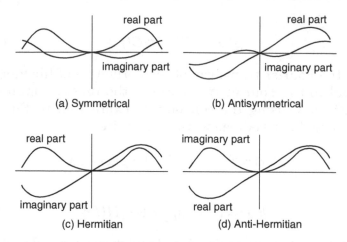

(a) Symmetrical   (b) Antisymmetrical

(c) Hermitian   (d) Anti-Hermitian

Figure 2.4   Possible symmetries of a function.

the *Hermitian property* of the spectrum of real functions. A few more properties of Fourier transforms, derived from their symmetry properties, include:

1. If the function $g(x)$ is complex — of the form $\exp i\phi(x)$, where $\phi(x)$ is positive for all values of x (the sign of the imaginary part is the same as the sign for the real part for all values of x) — then the spectral function $G(f)$ is different from zero only for positive values of f.

2. If the function $g(x)$ is complex — of the form $\exp i\phi(x)$, where $\phi(x)$ is negative for all values of x (the sign of the imaginary part is opposite the sign for the real part for all values of x) — then the spectral function $G(f)$ is different from zero only for negative values of f.

3. It is easy to show that for any complex function $g(x)$:

$$F\{g^*(x)\} = G^*(-f) \qquad (2.22)$$

where the symbol * stands for the complex conjugate.

A particular and important case is when the function $g(x)$ is real and we can write:

TABLE 2.3 Symmetry Properties of Fourier Transforms

$g(x)$		$G(f)$	
Real	Symmetrical	Real	Symmetrical
	Antisymmetrical	Imaginary	Antisymmetrical
	Asymmetrical	Complex	Hermitian
Imaginary	Symmetrical	Imaginary	Symmetrical
	Antisymmetrical	Real	Antisymmetrical
	Asymmetric	Complex	Anti-Hermitian
Complex	Symmetrical	Complex	Symmetrical
	Antisymmetrical	Complex	Antisymmetrical
	Hermitian	Real	Asymmetrical
	Anti-Hermitian	Imaginary	Asymmetrical
	Asymmetrical	Complex	Asymmetrical

$$G(f) = G^*(-f); \quad G(-f) = G^*(f) \tag{2.23}$$

which implies that

$$|G(f)| = |G^*(-f)| \tag{2.24}$$

From this expression, we may conclude that if the function $g(x)$ is real, as in any image to be digitized, the Fourier transform is Hermitian and that the Fourier spectrum (or magnitude) $|G(f)|$ is symmetrical.

2.4 THE CONVOLUTION OF TWO FUNCTIONS

The convolution operation of the two functions $g(x)$ and $h(x)$ is defined by:

$$g(x) * h(x) = \int_{-\infty}^{\infty} g(\alpha) h(x - \alpha) d\alpha \tag{2.25}$$

where the symbol * denotes the convolution operator. It may be seen that the convolution is commutative; that is,

$$g(x) * h(x) = h(x) * g(x) \tag{2.26}$$

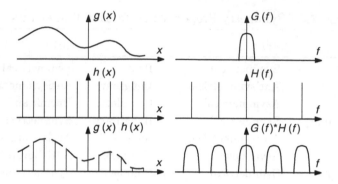

Figure 2.5   Product of a function $g(x)$ by a comb function $h(x)$ and the convolution of their Fourier transforms.

A property of the convolution operation is that the Fourier transform of the product of two functions is equal to the convolution of the Fourier transforms of the two functions:

$$F\{g(x)h(x)\} = G(f) * H(f) \tag{2.27}$$

or

$$F^{-1}\{G(f)H(f)\} = g(x) * h(x) \tag{2.28}$$

and, conversely, the Fourier transform of the convolution of two functions is equal to the product of the Fourier transforms of the two functions:

$$F\{g(x)h(x)\} = G(f) * H(f) \tag{2.29}$$

or

$$F^{-1}\{G(f) * H(f)\} = g(x)h(x) \tag{2.30}$$

Figure 2.5 shows the product of the function $g(x)$ and the comb function $h(x)$, as well as the convolution of the Fourier transforms of these functions. The convolution may be interpreted in several ways, and the following text provides two different models for such interpretation. One of these models is used more frequently in electronics, the other in optics, but they are equivalent.

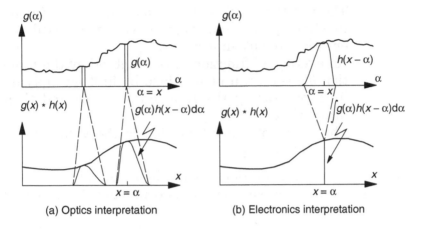

(a) Optics interpretation (b) Electronics interpretation

Figure 2.6 The convolution of two functions.

1. This interpretation of the convolution operation is typically used in optics to study the resolving power of optical instruments. It can be explained by the following four steps, as shown in Figure 2.6a:
   - The α-axis (object) is divided into many extremely narrow intervals of equal width $d\alpha$. The narrow interval at any position α is selected.
   - The function $h(x)$ is placed at the corresponding point $x = \alpha$ in the convolution space (image), without being reversed, to obtain the function $h(x - \alpha)$. The height is then made directly proportional to the value of $g(x)$ by multiplication of the two functions.
   - These two steps are repeated for all narrow intervals in the function space.
   - All of the $g(x)\, h(x - \alpha)d\alpha$ functions in the convolution space are added by integration.
2. The second interpretation is commonly used in electronics to study the signal distortion of electronic amplifiers. In this application, variable x is the time. This approach may be explained as follows (see Figure 2.6b):
   - A value of x is selected in the domain of the convolution (output signal).

- The function $h(\alpha)$ is placed at point $\alpha = x$ in the function space (input signal), with a reversed orientation, to obtain $h(x - \alpha)$.
- An average of function $g(\alpha)$, weighted by the function $h(x - \alpha)$, can be obtained by first multiplying function $g(\alpha)$ by the function $h(x - \alpha)$ and then integrating.
- The result of the integration is the value of the convolution at point x.

A property of the convolution is that the extent of the convolution is equal to the sum of the two function bases being convolved.

2.4.1   Filtering by Convolution

An important application of the convolution operation is low-pass, band-pass, or high-pass filtering of function $g(x)$ by means of a filter function $h(x)$. This filtering property of the convolution operation may be easily understood if we use Equations 2.27 and 2.25 to write:

$$\bar{g}(x) = F^{-1}\{G(x)H(x)\} = \int_{-\infty}^{\infty} g(\alpha)h(x-\alpha)\,d\alpha \qquad (2.31)$$

We see that the filtering or convolution operation is equivalent to multiplying the Fourier transform of the function to be filtered by the Fourier transform of the filtering function and then obtaining the inverse Fourier transform of the product. If the filtering function $h(x)$ has numerous low frequencies and no high frequencies, we have a low-pass filter. On the other hand, if the filtering function $h(x)$ has a large number of high frequencies and no low frequencies, we have a high-pass filter. This convolution process, with the associated low-pass filtering, is illustrated in Figure 2.6.

Let us consider the special case of the convolution of a sinusoidal real function $g(x)$ formed by the sum of a sine and a cosine function with filter function $h(x)$. Then, we obtain the filtered function $\bar{g}(x)$:

$$\bar{g}(x) = \int_{-\infty}^{\infty} (a\sin(2\pi nf\alpha) + b\cos(2\pi nf\alpha))\,h(x-\alpha)\,d\alpha \quad (2.32)$$

This expression, which is a function of x, must have a zero value for all values of x. The value of this function at the origin $(x = 0)$ is:

$$\bar{g}(0) = \int_{-\infty}^{\infty} (a_n\sin(2\pi nf\alpha) + b_n\cos(2\pi nf\alpha))\,h(-\alpha)\,d\alpha \quad (2.33)$$

The real sinusoidal function $g(x)$ with frequency f has two Fourier components, one with frequency f and the other with frequency $-f$. If only the first term (sine) is present in $g(x)$, then the signal is antisymmetric and the two Fourier components have the same magnitudes but opposite signs. In this case, if the signal is filtered with a filter function with symmetrical values at the frequency to be filtered, then we can see that the desired zero value is obtained at the origin but not at all values of x. If only the second term (cosine) is present in $g(x)$, then the signal is symmetrical and the two Fourier components have the same magnitudes and the same signs. In this case, if the signal is filtered with a filter function with antisymmetrical values at the frequency to be filtered, then the correct filtered value of zero is again obtained only at the origin.

In the most general case, when both the sine and cosine functions are present in $g(x)$, the magnitudes and signs of the two Fourier components may be different. Generally, the filtering function must have zero values at both Fourier components.

2.5 THE CROSS-CORRELATION OF TWO FUNCTIONS

The cross-correlation operation of the two functions $g(x)$ and $h(x)$ is similar to the convolution, and it is defined by:

$$g(x) \otimes h(x) = \int_{-\infty}^{\infty} g(\alpha)\,h(x+\alpha)\,d\alpha \quad (2.34)$$

where the symbol $\otimes$ denotes cross-correlation. This operation is not commutative but satisfies the relation:

$$g(x) \otimes h(x) = h(-x) \otimes g(-x) \qquad (2.35)$$

A property of the cross-correlation operation is that the Fourier transform of the product of the two functions is equal to the cross-correlation of the Fourier transforms:

$$F\{g(-x)h(x)\} = G(f) \otimes H(f) \qquad (2.36)$$

and, conversely, the Fourier transform of the cross-correlation is equal to the product of the Fourier transforms:

$$F\{g(x) \otimes h(x)\} = G(-f)H(f) \qquad (2.37)$$

The cross-correlation is related to the convolution by:

$$g(x) \otimes h(x) = g(-x) * h(x) \qquad (2.38)$$

As the convolution operation, the cross-correlation may be used to remove high-frequency Fourier components from a function $g(x)$ by means of a filter function $h(x)$.

2.6 SAMPLING THEOREM

Let us consider a band-limited real function $g(x)$ whose spectrum is $G(f)$. The width, Δf, of this spectrum is equal to the maximum frequency contained in the function. To sample the function $g(x)$ we need to multiply this function by the comb function $h(x)$, for which the spectrum $H(f)$ is also a comb function, as shown in Figure 2.5. The fundamental frequency of the comb function $h(x)$ is defined as the sampling frequency. A direct consequence of the convolution theorem is that the spectrum of this sampled function (a product of the two functions) is the convolution of the two Fourier transforms $G(f)$ and $H(f)$.

In Figure 2.7 we can see that, if the sampling frequency of the function $h(x)$ decreases, the spectral elements in the convolution of the functions $G(f)$ and $H(f)$ get closer to each other. If these spectral elements are completely separated

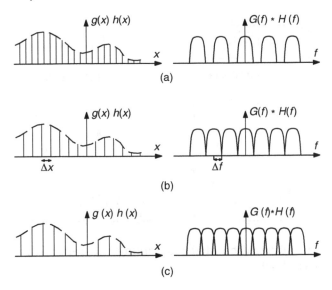

Figure 2.7 Sampling of a function with different sampling frequencies: (a) above the Nyquist limit, (b) just below the Nyquist limit, and (c) below the Nyquist limit.

without any overlapping, the inverse Fourier transform recovers the original function with full detail and frequency content. If the spectral elements overlap each other, as in Figure 2.7c, the process is not reversible. The original function may not be fully recovered after sampling if the spectral elements do overlap or even touch each other; thus, the sampling theorem requirements are violated when the spectral elements are just touching each other, as shown in Figure 2.7b.

The total width ($2\Delta f$) of the base of the spectral elements is smaller than twice the maximum frequency (f_{max}) present at the signal or function being sampled, as defined by its Fourier transform. On the other hand, the frequency separation between the peaks in the Fourier transform of the comb function is equal to the sampling frequency. Hence, the sampling frequency $f_S = 1/\Delta x$ must be greater than half the maximum frequency f_{max} contained in the signal or function to be sampled:

$$f_S > 2f_{max} \tag{2.39}$$

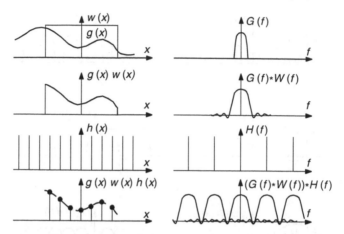

Figure 2.8 Illustration of the sampling theorem with a limiting aperture (window).

This condition is known as the *Whittaker–Shannon sampling theorem*, and the minimum sampling frequency is referred to as the *Nyquist frequency* (Nyquist, 1928). Alternatively, we can say that when a signal has been sampled the maximum frequency contained in this sampled signal is equal to half the sampling frequency. If the spectral elements overlap, recovery of the sampled function is not perfect, and a phenomenon known as *aliasing* occurs.

In this discussion we have assumed that the sampling function $h(x)$ extends from $-\infty$ to $+\infty$ and that the sampled function is band limited. In most practical cases, neither of these assumptions is true. If the sampling extends only from $-x_0$ to x_0, then for the sake of simplicity we may consider that the sampling points — that is, the function $h(x)$ — extend from $-\infty$ to $+\infty$ but that the function to be sampled, $g(x)$, is multiplied by a window function, $w(x)$, as shown in Figure 2.8. Then, by the convolution theorem, the spectrum of the product of these two functions is the convolution of its Fourier transforms. The Fourier transform of the window function is the sinc function, for which the spectrum extends from $-\infty$ to $+\infty$. Thus, the spectrum elements of the windowed sampled function necessarily have some overlap. The important conclusion here is

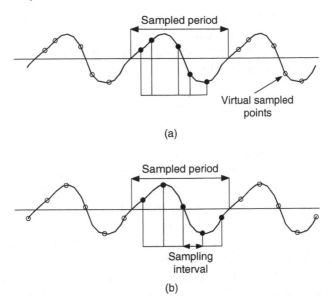

Figure 2.9 Sampling of a periodical function with a finite sampling interval.

that a bounded sampling function (or an interval-limited sampling function) is always imperfect, as perfect recovery of the function is not possible.

2.7 SAMPLING OF A PERIODICAL FUNCTION

In only one important case will limited sampling lead to perfect recovery of the function: when the function is periodic (not necessarily sinusoidal) and band limited (a highest order harmonic frequency must exist), with a fundamental spatial period equal to the length of the total sampling interval. If we assume that the function is periodic and band limited, then it may be represented by a Fourier series with a finite number of terms. Due to the periodicity of the function we may assume that the sampling pattern repeats itself outside the sampling interval, as shown in Figure 2.9. If the sampling points are equally spaced but not uniformly distributed in the interval (Figure 2.9a) and the sampling pattern is repeated, the entire distribution of virtual sampling points (empty points) is not uniform. Suppose, however, that the N sampling points are uniformly and equally spaced (Figure 2.9b) and that phases ϕ_n is given by:

$$\phi_n = \frac{2\pi(n-1)}{N} + \phi_0 \qquad (2.40)$$

where ϕ_0 is the phase at the first sampling point ($n = 1$). The virtual sampling points in the entire infinite interval will be equally distributed, and a sampling in an interval with length equal to the period of the fundamental is enough to obtain full recovery of the function. Of course, we are also assuming that the sampling frequency is greater than twice the maximum frequency contained in the function.

The advantage of extrapolating the function in this manner, outside the sampling interval, is that the sampling may be mathematically considered as extending to the entire interval from $-\infty$ to $+\infty$ and we can be sure that the sampling theorem is strictly satisfied.

An interesting example of a periodical and bandwidth-limited function is a pure sinusoidal function. If we sample a sinusoidal function, the sampling theorem requires a greater sampling frequency (equal is not acceptable) than twice the frequency of the sinusoidal function. Taking two sampling points in the period length makes the sampling frequency equal to twice the frequency of the sampled function. If the sampling interval is much larger than one period, we could sample with a frequency just slightly greater than this required minimum of two points per period; however, if the sampling interval is just one period (as in most phase-shifting algorithms), we need a minimum of three sampling points per period.

Figure 2.10a shows a sinusoidal signal sampled with a frequency (f_S) much higher than twice the frequency (f) of this signal. Figure 2.10b shows the sampling with three points per period. Figure 2.10c shows a smaller sampling frequency that still satisfies the sampling theorem requirements. Figure 2.10d illustrates a sampling frequency equal to two, just outside the sampling theorem requirements; we can see that the

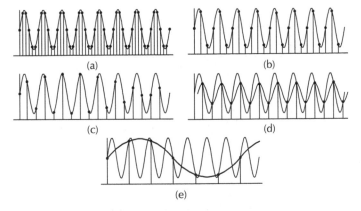

Figure 2.10 Sampling of a periodical function with a finite sampling interval: (a) frequency higher than twice the frequency of the function; (b) three points per period; (c) smaller sampling frequency, satisfying the sampling theorem; (d) sampling frequency equal to two; (e) sampling frequency lower than twice the frequency of the sinusoidal function.

function reconstruction can be achieved in several ways (two of which are illustrated here). Finally, Figure 2.10e shows a sampling frequency less than twice the frequency of the sinusoidal function, with the aliasing effect clearly shown. With aliasing, instead of obtaining a reproduction of the signal with frequency f, a false signal with a frequency of $f_S - f$ and the same phase at the origin as the signal appears. Because the requirements of the sampling theorem were violated, the frequency of this aliased wave is smaller than the signal frequency. Another way to visualize these concepts is by analyzing the same cases in the Fourier space, as shown in Figure 2.11. Each of these spectra corresponds to the same cases in Figure 2.10.

2.7.1 Sampling of a Periodical Function with Interval Averaging

We have studied the sampling of a periodical function using a detector that measures the signal at one value of the phase; however, most real detectors cannot measure the phase at one

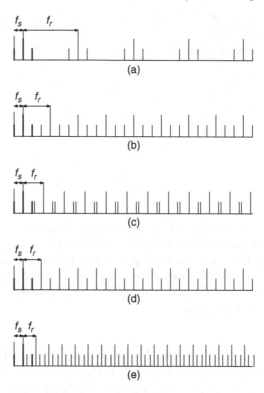

Figure 2.11 Spectra when sampling a periodical function with a finite sampling interval (as in Figure 2.10).

value of the phase but instead take the average value in one small phase interval. This may be the case in space signals as well as in time signals. In the case of a time-varying signal, as in phase-shifting interferometry, the phase may be continually changing while the measurements are being taken; thus, the number being read is the average of the irradiance during the time spent measuring. This method is frequently referred to as *bucket integration*.

In the case of a space-varying signal (such as when digitizing the image of sinusoidal interference fringes with a detector array), the detector may have a significant size compared to the separation between the detector elements. In this case, the measurements are also the average of the signal over the detector extension.

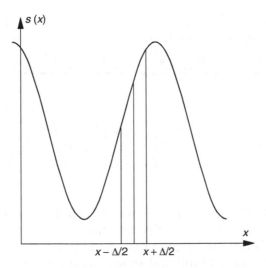

Figure 2.12 Signal averaging when measuring a sinusoidal signal in a phase interval from $-x_0/2$ to $x_0/2$.

Let us consider this signal averaging shown in Figure 2.12, where signal $s(x)$ is measured in an interval centered at x and extending from $x - x_0/2$ to $x + x_0/2$. Then, the average signal on this interval is given by:

$$\bar{s}(x) = \frac{\int_{-x_0/2}^{x_0/2} s(x)\,dx}{x_0} = \frac{\int_{-x_0/2}^{x_0/2} (a + b\cos x)\,dx}{x_0} \qquad (2.41)$$

thus, we obtain:

$$\bar{s}(x) = a + b\,\mathrm{sinc}(x_0/2)\cos x \qquad (2.42)$$

This result tells us that the effect of this signal averaging just reduces the contrast of the fringes with the filtering function sinc $(x_0/2)$. As it is to be expected, for an infinitely small averaging interval $(x_0 = 0)$ there is no reduction in contrast; however, for finite-size intervals, the contrast is reduced. The sinc function has zeros at $x_0 = 2m\pi$, where m is an integer. Thus, the first zero occurs at $x_0 = 2\pi$. If the sampling detectors

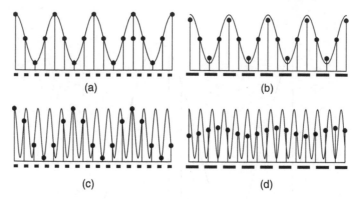

Figure 2.13 Contrast of a detected signal for a finite size of integration: (a) below the Nyquist limit and small integration interval; (b) below the Nyquist limit and large integration interval; (c) above the Nyquist limit and small integration interval, showing aliasing; and (d) below the Nyquist limit and large integration interval, showing reduction and inversion in the contrast.

have a size equal to its separation, so that no space exists between them (as in most practical charge-coupled device [CCD] detectors), this corresponds to half the sampling frequency allowed by the sampling theorem. In other words, when the signal frequency is increased, the Nyquist frequency is reached before the first zero of the contrast. Hence, at these values of x_0, when the averaging interval is a multiple of the wavelength of the signal (spatial or temporal), the contrast is reduced to zero and no signal is detected, but the DC component is detected. For averaging intervals between π and 2π, the contrast is reversed. These contrast changes are illustrated in Figure 2.13.

    When the signal is sampled at equally spaced intervals, there is an upper limit for the size of the averaging interval, when the averaging intervals just touch each other. Then, the averaging interval size is equal to the inverse of the sampling frequency; that is, $x_0 = 1/f_S$. With this detector, at the Nyquist limit (sampling frequency equal to twice the signal frequency) the integration interval is equal to half the period of the signal ($x_0 = \pi$) and the contrast reduction is $2/\pi = 0.6366$. The contrast is zero when the sampling frequency is equal to signal frequency f. In the digitization of images, this frequency-selective contrast reduction (filtering) is sometimes an advantage because it reduces the aliasing effect; however, in some interferometric applications, as described later in this book, the aliasing effect may be useful.

2.8 FAST FOURIER TRANSFORM

The numerical computation of a Fourier transform takes an extremely long time even for modern powerful computers. Several algorithms were designed by various authors early in the twentieth century, but they were not widely known. It was not until the work of J. W. Tukey and J. W. Cooley in the mid-1960s that one algorithm gained wide acceptance — the *fast Fourier transform* (FFT). Tukey devised an algorithm to compute the Fourier transform in a relatively short time by eliminating unnecessary calculations, and Cooley developed the required programming. Their work was not published, but it aroused enough interest that several researchers began using the algorithm. When R. L. Garwin was in need of this algorithm, he went to see Cooley to ask about his work. Cooley told him that he had not published it because he considered the algorithm to be quite elementary. Eventually, however, the Tukey–Cooley algorithm was, indeed, published and later came to be known as the fast Fourier transform. Explanations of this method can be found in numerous publications today (e.g., Brigham, 1974; Hayes, 1992). Code for programs using C language (Press et al., 1988) or Basic (Hayes, 1992) can also be found in the literature.

Because the Fourier transform is carried out by a computer, the function to be transformed must be sampled by means of a comb sampling function so the integral becomes a discrete sum. The *discrete Fourier transform* (DFT) pair is defined by:

$$G_k = \sum_{l=0}^{N-1} g_l e^{-i2\pi kl/N} \tag{2.43}$$

and

$$g_l = \frac{1}{N} \sum_{k=0}^{N-1} G_k e^{i2\pi kl/N} \tag{2.44}$$

The first expression may be written as:

$$G_k = \sum_{l=0}^{N-1} g_l W^{kl} \tag{2.45}$$

where

$$W = e^{-i2\pi/N} \tag{2.46}$$

We can see that the sampled function (g_l) to be Fourier transformed has a bounded domain contained in an array of N points. The Fourier transform (G_k) is calculated at another array of N points in the frequency space; thus, N multiplications must be carried out for each G_k. To calculate the entire Fourier transform set of numbers (G_k), N^2 multiplications are necessary; this is a huge number because the number of points N is generally quite a large number. This operation can be written in matrix notation (Iisuka, 1987) as:

$$
\begin{pmatrix} G_0 \\ G_1 \\ G_2 \\ \cdot \\ \cdot \end{pmatrix} =
\begin{pmatrix} W^0 & W^0 & W^0 & W^0 & \cdot & W^0 \\ W^0 & W^1 & W^2 & W^3 & \cdot & W^{N-1} \\ W^0 & W^2 & W^4 & W^6 & \cdot & W^{2(N-1)} \\ \cdot & \cdot & \cdot & \cdot & \cdot & \cdot \\ W^0 & W^{(N-1)} & \cdot & \cdot & \cdot & W^{(N-1)(N-1)} \end{pmatrix}
\begin{pmatrix} g_0 \\ g_1 \\ g_2 \\ \cdot \\ \cdot \end{pmatrix} \tag{2.47}
$$

Hence, the discrete Fourier transform may be regarded as a linear transform. If N points are to be sampled, then the transform has N points. The elements of the matrix are shown in Equation 2.47. This matrix has some interesting characteristics that may be used to reduce the time required for the matrix multiplication. Remember that the fast Fourier transform is simply an algorithm that reduces the number of operations, and note that the matrix in Equation 2.47 involves $N \times N$ multiplications and $N \times (N-1)$ additions.

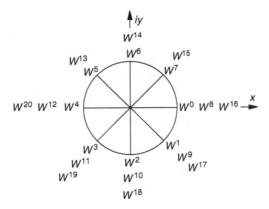

Figure 2.14 Phasor diagram representing values of W^{kl} for $N = 8$.

The values of W^{kl} may be represented in a phasor diagram in the complex plane as shown in Figure 2.14. All values fall in a unit circle, and we may see that we have only N different values. We may also notice that values at opposite sides of the circle differ only in their sign. Points symmetrically placed with respect to the x-axis have the same real part, and their imaginary parts differ only in sign. Points symmetrically placed with respect to the y-axis have the same imaginary parts and their real parts differ only in sign.

The key property that allows us to reduce the number of numerical operations when calculating this Fourier transform is that a discrete Fourier transform of length N can be expressed as the sum of two discrete Fourier transforms of length $N/2$. One of the two transforms is formed by the odd points and the other by the even points, as follows:

$$
\begin{aligned}
G_k &= \sum_{l=0}^{N-1} g_l e^{-i2\pi kl/N} \\
&= \sum_{l=0}^{N/2-1} g_{2l} e^{-i2\pi k(2l)/N} + \sum_{l=0}^{N/2-1} g_{2l+1} e^{-i2\pi k(2l+1)/N} \\
&= \sum_{l=0}^{N/2-1} g_{2l} e^{-i2\pi kl/(N/2)} + W^k \sum_{l=0}^{N/2-1} g_{2l+1} e^{-i2\pi kl/(N/2)}
\end{aligned}
\tag{2.48}
$$

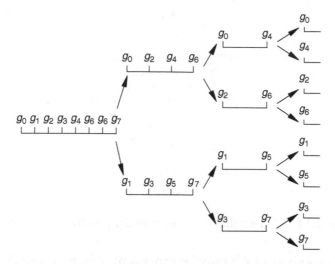

Figure 2.15   Fragmentation of a digitized signal with eight values in two parts in a successive manner to obtain eight single values.

where we have assumed N is even. This property is referred to as the *Danielson–Lanczos lemma*. Thus, we can also write:

$$G_k = G_k^{even} + W^k G_k^{odd} \qquad (2.49)$$

where each of these two Fourier transforms is of length $N/2$. So, now we have two linear transforms which are half the size of the original, and the total number of multiplications has been reduced to one fourth. This fragmentation procedure is known as *decimation*. After decimation, the smaller Fourier transforms are calculated and then a recombination of the results is performed to obtain the desired Fourier transform.

The wonderful thing is that this principle can be used recursively. It is only necessary that the number of points in each step is even. It is ideal when the total number of points is $N = 2^M$, where M is an integer. The result is that the number of multiplications has been reduced from N^2 to $N \log_2 N$.

As an example of how to find the fast Fourier transform, let us consider Figure 2.15, where we have a signal with eight digitized values (g_i). These values are divided into two groups, one with the odd sampled values and another with the even

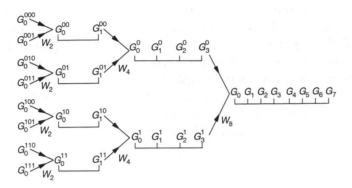

Figure 2.16 Calculation of the fast Fourier transform by grouping.

sampled values. Each of these groups is again divided into two, and so on, until we have eight groups with a single value.

The next step is to find the Fourier transform of each of the single values, which is trivial. Then, with the procedure described earlier, the Fourier transforms of larger groups of signal values are calculated until we obtain the desired Fourier transform at eight frequency values, as shown in Figure 2.16.

Figure 2.17 illustrates the positions of the sampling points in the space domain as well as the calculated points in the

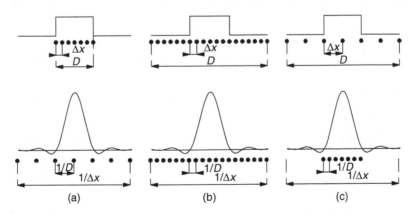

Figure 2.17 Location of sampling points in a transformed function and location of calculated points in the frequency space.

frequency domain for a rectangular function. It is interesting to note that if the sampling points are located only over the top of the rectangular function the calculated points do not have enough resolution to give the shape of the expected sinc function. A solution is to sample a larger space in the function domain with additional points, with zero values on both sides of the aperture. The details of the fast Fourier transform algorithms have been described by several authors — for example, Hayes (1992), Iisuka (1987), and Press et al. (1988).

REFERENCES

Bracewell, R.N., *The Fourier Transform and Its Applications*, 2nd ed., McGraw-Hill, New York, 1986.

Brigham, E.O., *The Fast Fourier Transform*, Prentice Hall, Englewood Cliffs, NJ, 1974.

Cooley, J.W. and Tukey, J.W., An algorithm for the machine calculation of complex Fourier series, *Math of Computation*, 19(90), 297–301, 1965.

Gaskill, J.D., *Linear Systems, Fourier Transforms, and Optics*, John Wiley & Sons, New York, 1978.

Gonzales, R.C. and Wintz, P., *Digital Image Processing*, 2nd ed., Addison-Wesley, Reading, MA, 1987.

Hayes, J., Fast Fourier transforms and their applications, in *Applied Optics and Optical Engineering*, Vol. XI, Wyant, J. C. and Shannon, R.R., Eds., Academic Press, New York, 1992.

Iisuka, K., *Optical Engineering*, 2nd ed., Springer-Verlag, Berlin, 1987.

Jain, A.K., *Fundamentals of Digital Image Processing*, Prentice Hall, Englewood Cliffs, NJ, 1989.

Nyquist, H., Certain topics in telegraph transmission theory, *AIEE. Trans.*, 47, 817–844, 1928.

Pratt, W.K., *Digital Image Processing*, John Wiley & Sons, New York, 1978.

Press, W.H., Flannery, B.P., Teukolsky, S.A., and Vetterling, W.T., *Numerical Recipes in C*, Cambridge University Press, Cambridge, U.K., 1988.

3

Digital Image Processing

3.1 INTRODUCTION

Digital image processing is a very important field by itself that has been treated in many textbooks (e.g., Pratt, 1978; Gonzales and Wintz, 1987; Jain, 1989) and chapter reviews (e.g., Morimoto, 1993). To digitize an image, it is separated into an array of small image elements called *pixels*. Each of these pixels has a different color and irradiance (gray level). The larger the number of pixels in an image, the greater the definition and sharpness of this image. Interferograms, as described in Chapter 1, may be analyzed using digital processing techniques. In this case, however, color information is not necessary, as is clearly illustrated in the images of the interferogram in Figure 3.1. The great advantage of digital image processing is that the image may be improved or analyzed using many different techniques, and these techniques may also be applied to the analysis of interferograms, as has been described by various authors for more than 20 years (see, for example, Kreis and Kreitlow, 1979). When digitizing an image, the gray levels (irradiance) are digitized and transformed into numbers by computer. These numbers are represented internally by binary numbers that have only ones and zeros and are called *bits*. A quantity written as a series of 8 bits is a *byte*. A quantity may be represented by 1, 2, or even

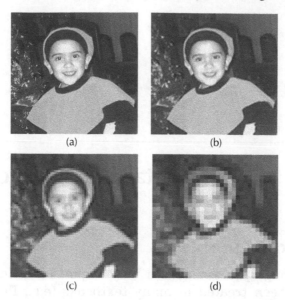

Figure 3.1   Digitized images with different pixel separations: (a) 256 × 256 pixels, (b) 128 × 128 pixels, (c) 64 × 64 pixels, and (d) 32 × 32 pixels.

3 bytes; thus, the total number of bits used to digitize an image represents the number of possible gray levels that may be used to represent the luminance level, as shown in Table 3.1.

3.2 HISTOGRAM AND GRAY-SCALE TRANSFORMATIONS

One of the most important properties of a digitized image is the relative population of gray levels. We may plot this information in a diagram where the x-axis represents the luminance of the pixel and the y-axis represents the number of pixels in the image with that value of the gray level. Such a diagram is referred to as a *histogram*. A gray level has a discrete quantized value that is determined by the number of bits representing it; thus, a histogram is not a continuous curve but a set of vertical line segments. Figure 3.2 shows a digitized

TABLE **3.1** Gray Levels According to the
Number of Bits

Number of Unsigned Bytes	Number of Bits	Number of Gray Levels
1	8	256
2	16	65,536

image and its histogram. The contrast of an image is reflected by its histogram, as shown in Figure 3.3, which uses the same image as in Figure 3.2 but with a much greater contrast, which can be seen in the histogram. It is interesting to note that the image of a digitized interferogram with perfectly sinusoidal fringes, without noise, has more dark and clear pixels than

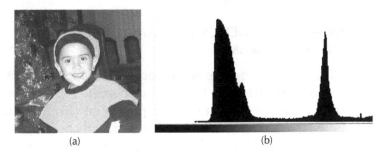

(a) (b)

Figure 3.2 (a) Digitized image; (b) its histogram.

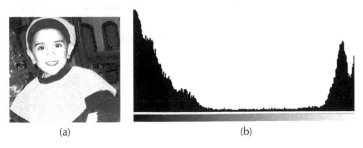

(a) (b)

Figure 3.3 (a) Increased contrast in a digitized image; (b) its modified histogram.

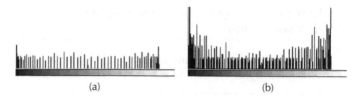

Figure 3.4 Histograms for two digitized interferograms: (a) with 20 pixels per fringe period, and (b) with 200 pixels per fringe period.

pixels with intermediate gray levels. A histogram has two maxima. The first corresponds to the gray level at the top of the clear fringes, and the second corresponds to the gray levels at the top of the dark fringes. If noise is present, the height of the first peak in the histogram is reduced. The aspect of the histogram depends on the number of pixels per fringe period, as shown in Figure 3.4.

3.3 SPACE AND FREQUENCY DOMAIN OF INTERFEROGRAMS

When digitizing or sampling an interferogram, the selection of the sampling points is extremely important, as indicated by a study on the effect of sampling points on the frequency domain by Womack (1983, 1984), who described the properties of this frequency domain of interferograms. Let us consider the interferogram of an aberrated wavefront with a large tilt (linear carrier), as shown in Figure 3.5a. Let us assume that the irradiance signal in this interferogram can be written as:

$$s(x, y) = a(x, y) + b(x, y) + \cos k[x \sin \theta - W(x, y)] \qquad (3.1)$$

This irradiance has been represented here by $s(x,y)$ instead of $I(x,y)$ so the Fourier transform becomes $S(f_x, f_y)$. The variable θ represents the tilt angle introducing the linear carrier, k is equal to $2\pi/\lambda$, and $W(x,y)$ is the wavefront deformation. We may also write this irradiance as:

$$s(x, y) = a(x, y) + b(x, y) \cos[2\pi f_0 x - k W(x, y)] \qquad (3.2)$$

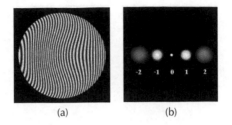

(a)   (b)

Figure 3.5 Interferogram and its frequency domain space image: (a) interferogram with tilt, and (b) spectrum. The second-order lobes are due to nonlinearities.

where f_0 is the spatial frequency introduced in the interferogram by the tilt. This expression may also be written as:

$$s(x,y) = a(x,y)\left[1+\frac{b(x,y)}{a(x,y)}\cos(2\pi f_0 x - kW(x,y))\right]$$

$$= a(x,y)\left[1+v(x,y)\cos(2\pi f_0 x - kW(x,y))\right]$$

(3.3)

where $v(x,y)$ is the fringe visibility. If we define the function $u(x,y)$, sometimes referred to as the *complex fringe visibility*, as:

$$u(x,y) = v(x,y)e^{-ikW(x,y)}$$

(3.4)

we obtain:

$$s(x,y) = a(x,y) + 0.5a(x,y)\left[\begin{array}{c}u(x,y)\exp(i2\pi f_0 x) \\ + u^*(x,y)\exp(-i2\pi f_0 x)\end{array}\right]$$

(3.5)

Then, using the convolution theorem and Equation 2.15, the Fourier transform of this function, $s(x,y)$, is:

$$S(f_x,f_y) = A(f_x,f_y) +$$

$$+ 0.5A(f_x,f_y) * \left[U(f_x - f_0,f_y) + U^*(-f_x - f_0,f_y)\right]$$

(3.6)

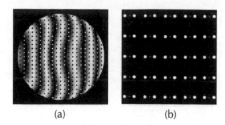

(a)   (b)

Figure 3.6   (a) Interferogram sampled with a rectangular array of points; (b) spectrum.

where the symbol * represents the convolution operation. Thus, this spectrum would be concentrated in three regions (lobes): a small one at the origin and two larger ones centered at f_0 and $-f_0$, with a radius equal to the frequency cut-off of $U(f)$.

The image in the frequency domain space (spectrum) of an interferogram without any tilt is a bright spot at the center in the frequency space. If tilt is added to the interferogram (Figure 3.5a), the spectrum splits in several orders (Figure 3.5b), but the three brightest components are the 0, −1, and +1 orders. The central bright peak is at the center, and the two smaller lobes correspond to the two first orders (−1, +1) on each side. If the tilt is increased, the separation between these lobes also increases.

If the interferogram is sampled with a rectangular array of points (Figure 3.6a), the spectrum looks like that shown in Figure 3.6b. To separate the different orders of diffraction and to be able to reconstruct the image of the interferogram, according to the sampling theorem the sampling point must have a spatial frequency higher than twice the maximum spatial frequency present in the interferogram.

3.4   DIGITAL PROCESSING OF IMAGES

In a digital image or interferogram, some types of spatial characteristics must sometimes be detected, reinforced, or eliminated, and some kinds of noise may have to be removed

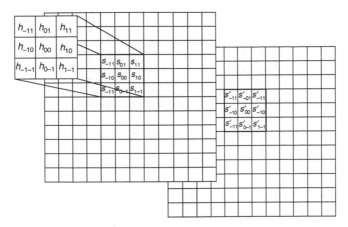

Figure 3.7 Image processing with window or mask.

using some type of averaging or spatial filtering. This section discusses the general procedures used in the digital processing of images, which is performed by means of a *window* or *mask* (also known as a *kernel*), represented by a matrix of $N \times N$ pixels. This mask is placed over the image to be processed, and each h_{nm} value in the mask is multiplied by the corresponding pixels with signal (gray level) s_{nm} in the image (Figure 3.7), and all these products are added to obtain the result s'_{00} as follows:

$$s'_{00} = \sum_{n=-M}^{M} \sum_{m=-M}^{M} h_{nm} s_{nm} \qquad (3.7)$$

where $M = (N - 1)/2$. The result (s') of this operation is used to define a new number to be inserted in the new processed image at the pixel corresponding to the center of the window. After this, the mask is moved to the next pixel in the image being processed, and the preceding operations are repeated for the new position. In this manner, the entire image is scanned. Following is a discussion of the primary image operations that can be performed.

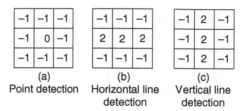

(a)   (b)   (c)
Point detection   Horizontal line   Vertical line
                   detection   detection

Figure 3.8   Masks for point and line detection.

3.4.1   Point and Line Detection

The simplest operation is detection of a pixel with a gray level that varies too greatly from the surrounding pixels. To do so, we take the average signal of eight pixels surrounding another one. If this average is very different from the signal at the pixel being considered, such a point has been identified. This operation may be carried out with the mask shown in Figure 3.8a. A point is said to be detected if:

$$s' > T \tag{3.8}$$

where T is a predefined threshold value. If s' is close to zero, the pixel is not different from the surrounding ones. A more complex operation is detection of a line. To detect a horizontal line, the average of the pixels above and below the line being considered are compared with the average of the pixels on the line. This is accomplished using the masks shown in Figures 3.8b and 3.8c. The criterion in Equation 3.8 is also used to determine if such a line has been detected.

3.4.2   Derivative and Laplacian Operators

The partial derivatives of the signal values with respect to x and y may be estimated if we calculate the difference in the signal values to two adjacent pixels:

$$\frac{\partial s}{\partial x} \propto s_{10} - s_{00} \tag{3.9}$$

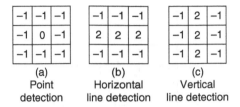

−1	−1	−1
−1	0	−1
−1	−1	−1

(a)
Point
detection

−1	−1	−1
2	2	2
−1	−1	−1

(b)
Horizontal
line detection

−1	2	−1
−1	2	−1
−1	2	−1

(c)
Vertical
line detection

Figure 3.9 Masks for evaluating derivatives: (a) Robert's operator, (b) Prewitt operator, and (c) Sobel operator.

The 2 × 2 Roberts masks (Figure 3.9a) can be used to evaluate the partial derivatives in the diagonal directions; however, an important problem with using these operators is their large susceptibility to noise so they are seldom used. The 3 × 3 Prewitt operators (Figure 3.9b) evaluate the partial derivatives in the x and y directions; they are less sensitive to noise than the Roberts operators because they take the average of three pixels in a line to evaluate these derivatives. The 3 × 3 Sobel operators (Figure 3.9c) also evaluate the partial derivatives in the x and y directions but they give more weight to the central points.

The Laplacian of a function s is given by:

$$\nabla^2 s = \frac{\partial^2 s}{\partial x^2} + \frac{\partial^2 s}{\partial y^2} \tag{3.10}$$

The value of the Laplacian is directly proportional to the average of the curvatures of function s in the directions x and y; this operator also is quite sensitive to noise. The 3 × 3 Laplacian operator is shown in Figure 3.10, and Figure 3.11 illustrates an interferogram processed with some of these operators.

3.4.3 Spatial Filtering by Convolution Masks

A filtering mask represents the filtering function $h(x,y)$ with a matrix of $N \times N$ pixels. As we have seen before in Chapter 2, a function may be filtered by convolving the function with a filter function. The Fourier transform of the filter function

0	−1	0
−1	4	−1
0	−1	0

Figure 3.10   Laplacian operator.

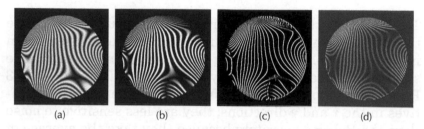

(a)   (b)   (c)   (d)

Figure 3.11   An interferogram processed by various operators: (a) original interferogram, (b) processed with a horizontal Sobel operator, (c) result after four passes with horizontal Sobel operator, and (d) after processing with the Laplacian.

is referred to as the *frequency response function* of the filter. The filtering function with a mask with $N \times N$ pixels may be written as:

$$h(x,y) = \sum_{n=-M}^{M} \sum_{m=-M}^{M} h_{nm}\delta(x - n\alpha, y - m\alpha) \qquad (3.11)$$

where $M = (N - 1)/2$. The Fourier transform (or frequency response) of this filter is:

$$H(f_x, f_y) = \sum_{n=-M}^{M} \sum_{m=-M}^{M} h_{nm} \exp{-i2\pi\alpha(nf_x + mf_y)} \qquad (3.12)$$

where α is the separation between two consecutive pixels; hence, we may write the sampling frequency as $f_S = 1/\alpha$.

The kernel or mask may be of any size $N \times N$. The larger the size, the greater the control over the functional form of the filter. This size must be decided based on the spatial frequencies in the image to be filtered, but a small 3×3 size is the most common. The mask may be asymmetrical or symmetrical. A symmetrical mask has a real Fourier transform and is thus referred to as a *zero phase mask*. In this case, we have $h_{-11} = h_{-1-1} = h_{1-1} = h_{11}$, $h_{-10} = h_{10}$, and $h_{0-1} = h_{01}$. Thus, in this particular case, we may write:

$$H(f_x, f_y) = h_{00} + 2h_{10}\cos\left(2\pi\frac{f_x}{fs}\right) + 2h_{01}\cos\left(2\pi\frac{f_y}{fs}\right) +$$

$$+4h_{11}\cos\left(2\pi\frac{f_x}{fs}\right)\cos\left(2\pi\frac{f_y}{fs}\right) \tag{3.13}$$

As pointed out before, when sampling a digital image it is assumed that it is band limited and that the conditions of the sampling theorem are not violated; hence, the maximum values that f_x and f_y may have are equal to half the sampling frequency. This filter function along the x-axis is:

$$H(f_x, 0) = h_{00} + 2h_{01} + 2(h_{10} + 2h_{11})\cos\left(2\pi\frac{f_x}{fs}\right) \tag{3.14}$$

The coefficients h_{nm} are frequently normalized so the filter frequency response at zero frequencies, $H(0,0)$, is equal to 1 in order to preserve the DC level of the image. In this case, we have:

$$H(0,0) = h_{00} + 2h_{10} + 2h_{01} + 4h_{11} = 1 \tag{3.15}$$

that is, the sum of all elements in the kernel should be equal to one. In some other kernels (for example, in the Laplacian), this sum of coefficients is made equal to zero to eliminate the DC level of the image. Examples of some common filtering masks are illustrated in Figure 3.12, and the frequency responses for some of these filters are shown in Figure 3.13. The frequency responses are plotted only up to the highest

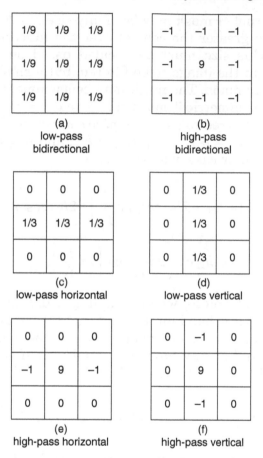

Figure 3.12   Some typical 3 × 3 kernels used to filter images.

frequency in the image, which is half the sampling frequency. For some of these filters, the response at some frequencies may become negative, so the contrast is reversed for these frequency components.

The main application of the low-pass filters is to reduce the noise level in an image. The low-pass kernel shown in Figure 3.12a is quite effective in reducing Gaussian noise, which affects the entire image randomly and seriously degrades its quality. The frequency response of this filter is shown in Figure 3.13a. We can see that the first zero of this

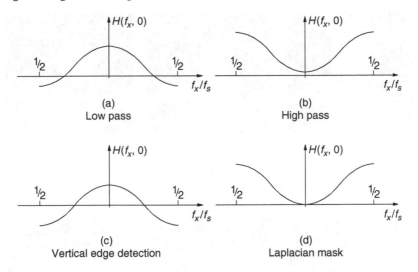

Figure 3.13 Frequency responses of some 3 × 3 kernels used to filter images.

filter is at approximately 0.31 of the sampling frequency. In other words, the period of the first zero is at 3.2 times the pixel separation, which is approximately the full mask size (3 pixels). A low-pass filter with its first zero at a lower spatial frequency requires a larger mask; thus, a rule of thumb is that the period of the first zero is about the mask size required.

Applying a low-pass filter reduces not only the noise but also the high-frequency content of the image. Another common consequence is that the image contrast is also reduced. The filter may be applied to the image several times to reduce the noise even more, but always at the expense of reducing the image sharpness. This is not the only type of noise that can affect an image, as *shot* or *binary noise* can affect isolated pixels having maximum brightness. This noise does not in general degrade the image definition, but it does produce the appearance of speckles. In such cases, the low-pass filter reduces the image definition without suppressing the binary noise.

A much better filter for reducing binary noise is the so-called *median filter*, which reduces binary noise without reducing the image definition. In the median filter, the value

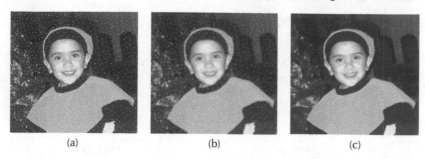

(a)   (b)   (c)

Figure 3.14 An image (a) with binary noise, (b) filtered with a low-pass filter, and (c) filtered with a median filter.

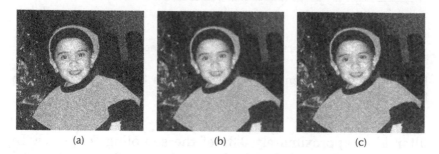

(a)   (b)   (c)

Figure 3.15 An image (a) with Gaussian noise, (b) filtered with a low-pass filter, and (c) filtered with a median filter.

to be inserted at the center of the kernel is not the average value of the surrounding pixels; instead, the median value of these pixels is taken. The median value is obtained by sorting the surrounding pixels in order of decreasing or increasing value, then the value of the pixel at the center is taken. If the kernel side is odd, as in the 3×3 example just considered, the number of pixels around the central one is even. In this case, the median is the average of the two pixels in the middle, after sorting. It is interesting to note that the median filter performs very poorly with Gaussian noise. Figures 3.14 and 3.15 show images with Gaussian and binary noise, respectively, and their filtered versions using these two noise filters. A high-pass filter is shown in Figure 3.12b and its frequency response in Figure 3.13b, and an example of filtering with this filter is provided in Figure 3.16.

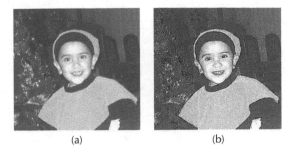

(a) (b)

Figure 3.16 (a) An image and (b) its filtered version using a high-pass filter.

3.4.4 Edge Detection

It is possible to detect fringe edges by means of a derivative, as shown in Figure 3.17, where the location of the edge is defined by the points with maximum slopes. At the maximum slope locations, the second derivative is zero, as shown in the same figure. We have seen in Chapter 2 that the derivative of a function may be found by convolving it with a filtering function for which the Fourier transform is linear with the frequency. This is possible only for a large mask; however, as we have already seen, a good approximation may be obtained with some 3×3 masks, in which case the edges can de detected by calculating the partial derivatives in order to obtain the gradient, defined by a vector with the following two components:

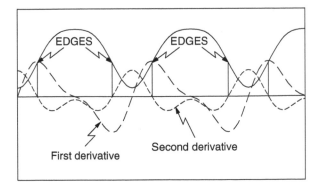

Figure 3.17 Edge detection with first and second derivatives.

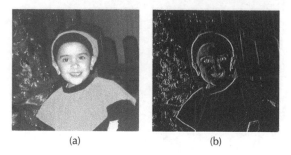

(a)   (b)

Figure 3.18 (a) An image and (b) its filtered version using an edge-detection filter.

$$\nabla s = \left(\frac{\partial s}{\partial x}, \frac{\partial s}{\partial y} \right) \tag{3.16}$$

The edges are located where the gradient has a maximum value, with an orientation perpendicular to the gradient. The Laplacian is not often used for edge detection due to its large sensitivity to noise; however, it can be useful when determining which side of the edge is the dark or clear zone. Figure 3.18 shows an example of edge detection.

3.4.5 Smoothing by Regularizing Filters

We have seen how we can use small convolution matrices to filter images. In fringe analysis, we often need to apply a low-pass filter to a fringe pattern that has a finite extension. This finite extension may be due to the pupil of the optical instrument under analysis. The main drawback of using low-pass convolution filters is that at the edges of the fringe pattern the fringe pattern is mixed with the illumination background. In other words, cross talk occurs at the fringe boundary between the background illumination and the fringe pattern which causes problems for phase detection near the boundary. The phase distortion at the edge introduced by a convolution filter may be very important when testing, for example, a large telescope mirror.

A filtering method that alleviates this cross-talk problem uses the so-called regularized filters (Marroquin, 1993). These filters are obtained as minimizers of quadratic cost functionals. The basic principle behind those filters is to assume that neighboring pixels of the filtered image must have similar values while the processed value still resembles the raw image data; that is, large changes among neighboring pixels are penalized. A merit function (U) may be defined as:

$$U = \sum_{i,j} \begin{bmatrix} \left(s'_{i,j} - s_{i,j}\right)^2 m_{i,j} + \eta_x\left(s'_{i,j} - s'_{i-1,j}\right)^2 m_{i,j}m_{i-1,j} \\ +\eta_y\left(s'_{i,j} - s'_{i,j-1}\right)^2 m_{i,j}m_{i,j-1} \end{bmatrix} \quad (3.17)$$

where the field signal ($s_{i,j}$) is the image being filtered and $s'_{i,j}$ is the filtered field signal. The mask field ($m_{i,j}$) is equal to one in the region of valid image data and zero otherwise. The first term in the quadratic merit function defined by this expression is fidelity to the observed term. The constants η_x and η_y penalize large gray-level changes of the filtered field signals ($s'_{i,j}$) in the i and j directions, respectively. We need to specify a mask field ($m_{i,j}$) over the image being filtered by setting on the valid region a value $m_{i,j} = 1$ and on the background a value $m_{i,j} = 0$. This field mask, therefore, represents the region where we want to filter the field $s_{i,j}$ to obtain a filtered field ($s'_{i,j}$). The filtered field, then, will be the one that minimizes the above cost functional for each pixel. This field may be found by deriving the cost functional (U) with reference to the filtered field ($s'_{i,j}$) and making this derivative equal to zero; that is,

$$\frac{\partial U}{\partial s'_{i,j}} = \left(s'_{i,j} - s_{i,j}\right)m_{i,j} + \eta_x \begin{bmatrix} \left(s'_{i,j} - s'_{i-1,j}\right)m_{i,j}m_{i-1,j} \\ -\left(s'_{i+1,j} - s'_{i,j}\right)m_{i+1,j}m_{i,j} \end{bmatrix} +$$

$$+ \eta_y \begin{bmatrix} \left(s'_{i,j} - s'_{i,j-1}\right)m_{i,j}m_{i,j-1} \\ -\left(s'_{i,j} - s'_{i,j-1}\right)m_{i,j}m_{i,j-1} \end{bmatrix} \quad (3.18)$$

$$= 0$$

This expression represents a linear set of simultaneous equations that must be solved for the $s'_{i,j}$ field. One simple iterative method that can be used to solve Equation 3.18, thus minimizing the merit function, utilizes gradient descent:

$$s'^{k+1}_{i,j} = s'^{k}_{i,j} - \tau \frac{\partial U}{\partial s'_{i,j}} \qquad (3.19)$$

where τ is a damping parameter. Coding this equation into a computer is very simple, but this is not a very efficient method. We may instead use the conjugate gradient.

The Fourier method can also be used to analyze this kind of filter. The Fourier method of analyzing these filters assumes that the region of valid image data is very large; that is, the indicating mask field $(m_{i,j})$ is equal to one over the entire (i,j) plane. With this in mind, Equation 3.18 may be rewritten as:

$$\frac{\partial U}{\partial s'_{i,j}} = s'_{i,j} - s_{i,j} + \eta_x \left[-s'_{i-1,j} + 2s'_{i,j} - s'_{i+1,j} \right] +$$

$$+ \eta_y \left[-s'_{i,j-1} + 2s'_{i,j} - s'_{i,j+1} \right] \qquad (3.20)$$

Taking the Fourier transform on both sides of Equation 3.20, we may obtain the frequency response of the system as:

$$H(\omega) = \frac{F\{s'_{i,j}\}}{F\{s_{i,j}\}} = \frac{1}{1 + 2\eta_x \left[1 - \cos(\omega_x) \right] + 2\eta_y \left[1 - \cos(\omega_y) \right]} \qquad (3.21)$$

This transfer function represents a low-pass filter with bandwidth controlled by the parameters constants η_x and η_y.

3.5 SOME USEFUL SPATIAL FILTERS

We will now describe some of the filters most commonly used in interferogram analysis and their associated properties.

3.5.1 Square Window Filter

One common filter function is a square function, with width x_0 and defined by:

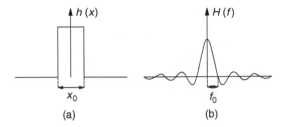

Figure 3.19 (a) One-dimensional square filter and (b) its spectrum.

$$h(x) = 1.0 \quad \text{for } |x| < \frac{x_0}{2} \tag{3.22}$$
$$ = 0 \quad \text{elsewhere}$$

The spectrum of this filter (Figure 3.19a) is the sinc function (Figure 3.19b) given by:

$$H(f) = \frac{\sin(\pi f x_0)}{\pi f x_0} = \text{sinc}(\pi f x_0) \tag{3.23}$$

The first zero of the spatial frequency is for the frequency f_0 given by:

$$f_0 = \frac{1}{x_0} \tag{3.24}$$

This filter is equivalent to averaging the irradiance over all pixels in a window 1 pixel high by N pixels wide. This width is selected so that the row of N pixels just covers the window width (x_0) defined by the desired low-pass cutting point (f_0) for the spatial frequency. In other words, the length of the filtering window should be equal to the period of the signal to be filtered out. The height of the first secondary (negative) lobe is equal to 0.2172 times the height of the main lobe (central peak); hence, the amplitude of this secondary maximum is 7.63 decibels (dB) below the central peak. We may also use a window with a sinc profile, in which case the spectrum would be a square function.

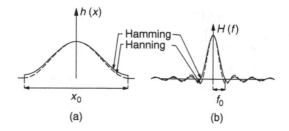

Figure 3.20 (a) Hamming and Hanning filters and (b) their Fourier transforms.

3.5.2 Hamming and Hanning Window Filters

The square filter just described is not the ideal because it leaves some high frequencies unfiltered due to the secondary maxima in the spectrum of the sinc function. A better filtering function is the Hamming function, defined by:

$$h(x) = 0.54 + 0.46\cos\frac{2\pi x}{x_0} \quad \text{for } |x| < \frac{x_0}{2}$$
$$= 0 \qquad\qquad\qquad\qquad \text{elsewhere} \tag{3.25}$$

This function and its spectrum are illustrated in Figure 3.20. The Fourier transform of this filter is given by:

$$H(f) = 1.08\,\text{sinc}(\pi f x_0) + 0.23\,\text{sinc}(\pi f x_0 + \pi) +$$
$$+ 0.23\,\text{sinc}(\pi f x_0 - \pi) \tag{3.26}$$

The first zero for the spatial frequency of this filter is:

$$f_0 = \frac{1}{2x_0} \tag{3.27}$$

The height of the first secondary lobe (negative) is equal to 0.0063 times the height of the main lobe, or 22 dB down, which is a much lower value than for the square filter.

The Hanning filter is very similar to the Hamming filter and is defined by:

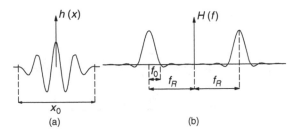

Figure 3.21 (a) Cosinusoidal window filter and (b) its spectrum.

$$h(x) = 0.5\left(1 + \cos\frac{2\pi x}{x_0}\right) \quad \text{for } |x| < \frac{x_0}{2}$$
$$= 0 \qquad\qquad\qquad \text{elsewhere} \tag{3.28}$$

This function and its spectrum are illustrated in Figure 3.20. The Fourier transform of this filter is given by:

$$H(f) = 1.00\,\text{sinc}(\pi f x_0) + 0.25\,\text{sinc}(\pi f x_0 + \pi) +$$
$$+ 0.25\,\text{sinc}(\pi f x_0 - \pi) \tag{3.29}$$

The difference between the Hamming and Hanning filters is the relative height of the secondary lobes with respect to the main lobe and the main lobe widths.

3.5.3 Cosinusoidal and Sinusoidal Window Filters

These are not low-pass but band-pass filters. The cosinusoidal filter may be expressed as the product of a Hamming filter and a cosinusoidal function (Figure 3.21):

$$h(x) = \left(0.54 + 0.46\cos\frac{2\pi x}{x_0}\right)\cos(2\pi f_R) \quad \text{for } |x| < \frac{x_0}{2}$$
$$= 0 \qquad\qquad\qquad\qquad\qquad \text{elsewhere} \tag{3.30}$$

The half-width of each band is the same as in the Hamming filter, and their separation from the origin is equal to f_R. The disadvantage of this filter is that it has two symmetrical pass

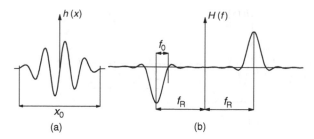

Figure 3.22   Sinusoidal window filter and its spectrum.

bands; hence, one of the sidebands cannot be isolated. The solution is to complement its use with a sinusoidal filter, defined by:

$$h(x) = \left(0.54 + 0.46\cos\frac{2\pi x}{x_0}\right)\sin(2\pi f_R) \quad \text{for } |x| < \frac{x_0}{2}$$
$$= 0 \qquad\qquad\qquad\qquad\qquad\qquad \text{elsewhere}$$
(3.31)

This filter has a spectrum as shown in Figure 3.22, where we can see that the two pass bands now have opposite signs. Any of the sidebands may be isolated by using a combination of both filters. The combination of these two filters is known as a *quadrature filter*.

3.6   EXTRAPOLATION OF FRINGES OUTSIDE OF THE PUPIL

In order to avoid some errors in phase detection, as suggested by Roddier and Roddier (1987), the Gerchberg (1974) method may be used to extrapolate the fringes in interferograms with a large tilt (spatial carrier) outside the pupil boundary. Let us assume that the irradiance signal in the interferogram with a large spatial carrier can be written as:

$$s(x, y) = p(x, y)a(x, y)\left[1 + v(x, y)\cos(2\pi f_0 \cdot x - kW(x, y))\right] \quad (3.32)$$

where $p(x,y)$ is the domain on which the interferogram extends, as follows:

$$p(x,y) = 1; \quad \text{inside the pupil}$$
$$p(x,y) = 0; \quad \text{outside the pupil} \tag{3.33}$$

Now, we can define the continuum as the interferogram irradiance when there are no fringes which is equal to $a(x,y)$. This continuum may be measured by several different procedures, as described by Roddier and Roddier (1987). If we divide the irradiance by the continuum and subtract the pupil domain function, we obtain:

$$g(x,y) = \frac{s(x,y)}{a(x,y)} - p(x,y) \tag{3.34}$$
$$= p(x,y)v(x,y)\cos(2\pi f_0 \cdot x - kW(x,y))$$

If we use the complex fringe visibility, $u(x,y)$, as defined in Equation 3.4, we obtain:

$$g(x,y) = \frac{p(x,y)}{2}\begin{bmatrix} u(x,y)\exp(i2\pi f_0 x) \\ + u^*(x,y)\exp(-i2\pi f_0 x) \end{bmatrix} \tag{3.35}$$

The Fourier transform of function $g(x,y)$, using the convolution theorem in Equation 2.30, is:

$$G(f_x, f_y) = 0.5P(f_x, f_y) * \left[U(f_x - f_0, f_y) + U^*(-f_x - f_0, -f_y) \right] \tag{3.36}$$

Thus, if the interferogram has no pupil boundaries, this spectrum would be concentrated in two circles with radii equal to the frequency cut-off of $U(f)$ centered at f_0 and $-f_0$. Due to the circular boundary of the pupil, these circles increase in size as the pupil size decreases. Extrapolation of the fringes is easily achieved if the size of these two spots is reduced by cutting them around and then taking the inverse Fourier transform. This cut, however, distorts the fringes a little. The original fringe pattern inside the pupil area is recovered by inserting it back into the extrapolated fringe pattern. This process is repeated iteratively several times. This algorithm to extrapolate the fringes outside of the boundary of the pupil

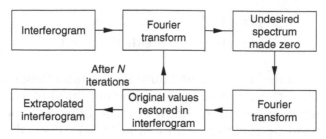

Figure 3.23 Algorithm used to extrapolate the fringes in an interferogram.

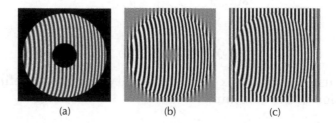

Figure 3.24 (a) Interferogram and its extrapolated interferogram using Gerchberg method and filtering with a Gaussian filter; (b) after 10 passes; (c) after 60 passes.

is illustrated in Figure 3.23, and Figure 3.24 provides an example of fringe extrapolation using this method. If the interferogram has no noise and the interferogram boundary is well defined, this algorithm works quite well, producing clean and continuous fringes. An improved version of this algorithm for use when some noise is present was proposed by Kani and Dainty (1988).

3.7 LIGHT DETECTORS USED TO DIGITIZE IMAGES

Modern instrumentation to digitize images is of many different types and is rapidly evolving and changing, and a description of these instruments is bound to be obsolete in a relatively short time; nevertheless, a brief overview may be useful for

Figure 3.25   Television charge-coupled devices (CCDs).

people beginning to work in the field of interferogram analysis. Microcomputer systems for the acquisition and processing of interferogram video images can have many different configurations, one of which was described by Oreb et al. (1982).

3.7.1   Image Detectors and Television Cameras

Image detectors vary, depending on several factors such as wavelength, resolution, or price. For example, Stahl and Koliopoulos (1987) reported the use of pyroelectric vidicons to detect interferograms produced with infrared light. Prettyjohns (1984) described the use of charge-coupled device (CCD) arrays. A television camera is one of the most commonly used image detectors for digitizing interferograms (Hariharan, 1985). The most important characteristic of such an application is the resolving power.

The typical image detector is a charge-coupled device, illustrated in Figure 3.25 and described extensively in the scientific literature (e.g., Tredwell, 1995). Among the many different television systems are the National Television Systems Committee (NTSC) and the Electronics Industries Association (EIA) systems, which are used in the United States, Canada, Mexico, and Japan. The phase alternating line (PAL) system is used in Germany, the United Kingdom, and parts of Europe, South America, Asia, and Africa. The *Sequential Couleur à Mémoire* (SECAM) system is used in France, Eastern Europe, and Russia. Table 3.2 shows the typical image resolutions for these three systems.

The image is formed by a series of horizontal lines. A complete scan of an image is called a *frame*. Frequently, to avoid flickering, the odd-numbered lines are scanned first and

TABLE 3.2   Image Resolution in Vertical Lines for the Main
Television Systems

Resolution	System			
	NTSC	EIA	PAL	SECAM
Vertical	340	340	400	400
Horizontal	330	360	390	470

then the even-numbered lines, in an alternating manner (Figure 3.26). The set of all odd-numbered lines is the *odd field*, and the set of all even-numbered lines is the *even field*. This manner of scanning is referred to as *interlaced scanning*. The total number of lines per frame is 525 in the NTSC system. In interlaced scanning, each of the two alternating fields has 263.5 lines.

Not all lines in the frame contribute to the image. Approximately 41 lines are blanked out because they are either retraced lines or are at the extreme top or bottom of the frame. Subtracting these lines from the total number in the entire frame, we are left with about 484 visible lines. The aspect ratio of a standard television image is 4:3 (1.33:1); however, broadcast television images have an aspect ratio of 1.56:1, which is based on an unofficial standard for professional digital television equipment (Figure 3.27).

The main characteristics of the two main television systems, NTSC and PAL, are provided in Table 3.3. The vertical resolution depends on the number of scanning lines, and a line covers a row of pixels on the CCD, as illustrated in Figure 3.28; hence, a CCD array must have 485 pixels or more in the

Figure 3.26   Interlaced lines in a television frame.

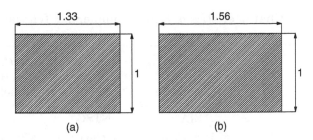

(a)   (b)

Figure 3.27 Aspect ratios in a television frame: (a) standard television image; and (b) broadcast television image.

TABLE 3.3 Characteristics of NTSC and PAL Systems

	NTSC	PAL
Field rate	60 Hz	50 Hz
Number of lines	525	625
Number of active lines	480	576
Time per line	63.49 μs	64 μs
Video bandwidth	4.5 MHz	5.5 MHz

vertical direction. The maximum vertical resolution, then, is 486 television lines. The signals from each row (image line) in the CCD detector are transformed into an analog signal. The horizontal detail (i.e., the number of image elements in the horizontal line) is defined by the bandwidth of the television signal, which is approximately 4.0 MHz, but it may vary, as shown in Table 3.3. If the horizontal resolution is equal to the

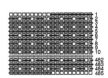

Figure 3.28 Scanning the image from a CCD detector in a television camera. Continuous odd-numbered lines show the first field, while dotted even-numbered lines show the second field.

TABLE 3.4　Characteristics of Some Commercial Television
Cameras

Specifications	Monochrome	Color	Color (High Resolution)
Signal format	EIA	NTSC	NTSC
Horizontal resolution	570 television lines	330 television lines	470 television lines
Picture elements	768 H × 494 V	510 H × 492 V	768 H × 494 V
Sensing area (Hmm × Vmm)	6.2 × 4.6	6.2 × 4.6	6.3 × 4.7
Interlaced	Optional	Yes	Yes

vertical resolution, we say that the horizontal resolution is equal to 484 television lines; however, because the aspect ratio is equal to 4:3, the horizontal resolution is equivalent to having $(484 \times 4)/3 = 645$ lines. The horizontal resolution specified in television lines is variable, depending on the number of pixels on the CCD. The frequency bandwidth in the electronics of a camera is constructed to fit the horizontal resolution of the CCD detector; thus, the horizontal resolution may be higher than the vertical resolution. Table 3.4 shows the resolution characteristics for some commercial television cameras.

In color television cameras, dichroic red–green–blue (RGB) color filters are built on each element of the CCD array. Because each element contains only one of these colors, the effective resolution in a color camera is lower than that of a black-and-white camera. Some expensive cameras use three CCD detectors to improve the image characteristics.

Television cameras for scientific applications may utilize systems different from NTSC or any other commercial systems, and their resolution may generally be higher. Television cameras are either analog or digital. Analog cameras work in a manner similar to NTSC cameras, but they may have more scanning lines and a larger bandwidth to increase their resolution. Digital cameras, on the other hand, do not transform

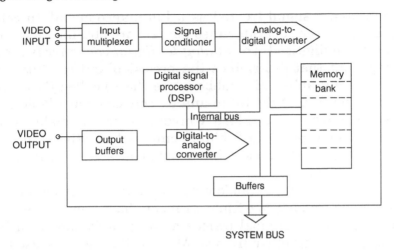

Figure 3.29 Block diagram of a typical frame grabber.

the signals from each row in the CCD detector into analog signals; instead, the signal from each element (pixel) in the detector is directly read and transmitted to the receiver or computer.

3.7.2 Frame Grabbers

When an analog camera is used to sample the image to be digitized, an electronic circuit has to be used to convert the analog signals from each line in the image into digital signals for each pixel image. This analog-to-digital converter is referred to as a *frame grabber*. Frame grabbers are usually located inside the computer, although some models are external modules that connect to a computer port. A typical frame grabber has one or more of the following components (Figure 3.29).

The *input multiplexer* selects from several available inputs, some with different specifications (RGB, composite video, S-video), into a single input channel. The *signal conditioner* adjusts the input signal to a level compatible with the analog-to-digital converter. For monochrome frame grabbers, the chroma signal is removed to avoid having the chrominance signal treated as a luminance signal. In color grabbers, three separate video signals are obtained for each color to be digitized. The *analog-to-digital converter* is a key component that determines the precision and resolution of the entire grabber. All grabbers use the so-called *flash converter*, the fastest digital-to-analog converter available and the most expensive. Flash converters are available with lower resolution (6 to 8 bits), compared to other kinds of converters, as their most important characteristic is speed of conversion.

Image memory is random-access memory used for storing a digitized frame. Some frame grabbers have enough memory to store several original frames as well as frames resulting from processing other frames. Most of the memory used in frame grabbers is double-port memory, which allows simultaneous reading and writing at different memory locations. The data can be written while being displayed. Color and high-resolution grabbers require a large amount of memory. Some grabbers include a *digital signal processor* (DSP) to perform dedicated high-speed calculations. In other cases, the grabber is connected to an external array or a high-speed processor board. A *digital-to-analog converter* translates the digital image to an analog signal for display. The rate at which the data are converted defines the output format. By selecting a window from the original data and by adjusting the reading rate, a grabber may be used for format conversion.

The least expensive grabbers usually work at standard television rates. Some more expensive handle nonstandard rates, including slow-scan, line-scan, high-resolution, or custom-defined formats. Grabbers are available commercially for several computer architectures, such as PC bus, EISA, VMEbus, and microVAX, among others. The software to be used determines the selection of a frame grabber, as does hardware compatibility. Many grabbers are sold with bundled software (e.g., drivers, demos), and a variety of image processing software is widely available.

REFERENCES

Gerchberg, R.W., Super-resolution through error energy reduction, *Opt. Acta*, 21, 709–720, 1974.

Gonzales, R.C. and Wintz, P., *Digital Image Processing*, 2nd ed., Addison–Wesley, Reading, MA, 1987.

Hariharan, P., Quasi-heterodyne hologram interferometry, *Opt. Eng.*, 24, 632–638, 1985.

Jain, A.K., *Fundamentals of Digital Image Processing*, Prentice Hall, Englewood Cliffs, NJ, 1989.

Kani, L.M. and Dainty, J.C., Super-resolution using the Gerchberg algorithm, *Opt. Commun.*, 68, 11–15, 1988.

Kreis, T. and Kreitlow, H., Quantitative evaluation of holographic interference patterns under image processing aspects, *Proc. SPIE*, 210, 196–202, 1979.

Kuan, D.T., Sawchuk, A.A., Strand T.C., and Chavel, P., Adaptive restoration of images with speckle, *Proc. SPIE*, 359, 28–38, 1982.

Marroquin, J.L., Deterministic interactive particle models for image processing and computer graphics, *Comput. Vision Graphics Image Process.*, 55, 408–417, 1993.

Morimoto, Y., Digital image processing, in *Handbook of Experimental Mechanics*, Kobayashi, A. S., Ed., VHC Publishers, New York, 1993.

Oreb, B.F., Brown, N., and Hariharan, P., Microcomputer system for acquisition and processing of video data, *Rev. Sci. Instrum.*, 53, 697–699, 1982.

Pratt, W.K., *Digital Image Processing*, John Wiley & Sons, New York, 1978.

Prettyjohns, K.N., Charge-coupled device image acquisition for digital phase measurement interferometry, *Opt. Eng.*, 23, 371–378, 1984.

Roddier, C. and Roddier, F., Interferogram analysis using Fourier transform techniques, *Appl. Opt.*, 26, 1668–1673, 1987.

Stahl, H.P. and Koliopoulos, C.L., Interferometric phase measurement using pyroelectric vidicons, *Appl. Opt.*, 26, 1127–1136, 1987.

Tredwell, T.J., Visible array detectors, in *Handbook of Optics*, 2nd ed., Vol. I, Bass, M., Ed., Optical Society of America, Washington, D.C., 1995.

Womack, K.H., A frequency domain description of interferogram analysis, *Proc. SPIE*, 429, 166–173, 1983.

Womack, K.H., Frequency domain description of interferogram analysis, *Opt. Eng.*, 23, 396–400, 1984.

4

Fringe Contouring and Polynomial Fitting

4.1 FRINGE DETECTION USING MANUAL DIGITIZERS

If a large tilt is introduced in a Twyman–Green type interferometer of a perfectly flat wavefront interfering with a reference flat wavefront, the fringes will look straight, parallel, and equidistant. If the wavefront under analysis is not flat, the fringes are curved, not straight. These fringes are called *equal-thickness fringes* because they represent the locus of the points with constant wavefront separation. The wavefront deformations may be easily estimated from a visual examination of their deviation from straightness. If the maximum deviation of a fringe from its ideal straight shape is Δx and the average separation between the fringes is equal to s, then its wavefront deviation (in wavelengths) from flat is equal to $\Delta x/s$.

This visual method gives us a precision that greatly depends on the skills of the person making the measurements. In the best case, we can probably approximate $\lambda/20$; norms have been established for defining and classifying visually detected errors (Boutellier and Zumbrunn, 1986). Even image quality can be determined from manual measurements in an interferogram (Platt et al., 1978). Some measuring devices were proposed to aid in this fringe measurement (Dyson, 1963; Dew, 1964; Zanoni, 1978), and this procedure is still used in many manufacturing facilities, which use test plates as references.

The simplest interferometric quantitative analysis method involves visually identifying and then tracking fringes in an interferogram. In this method, a photograph of the interferogram is taken and then a digitizing tablet is used to enter into the computer the x,y coordinates of some selected points on the interferogram located on the peak of the fringes. In contrast, Kingslake (1926–1927) computed the primary aberration coefficients by measuring a few points on the fringe peaks in an interferogram.

Alternatively, to avoid the need for a photograph, the image of an interferogram can be captured with a television camera and displayed on a computer screen, where the peaks of the fringes can be manually sampled (Augustyn et al., 1978; Augustyn, 1979a,b). When the image is digitized with a television camera, mechanical vibrations may introduce errors, but some methods are available to reduce these errors (Crescentini and Fiocco, 1988; Crescentini, 1988).

For manual sampling, the fringes are assigned consecutive numbers that increase by one from one fringe to the next. This number is the interference order number (m). A tilt that is large enough to eliminate closed fringes presents no problem. Every time a point on top of a fringe is selected, the x and y coordinates are read by the graphic tablet or computer and an order number (n) is assigned. This number is entered by the computer operator each time a new fringe is beginning to be measured. The wavefront deformation, $W(x,y)$, at the sampled points on top of the fringes is:

$$W(x, y) = m\lambda \qquad (4.1)$$

The value of n may differ from the real number m by a constant quantity at all measurements, but this is not important. It is more important to know in which direction the number m must increase; otherwise, the sign of the wavefront deformations will be undetermined. It is impossible to determine in which direction the fringe order number increases

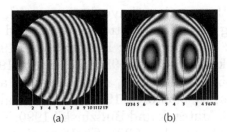

Figure 4.1 Sampling fringe positions at some points and assigning order numbers in an interferogram: (a) open fringes, and (b) closed-loop fringes.

from a single picture of the interference pattern, unless the sign of any of the component aberrations is known. For example, it would be sufficient if the sign of the tilt is known. This sign has to be determined when adjusting the interferometer to take the interferogram picture.

If some of the fringes form closed loops, the order number assignment is a little more difficult but not impossible if carefully done (Figure 4.1). Many systems have been developed to perform semiautomatic analyses of fixed interferograms of pictures or in real time (Jones and Kadakia, 1968; Augustyn, 1979a,b; Moore, 1979; Womack et al., 1979; Cline et al., 1982; Trolinger, 1985; Truax and Selberg, 1986/87; Truax, 1986; Vrooman and Maas; 1989). Reviews on the problems associated with the automatic analysis of fringes have been published by several authors (e.g., Reid, 1986/87, 1988; Choudry, 1987).

4.2 FRINGE TRACKING AND FRINGE SKELETONIZING

The next stage in the automation process is detecting the fringes, assigning order numbers by reading the interferogram image with a two-dimensional light detector or television camera, and computer analysis of the image. The objective here is to locate the fringe maxima or minima by searching with algorithms based on line tracking, threshold comparison, or adaptive binarization. Automatic location of the fringe maxima has been available since the end of the 1970s (e.g., Hot and Durou,

1979). When the maxima have been located, a subsequent fringe thinning or skeletonization is performed (Tichenor and Madsen, 1978; Schluter, 1980; Becker et al., 1982; Yatagai et al., 1982b; Nakadate et al., 1983; Robinson, 1983a,b; Becker and Yung, 1985; Button et al., 1985; Osten et al., 1987; Eich-horn and Osten, 1988; Gillies, 1988; Hunter et al., 1989a,b; Liu and Yang, 1989; Matczak and Budzinski, 1990; Yan et al., 1992; Huang, 1993; He et al., 1999). Skeletonizing is based on a search of local irradiance peaks by segmentation algorithms based on adaptive thresholds, gradient operators, piecewise approximations, thinning procedures, or spatial frequency filtering. The result is a skeleton of the interferogram formed by lines one pixel wide.

Servin et al. (1990) described a technique they refer to as *rubber band* to find the shape of a fringe. The method is based on a set of points linked together in a way similar to a rubber band that attracts these points to a local maximum of the fringe.

Before sampling the fringes it is useful to add a tilt to the interferogram. This tilt straightens the fringes and reduces the fringe spacing, making it more uniform. Another benefit of the tilt is that it makes fringe measurement and order identification easier. Wide spacing between fringes increases accuracy when locating the top of the fringe. On the other hand, a large number of fringes increases the number of fringes that must be sampled and hence the amount of measured information, so it is desirable to determine an optimum intermediate tilt. For the case of digital sampling, Macy (1983) and Hatsuzawa (1985) used a two-dimensional light detector array to determine that the optimum fringe spacing is that which produces a fringe separation of about four pixels.

The fringe analysis procedure can be summarized as follows (Reid, 1986/87, 1988):

1. Spatial filtering of the image
2. Identification of fringe maxima
3. Assignment of order number to fringes
4. Interpolation of results between fringes.

The next few sections examine these steps in some detail.

4.2.1   Spatial Filtering of the Image

Spatial filtering is used to reduce the noise. This noise reduction can be performed in several different ways (Varman and Wykes, 1982). If the spatial frequency of the noise is higher than that of the fringes, low-pass filtering is appropriate. When the spatial frequency of the noise is much lower than that of the fringes (for example, due to an uneven illumination), high-pass filtering can improve the fringe contrast. A more difficult situation arises when the spatial frequency of the noise is similar to that of the fringes. Sometimes, the noise is fixed to the aperture (for example, due to diffracting particles in the interferometer components); in this case, we can take a second interferogram after moving the fringes and changing the optical path difference (OPD) by $\lambda/2$, so the two interferograms are complementary (i.e., a dark fringe in one pattern corresponds to a clear fringe in the other) (Kreis and Kreitlow, 1983). If we subtract one fringe pattern from the other, the fixed noise will be greatly reduced.

4.2.2   Identification of Fringe Maxima

Skeletonizing techniques detect the fringe peaks on the entire area of the digitized interferogram. Many different methods may be used to detect the fringe peaks. Schemm and Vest (1983) reduced the noise and located the fringe peaks using nonlinear regression analysis with a least-squares fit of the irradiance measurements in a small region to a sinusoid function. Snyder (1980) plotted the fringe profiles in a direction perpendicular to the fringes by first smoothing and reducing the data using an adaptive digital filter that located the symmetry points of the fringe pattern. Yi et al. (2002) used a least-squares fitting to find the maxima of the fringes. Mastin and Ghiglia (1985) skeletonized fringe patterns by using the fast Fourier transform and then locating the dominant spatial frequency in the vicinity of each fringe and also by using a set of logical transformations in the neighborhood of a fringe peak. Zero crossing algorithms have also been used (Gasvik, 1989).

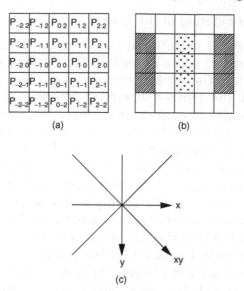

Figure 4.2 Yatagai matrix to find fringe maxima (see text).

These peaks can also be detected using a matrix of 5×5 pixels (Figure 4.2), as proposed by Yatagai et al. (1982b). Assume that the matrix in Figure 4.2a is centered on top of a vertical fringe. Then, the average values of the irradiance in the shaded pixels in Figure 4.2b will be smaller than the average values of the irradiance in the pixels with dots. The same principle can be applied to horizontal fringes and inclined fringes (Figure 4.2c). Thus, the conditions for detecting a fringe maxima are:

$$P_{00} + P_{0-1} + P_{01} = P_{-21} + P_{-20} + P_{-2-1} \tag{4.2}$$

and

$$P_{00} + P_{0-1} + P_{01} = P_{21} + P_{20} + P_{2-1} \tag{4.3}$$

in the x direction;

$$P_{00} + P_{-10} + P_{10} = P_{-1-2} + P_{0-2} + P_{1-2} \tag{4.4}$$

and

$$P_{00} + P_{-10} + P_{10} = P_{-12} + P_{02} + P_{12} \tag{4.5}$$

in the y direction;

$$P_{00} + P_{-1-1} + P_{11} = P_{-22} + P_{-21} + P_{-12} \qquad (4.6)$$

and

$$P_{00} + P_{-1-1} + P_{11} = P_{2-2} + P_{2-1} + P_{1-2} \qquad (4.7)$$

in the x,y direction;

$$P_{00} + P_{-11} + P_{1-1} = P_{22} + P_{21} + P_{12} \qquad (4.8)$$

and

$$P_{00} + P_{-11} + P_{1-1} = P_{-2-2} + P_{-2-1} + P_{-1-2} \qquad (4.9)$$

in the $-x,y$ direction.

When at least two of these conditions are satisfied, the point is assumed to be on top of a fringe. Figure 4.3 shows an example of fringe skeletonizing using this method. Yu et al. (1994) showed that, if the interferogram illumination has a strong modulation (for example, if a large-aperture Gaussian beam is used), the central peak of the fringes shifts laterally a small amount. This shift is greater where a larger slope of the interferogram illumination exists. The extracted skeletons may contain many disconnections, so the next step is to localize these and make some corrections. Many sophisticated methods have been devised to perform this operation (Becker et al., 1982). For simple interferograms with low noise and good contrast, the matrix operators described in Chapter 3 can be used.

4.2.3 Assignment of Order Number to Fringes

The assignment of order numbers to the fringes is an extremely important step. A mistake in just one of the fringes can lead to significant errors when calculating the wavefront deformation. This step can be made quite simple if a large amount of tilt is introduced to eliminate closed fringes (Hovanesian and Hung, 1990). In this case, the order number increases monotonically from one fringe to the next. Sometimes, however, when such a large tilt is not possible or practical, we can use two interferograms taken with different colors or with slightly different optical path differences (Livnat et al., 1980). Such an approach is equivalent to methods used in optical shops where

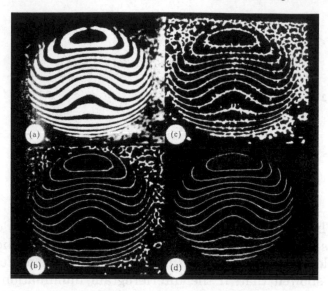

Figure 4.3 Skeletonizing and thinning of interferometric fringes: (a) original interferogram, (b) result after detecting peaks in one direction, (c) result after detecting peaks in two orthogonal directions, and (d) thinned skeletons with noise outside of pupil being removed. (Adapted from Yatagai, T., in *Interferogram Analysis, Digital Fringe Pattern Measurement Techniques*, Robinson, D.W. and Reid, G.T., Eds., Institute of Physics, Philadelphia, PA, 1993.)

test plates are used to determine if a surface is concave or convex with respect to the test plate (Mantravadi et al., 1992). Hovanesian and Hung (1990) studied three similar methods to identify the fringe order number.

Trolinger (1985) discussed the problems of a completely automatic fringe analysis, and frequently, when an automatic method is difficult, the order number must still be determined by visual observation of the fringes, in which case interactive procedures are convenient. These semiautomatic algorithms allow the operator to interact with the computer during the interferogram processing. Yatagai et al. (1982b) reported an interactive system for analyzing interferograms in which operators used a light pen to indicate their decisions. Funnell (1981) developed an interactive system in which the operator helped the machine with fringe identification by using keyboard commands. Still another interactive system was reported by Yatagai et al. (1984) to test the flatness of very large integrated circuit wafers. Finally, Parthiban and Sirohi (1989) constructed an interactive system in which the operator helped the machine identify fringe order numbers using a gray-scale coding with different colors for the fringes. The problem of fringe number identification may be simplified if some *a priori* information is known (Robinson, 1983a). A clear example is when we know in advance that the fringes are circular.

4.3 GLOBAL POLYNOMIAL INTERPOLATION

When the values of the wavefront deformations have been determined for many points over the interferogram, an interpolation between the points must be made in order to estimate the complete wavefront shape. This interpolation is accomplished by the use of a two-dimensional function. This is a global interpolation, because a single analytical function is used to represent the wavefront for the entire interferogram. To perform a global interpolation, the polynomials used most frequently are the Zernike polynomials (Malacara et al., 1976, 1987, 1990; Loomis, 1978; Plight, 1980; Swantner and Lowrey, 1980; Wang and Silva, 1980; Mahajan, 1981, 1984; Kim, 1982; Malacara, 1983; Hariharan et al., 1984; Kim and Shannon, 1987; Prata and Rusch, 1989; Malacara and DeVore, 1992).

Because the pupil of optical systems is frequently circular, it seems logical to express this two-dimensional function in polar coordinates, as follows:

$$x = \rho \sin\theta \qquad (4.10)$$

and

$$y = \rho \cos\theta \qquad (4.11)$$

where angle θ is measured with respect to the y-axis (Figure 4.4).

The wavefront deformations can be represented by many types of two-dimensional analytical functions, but the most commonly used are the Zernike polynomials. When the fit is

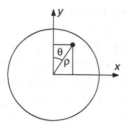

Figure 4.4   Polar coordinates used for two-dimensional polynomials.

not perfect, we define the fit variance, σ_f^2, as the difference between the actual sampled wavefront, W', and the analytical wavefront, $W(\rho,\theta)$, as follows:

$$\sigma_f^2 = \frac{1}{\pi} \int_0^1 \int_0^{2\pi} \left(W' - W(\rho,\theta)\right)^2 \rho \, d\rho \, d\theta \qquad (4.12)$$

The normalizing factor in front of the integral is $1/\pi$. If the fit variance is zero, the analytic function is an exact representation of the wavefront.

Sometimes it is also important to specify the mean wavefront deformation (W_{av}) including the normalizing factor, which is defined by:

$$W_{av} = \frac{1}{\pi} \int_0^1 \int_0^{2\pi} W(\rho,\theta) \rho \, d\rho \, d\theta \qquad (4.13)$$

Wavefront deformations are nearly always measured with respect to a close spherical reference. This spherical reference is defined by the position of the center of curvature and the radius of curvature.

The average wavefront deviations with respect to the spherical reference is the variance (σ_w^2), defined as:

$$\sigma_w^2 = \frac{1}{\pi} \int_0^1 \int_0^{2\pi} \left(W(\rho,\theta) - W_{av}\right)^2 \rho \, d\rho \, d\theta$$

$$= \frac{1}{\pi} \int_0^1 \int_0^{2\pi} W^2(\rho,\theta) \rho \, d\rho \, d\theta - W_{av}^2 \qquad (4.14)$$

which is frequently referred to as the root mean square (*rms*) value of the wavefront deformations. The reference spherical wavefront may be defined with any value of the radius of curvature (piston term) without modifying the position of the center of curvature. Nevertheless, the value of the wavefront variance may be affected by this selection, because the average wavefront is also affected. A convenient way to eliminate this problem is to select the reference sphere, when defining the wavefront variance, as the one with the same position as the mean wavefront deformation. This is why we subtract W_{av} in this expression.

4.3.1 Zernike Polynomials

The Zernike polynomials have unique and desirable properties that are derived from their orthogonality. These polynomials have been described in many places in the literature (e.g., Zernike, 1934, 1954; Bathia and Wolf, 1952, 1954; Born and Wolf, 1964; Barakat, 1980; Malacara and DeVore, 1992; Wyant and Creath, 1992), and a brief review is made here. The Zernike polynomials, $U(\rho,\theta)$, written in polar coordinates, are orthogonal in the unit circle in a continuous fashion (exit pupil with radius one) with the condition:

$$\int_0^1 \int_0^{2\pi} U_n^l(\rho,\theta) U_{n'}^{l'}(\rho,\theta) \rho \, d\rho \, d\theta = \frac{\pi}{2(n+1)} \delta_{nn'} \delta_{ll'} \qquad (4.15)$$

where $\rho = S/S_{max}$ is the normalized radial coordinate, with S being the non-normalized radial coordinate. The Kronecker delta ($\delta_{nn'}$) is zero if n is different from n'. The Zernike polynomials are represented with two indices (n and l) because they are dependent on two coordinates. Index n is the degree of the radial polynomial, and l is the angular dependence index. The numbers n and l are both even or both odd, making $n - l$ always even. There are $(1/2)(n + 1)(n + 2)$ linearly independent polynomials $U_n^l(\rho,\theta)$ of degree $\leq n$, one for each pair of numbers n and l.

The polynomials can be separated into two functions, one depending only on radius ρ and the other depending only on angle θ, thus obtaining:

$$U_n^l(\rho,\theta) = R_n^l(\rho)\begin{bmatrix} \sin \\ \cos \end{bmatrix} l\,\theta$$

$$= U_n^{n-2m}(\rho,\theta) = R_n^{n-2m}(\rho)\begin{bmatrix} \sin \\ \cos \end{bmatrix} (n-2m)\theta$$

(4.16)

where the sine function is used when $n - 2m > 0$ (antisymmetric functions), and the cosine function is used when $n - 2m \leq 0$ (symmetric functions). The degree of the radial polynomial $R_n^l(\rho)$ is n and $0 \leq m \leq n$. It can be shown that $|l|$ is the minimum exponent of the polynomials R_n^l. The radial polynomial is given by:

$$R_n^{n-2m}(\rho) = R_n^{-(n-2m)}(\rho) = \sum_{s=0}^{m} (-1)^s \frac{(n-s)!}{s!(m-s)!(n-m-s)!} \rho^{n-2s} \quad (4.17)$$

All Zernike polynomials, $U_n(\rho)$, may be ordered with a single index, r, defined by:

$$r = \frac{n(n+1)}{2} + m + 1 \quad (4.18)$$

Table 4.1 shows the first 15 Zernike polynomials. Kim and Shannon (1987) developed isometric plots for the first 37 Zernike polynomials, some of which are shown in Figure 4.5.

Triangular and ashtray astigmatisms may be visualized as the shape that a flexible disc adopts when supported on top of three or four points equally distributed around the edge. It should be pointed out that these polynomials are orthogonal only if the pupil is circular, without any central obscurations.

Any continuous wavefront shape, $W(x,y)$, may be represented by a linear combination of the Zernike polynomials:

TABLE 4.1 First Fifteen Zernike Polynomials

n	m	r	Zernike Polynomial	Meaning
0	0	1	1	Piston term
1	0	2	$\rho \sin\theta$	Tilt about x-axis
1	1	3	$\rho \cos\theta$	Tilt about y-axis
2	0	4	$\rho^2 \sin(2\theta)$	Astigmatism with axis at $\pm 45°$
2	1	5	$2\rho^2 - 1$	Defocusing
2	2	6	$\rho^2 \cos(2\theta)$	Astigmatism, axis at $0°$ or $90°$
3	0	7	$\rho^3 \sin(3\theta)$	Triangular astigmatism, base on x-axis
3	1	8	$(3\rho^3 - 2\rho)\sin\theta$	Primary coma along x-axis
3	2	9	$(3\rho^3 - 2\rho)\cos\theta$	Primary coma along y-axis
3	3	10	$\rho^3 \cos(3\theta)$	Triangular astigmatism, base on y-axis
4	0	11	$\rho^4 \sin(4\theta)$	Ashtray astigmatism, nodes on axes
4	1	12	$(4\rho^4 - 3\rho^2)\sin(2\theta)$	
4	2	13	$64\rho^4 - 6\rho^2 + 1$	Primary spherical aberration
4	3	14	$(4\rho^4 - 3\rho^2)\cos(2\theta)$	
4	4	15	$\rho^4 \cos(4\theta)$	Ashtray astigmatism, crests on axis

$$W(\rho,\theta) = \sum_{n=0}^{k}\sum_{m=0}^{n} A_{nm}U_{nm}(\rho,\theta)$$

$$= \sum_{r=0}^{L} A_r U_r(\rho,\theta)$$

(4.19)

If the maximum power is L, coefficients A_r can be found by any of several possible procedures — for example, by requiring that the fit variance defined is minimized.

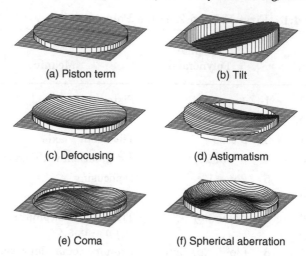

(a) Piston term   (b) Tilt

(c) Defocusing   (d) Astigmatism

(e) Coma   (f) Spherical aberration

Figure 4.5 Isometric plots for some Zernike polynomials.

4.3.2 Properties of Zernike Polynomials

The advantage of expressing the wavefront by a linear combination of orthogonal polynomials is that the wavefront deviation represented by each term is a best fit (minimum-fit variance) with respect to the actual wavefront. Any combination of these terms must also be a best fit. Each Zernike polynomial is obtained by adding to each type of aberration the proper amount of piston, tilt, and defocusing so that the *rms* value (σ_w^2), for each Zernike polynomial is minimized. To illustrate this with an example, let us consider a spherical aberration polynomial, where we see that a term $+ 1$ (piston term) and a term $- 6\rho^2$ (defocusing) have been added to the spherical aberration term, $6\rho^4$. These additional terms minimize the *rms* deviation of spherical aberration with respect to a flat wavefront. The practical consequence of the orthogonality of the Zernike polynomials is that any aberration terms, such as defocusing or tilt, may be added or subtracted from the wavefront function, $W(x,y)$, without losing the best fit to the data points.

Using the orthogonality condition, the mean wavefront deformation for each Zernike polynomial may be shown to be:

$$W_{av} = \frac{1}{\pi} \int_0^1 \int_0^{2\pi} U_r(\rho,\theta)\rho \, d\rho \, d\theta$$

$$= \frac{1}{2}; \quad \text{if } r = 1 \tag{4.20}$$

$$= 0; \quad \text{if } r > 1$$

This means that the mean wavefront deformation is zero for all Zernike polynomials, with the exception of the piston term; thus, the wavefront variance, is given by:

$$\sigma_W^2 = \frac{1}{2}\sum_{r=1}^{L} \frac{A_r^2}{n+1} - W_{av}^2$$

$$= \frac{1}{2}\sum_{r=2}^{L} \frac{A_r^2}{n+1} \tag{4.21}$$

where n is related to r by:

$$n = \text{next integer greater than } \frac{-3+[1+8r]^{1/2}}{2} \tag{4.22}$$

4.3.3 Least-Squares Fit to Zernike Polynomials

The analytic wavefront in terms of Zernike polynomials may be obtained using a two-dimensional least-squares fit (Malacara et al., 1990; Malacara and DeVore, 1992). If we have N measured points with coordinates (ρ_n,θ_n) and values W_n', measured with respect to a close analytical function, $W(\rho,\theta)$, then the discrete variance (v^2) is defined by:

$$v^2 = \frac{1}{N}\sum_{n=1}^{N}\left[W_n - W(\rho_n,\theta_n)\right] \tag{4.23}$$

The best least-squares fit to the function $W(\rho,\theta)$ is defined when the analytical function is chosen so this variance is a minimum with respect to the parameters of this function. We can see that the discrete variance S^2 and variance σ_f^2 are the same if the number of points is infinite, and they are uniformly distributed on the sampling region (aperture of the interferogram).

Let us now consider the analytical function $W(\rho,\theta)$ when it is a linear combination of some predefined polynomials, $V(\rho,\theta)$:

$$W(\rho_n,\theta) = \sum_{r=1}^{L} B_r V_r(\rho_n,\theta_n) \tag{4.24}$$

In order to have the best fit, we require that

$$\frac{\partial v}{\partial B_p} = 0 \tag{4.25}$$

where $p = 1, 2, 3, ..., L$. We then obtain the following system of L linear equations:

$$\sum_{r=1}^{L} B_r \sum_{n=1}^{N} V_r(\rho_n,\theta_n) V_p(\rho_n,\theta_n) = \sum_{n=1}^{N} W_n' V_p(\rho_n,\theta_n) = 0 \tag{4.26}$$

The matrix of this linear system of equations becomes diagonal if the polynomials V_r satisfy the condition that

$$\sum_{n=1}^{N} V_r(\rho_n,\theta_n) V_p(\rho_n,\theta_n) = \left(\sum_{n=1}^{N} V_n^2(\rho_n,\theta_n) \right) \delta_{rp} \tag{4.27}$$

This expression means that the polynomials V_r are orthogonal on the discrete base of the measured data points, as opposed to the Zernike polynomials, which are orthogonal in a continuous manner; that is, they are not orthogonal in the unitary circle, as the Zernike polynomials are.

The solution to the system of equations then becomes:

$$B_p = \frac{\displaystyle\sum_{n=1}^{N} W_n' V_p(\rho_n,\theta_n)}{\displaystyle\sum_{n=1}^{N} V_n^2(\rho_n,\theta_n)} \tag{4.28}$$

The polynomials V_p are not the Zernike polynomials U_p, but they approach them when the number of sampling points is extremely large and they are uniformly distributed on the unitary circle. The most important and useful property of orthogonal polynomials, as was pointed out earlier, is that when a least-squares fit is made any polynomial in the linear combination can be taken out without losing the best fit. Hence, it is more convenient to use V_p instead of U_p to make the wavefront representation. If desired, these polynomials can later be transformed into Zernike polynomials. A small problem, however, is that, because the locations of sampling points are different for different interferograms, the polynomials V_p are not universally defined, so they must be found for every particular case by a process referred to as *Gram–Schmidt orthogonalization*.

4.3.4   Gram–Schmidt Orthogonalization

The desired polynomials, orthogonal in the datapoint base, can be found as a linear combination of the Zernike polynomials:

$$V_r(\rho,\theta) = U_r + \sum_{s=1}^{r-1} D_{rs} V_s(\rho,\theta) \qquad (4.29)$$

where $r = 1, 2, 3, ..., L$. Now, using the orthogonality property and summing for all data points, we obtain for all values of r different from p:

$$\sum_{n=1}^{N} V_r(\rho_n,\theta_n) V_p(\rho_n,\theta_n) = \sum_{n=1}^{N} U_r(\rho_n,\theta_n) V_p(\rho_n,\theta_n) +$$
$$\qquad (4.30)$$
$$+ D_{rp} \sum_{n=1}^{N} V_p^2(\rho_n,\theta_n)$$

Thus, D_{rp} can be written as:

$$D_{rp} = \frac{\sum\limits_{n=1}^{N} U_r(\rho_n, \theta_n) V_p(\rho_n, \theta_n)}{\sum\limits_{n=1}^{N} V_p^2(\rho_n, \theta_n)} \qquad (4.31)$$

where $r = 2, 3, 4, ..., L$, and $p = 1, 2, ..., r - 1$. These coefficients give us the desired orthogonal polynomials. Factors affecting the accuracy of global interpolation using Zernike polynomials were studied by Wang and Ling (1989).

4.4 LOCAL INTERPOLATION BY SEGMENTS

A set of data points may be fitted to a polynomial, as we have seen in last section. This approach, however, has some problems, perhaps the most important being that, when the number of sampling points is large, the fit tends to have many oscillations and to deviate strongly at the edges, as illustrated in Figure 4.6. Global and local fitting of interferograms has been studied and compared by several researchers (e.g., Roblin and Prévost, 1978; Hayslett and Swantner, 1978, 1980; Freniere et al., 1979, 1981).

Local interpolation can be performed by several possible methods. The simplest one is Newton trapezoidal interpolation, but frequently better approximations are necessary. The three procedures most commonly used, then, are (Mieth and Osten, 1990):

1. One-dimensional spline interpolation
2. Two-dimensional bilinear interpolation
3. Triangular interpolation

A spline is a mechanical device, made of flexible material, that is used by draftsmen to draw curves. In mathematics, however, a spline is also an extension-limited piece of curve that may be used to represent a small interval in the set of points to be interpolated. The theory of splines has been treated in several books (e.g., Lancaster and Salkauskas, 1986). This method has the great advantage of providing greater control

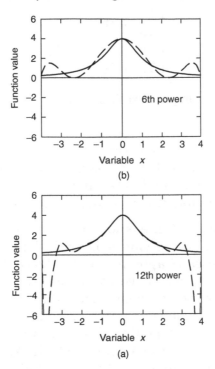

Figure 4.6 Errors in curve fitting for several polynomial degrees.

over the quality of the interpolation, as we proceed segment by segment to construct an entire curve. The problem, however, is that no single analytical representation exists for the entire curve. The points to be joined by splines are called *knots*. When the knots are connected with a straight line, the spline is *linear*. Additionally, at two consecutive knots joined by a spline, we must satisfy at least one of the two following conditions:

1. To have the same slope (first derivative) at the common knot. This condition can be satisfied with a third-degree polynomial, and the spline is *cubic*.
2. To have the same curvature (second derivative) at the common knot; under certain conditions, this criterion can also be satisfied with a cubic spline.

Figure 4.7 An example of spline fitting.

In interferometric data fitting, the cubic spline is a most popular and useful tool. To construct a cubic spline, the first derivative (slope) at the knots must be continuous; however, we have two possible ways to construct this spline:

1. The slope at the knots is calculated first, and the choice of these slopes is critical to the final result. One possible approach is to choose the slope of the second-degree curve (parabola) that passes through the point being considered and the two points on each side. The slopes at the extremes are those of the straight lines joining the first two and the last two points. When the slopes at all the knots are defined, the cubic spline may be calculated.

2. Another possibility is not to define the slope values at each knot; it is only required that they be continuous. We use this extra degree of freedom to require that the curvatures (second derivatives) are also continuous at the knots. In this case, we have a *classic cubic spline*. We only have to define the slopes or the curvatures at the first knot and at the last knot. If we define these curvatures as zero, we have a *natural cubic spline*. Figure 4.7 shows an example of a spline fitting.

Press et al. (1988) provided an algorithm in C to calculate the classic spline and the algebraic expressions to calculate the splines for interpolation of an array of points (y_i, x_i) with $x_1 < x_2 < \ldots < x_N$.

In addition to the point coordinates we must also supply the program with the values of the slopes at the beginning and at the end of the array. This procedure begins with solving a system of N linear equations with N unknowns. The first $N - 2$ equations are:

$$\frac{x_j - x_{j-1}}{6} y''_{j-1} + \frac{x_{j+1} - x_{j-1}}{3} y''_j + \frac{x_{j+1} - x_j}{6} y''_{j+1} =$$

$$= \frac{y_{j+1} - y_j}{x_{j+1} - x_j} - \frac{y_j - y_{j-1}}{x_j - x_{j-1}}; \quad j = 2,\ldots,(N-1) \tag{4.32}$$

where the unknowns (y'') are second derivatives at each of the knots. Two other equations necessary to solve this system are:

$$y''_1 = 0$$

$$y''_N = 0 \tag{4.33}$$

if the natural cubic spline is desired. Alternatively, we may set both of the first derivatives at the beginning and the end of the array of points to the desired values and use the following two equations:

$$y'_1 = \frac{y_2 - y_1}{x_2 - x_1} - \frac{3A_1^2 - 1}{6}(x_2 - x_1)y''_1 - \frac{3B_1^2 - 1}{6}(x_2 - x_1)y''_2 \tag{4.34}$$

with

$$A_1 = \frac{y_2 - x_1}{x_2 - x_1}; \quad B_1 = 1 - A_1 \tag{4.35}$$

and

$$y'_n = \frac{y_N - y_{N-1}}{x_N - x_{N-1}} - \frac{3A_N^2 - 1}{6}(x_N - x_{N-1})y''_{N-1} -$$

$$- \frac{3B_N^2 - 1}{6}(x_N - x_{N-1})y''_N \tag{4.36}$$

with

$$A_N = \frac{y_N - x_N}{x_N - x_{N-1}}; \quad B_N = 1 - A_N \tag{4.37}$$

In two dimensions, a similar approach can be used with bicubic splines.

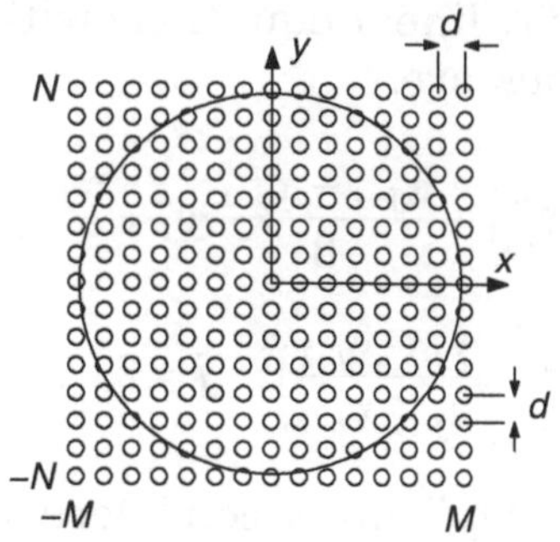

Figure 4.8 Sampling a wavefront with a two-dimensional array of Gaussians.

4.5 WAVEFRONT REPRESENTATION BY AN ARRAY OF GAUSSIANS

Frequently, the description of a wavefront shape can be inaccurate when using a polynomial representation if sharp local deformations are present. The most important errors in the analytical representation occur at these sharp deformations and near the edge of the pupil. An analytical representation by means of a two-dimensional array of Gaussians may be more accurate, as described by Montoya-Hernández et al. (1999). Let us consider a two-dimensional array of $(2M + 1) \times (2N + 1)$ Gaussians with separation d (Figure 4.8). The height (w_{nm}) of each Gaussian in the array is adjusted to obtain the desired wavefront shape, $W(x,y)$, with the expression:

$$W(x,y) = \sum_{m=-M}^{M} \sum_{n=N}^{N} w_{nm} e^{-\left((x-md)^2+(y-nd)^2\right)/\rho^2} \tag{4.38}$$

The spatial frequency content of this wavefront is represented by the Fourier transform $F\{W(x,y)\}$ of the function $W(x,y)$ as follows:

$$F\{W(x,y)\} = \pi\rho^2 e^{\left[-\pi^2\rho^2\left(f_x-f_y\right)\right]} \sum_{m=-M}^{M} \sum_{n=N}^{N} w_{nm} e^{-i2\pi d\left(mf_x-nf_y\right)} \tag{4.39}$$

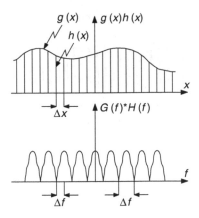

Figure 4.9 Sampling a two-dimensional function with a comb function.

Two important parameters to be determined are the separation (d) and width (ρ) of the Gaussians. To determine these quantities, let us consider a one-dimensional function, $g(x)$, which is sampled by a comb function, $h(x)$, as shown in Figure 7.9a. We assume that function $g(x)$ is band limited, with a maximum spatial frequency (f_{max}). To satisfy the sampling theorem, the comb sampling frequency should be smaller than half of f_{max}. Function $g(x)$ can then be reconstructed.

From the convolution theorem we know that the Fourier transform of the product of two functions is equal to the convolution of the Fourier transforms of the two functions:

$$\mathrm{F}\{g(x)h(x)\} = G(f) * H(f) \qquad (4.40)$$

We can see in Figure 4.9b that in the Fourier or frequency space an array of lobes represents the Fourier transforms of the sampled function. If the sampling frequency is higher than $2f_{max}$, the lobes are separated without any overlapping. Ideally, they should just touch each other. The function $g(x)$ is well represented only if all lobes in the Fourier space are filtered out with the exception of the central lobe. To perform the necessary spatial filtering, the comb function is replaced by an array of Gaussians, as shown in Figure 4.10a. In the Fourier

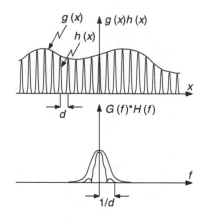

Figure 4.10 Sampling a one-dimensional function with an array of Gaussians.

space, the Fourier transform of these Gaussians appears as a modulating envelope that filters out the undesired lobes (Figure 4.10b). To obtain good filtering, the Gaussians should have a width (ρ) approximately equal to the array separation (d).

The remaining parameter to be determined is the Gaussian height (w_{nm}). This can be done using an iterative procedure. To obtain the wavefront deformation at a given point, it is not necessary to evaluate all the Gaussian heights, as the contributions of the Gaussians decay very quickly with their distance from that point. The height of each Gaussian is adjusted until the function $g(x)$ has the desired value at that point. A few iterations are sufficient to obtain a good fitting.

REFERENCES

Augustyn, W.H., Automatic data reduction of both simple and complex interference patterns, *Proc. SPIE*, 171, 22–31, 1979a.

Augustyn, W.H., Versatility of a microprocessor-based interferometric data reduction system, *Proc. SPIE*, 192, 128–133, 1979b.

Augustyn, W.H., Rosenfeld, A.H., and Zanoni, C.A., An automatic interference pattern processor with interactive capability, *Proc. SPIE*, 153, 146–155, 1978.

Barakat, R., Optimum balanced wave-front aberrations for radially symmetric amplitude distributions: generalizations of Zernike polynomials, *J. Opt. Soc. Am.*, 70, 739–742, 1980.

Bathia, A.B. and Wolf, E., The Zernike circle polynomials occurring in diffraction theory, *Proc. Phys. Soc.*, B65, 909–910, 1952.

Bathia, A.B. and Wolf, E., On the circle polynomials of Zernike and related orthogonal tests, *Proc. Cambridge Phil. Soc.*, 50, 40–48, 1954.

Becker, F., Zur Automatischen Auswertung von Interferogrammen, *Mitteilungen aus der Max-Planck-Institut fuer Stroemungsforschung*, Nr. 74, 1982.

Becker, F. and Yung, Y.H., Digital fringe reduction techniques applied to the measurement of three-dimensional transonic flow fields, *Opt. Eng.*, 24, 429–434, 1985.

Becker, F., Maier, G.E.A., and Wegner, H., Automatic evaluation of interferograms, *Proc. SPIE*, 359, 386–393, 1982.

Born, M. and Wolf, E., *Principles of Optics*, Pergamon Press, New York, 1964.

Button, B.L., Cutts, J., Dobbins, B.N., Moxon, J.C., and Wykes, C., The identification of fringe positions in speckle patterns, *Opt. Laser Technol.*, 17, 189–192, 1985.

Boutellier, R. and Zumbrunn, R., Digital interferogram analysis and DIN norms, *Proc. SPIE*, 656, 128–134, 1986.

Choudry, A., Automated fringe reduction analysis, *Proc. SPIE*, 816, 49–55, 1987.

Cline, H.E., Holik, A.S., and Lorensen, W.E., Computer-aided surface reconstruction of interference contours, *Appl. Opt.*, 21, 4481–4488, 1982.

Crescentini, L., Fringe pattern analysis in low-quality interferograms, *Appl. Opt.*, 28, 1231–1234, 1988.

Crescentini, L. and Fiocco, G., Automatic fringe recognition and detection of subwavelength phase perturbations with a Michelson interferometer, *Appl. Opt.*, 27, 118–123, 1988.

Dew, G.D., A method for the precise evaluation of interferograms, *J. Sci. Instrum.*, 41, 160–162, 1964.

Dyson, J., The rapid measurement of photographic records of inter-ference fringes, *Appl. Opt.*, 2, 487–489, 1963.

Eichhorn, N. and Osten, W., An algorithm for the fast derivation of the line structures from interferograms, *J. Mod. Optics*, 35, 1717–1725, 1988.

Freniere, E.R., Toler, O.E., and Race, R., Interferogram evaluation program for the HP-9825A calculator, *Proc. SPIE*, 171, 39–42, 1979.

Freniere, E.R., Toler, O.E., and Race, R., Interferogram evaluation program for the HP-9825A calculator, *Opt. Eng.*, 20, 253–255, 1981.

Funnell, W.R.J., Image processing applied to the interactive analysis of interferometric fringes, *Appl. Opt.*, 20, 3245–3249, 1981.

Gasvik, K.J., Fringe location by means of a zero crossing algorithm, *Proc. SPIE*, 1163, 64–70, 1989.

Gillies, A.C., Image processing approach to fringe patterns, *Opt. Eng.*, 27, 861–866, 1988.

Hariharan, P., Oreb, B.F., and Wanzhi, Z., Measurement of aspheric surfaces using a microcomputer-controlled digital radial-shear interferometer, *Optica Acta*, 31, 989–999, 1984.

Hatsuzawa, T., Optimization of fringe spacing in a digital flatness test, *Appl. Opt.*, 24, 2456–2459, 1985.

Hayslett, C.R. and Swantner, W.H., Mathematical methods for deriv-ing wavefronts from interferograms, in *Optical Interferograms: Reduction and Interpretation*, Guenther, A.H. and Liedbergh, D.H., Eds., ASTM Symposium, Tech. Publ. 666, American Soci-ety for Testing and Materials, West Conshohocken, PA, 1978.

Hayslett, C.R. and Swantner, W.H., Wave-front derivation from interferograms by three computer programs, *Appl. Opt.*, 19, 3401–3406, 1980.

He, R., Yan, H., and Hu., J., Skeletonization algorithm based on cross segment analysis, *Opt. Eng.*, 38, 662–671, 1999.

Hot, J.P. and Durou, C., System for the automatic analysis of inter-ferograms obtained by holographic interferometry, *Proc. SPIE*, 210, 144–151, 1979.

Hovanesian, J. Der and Hung, Y.Y., Fringe analysis and interpretation, *Proc SPIE*, 1121, 64–71, 1990.

Huang, Z., Fringe skeleton extraction using adaptive refining, *Opt. Lasers Eng.*, 18, 281–295, 1993.

Hunter, J.C., Collins, M.W., and Tozer, B.A., An assessment of some image enhancement routines for use with an automatic fringe tracking programme, *Proc. SPIE*, 1163, 83–94, 1989a.

Hunter, J.C., Collins, M.W., and Tozer, B.A., A scheme for the analysis of infinite fringe systems, *Proc. SPIE*, 1163, 206–219, 1989b.

Jones, R.A. and Kadakia, P.L., An automated interferogram analysis, *Appl. Opt.*, 7, 1477–1481, 1968.

Kingslake, R., The analysis of an interferogram, *Trans. Opt. Soc.*, 28, 1, 1926–1927.

Kim, C.-J., Polynomial fit of interferograms, *Appl. Opt.*, 21, 4521–4525, 1982.

Kim, C.-J. and Shannon, R., Catalog of Zernike polynomials, in *Applied Optics and Optical Engineering*, Vol. 10, Shannon, R. and Wyant, J.C., Eds., Academic Press, New York, 1987.

Kreis, T.M. and Kreitlow, H., Quantitative evaluation of holographic interferograms under image processing aspects, *Proc. SPIE*, 210, 2850–2853, 1983.

Lancaster, P. and Salkauskas, K., *Curve and Surface Fitting: An Introduction*, Academic Press, San Diego, CA, 1986.

Liu, K. and Yang, J.Y., New method of extracting fringe curves from images, *Proc. SPIE*, 1163, 71–76, 1989.

Livnat, A., Kafri, O., and Erez, G., Hills and valleys analysis in optical mapping and its application to moiré contouring, *Appl. Opt.*, 19, 3396–3400, 1980.

Loomis, J.S., A computer program for analysis of interferometric data, in *Optical Interferograms: Reduction and Interpretation*, Guenther, A.H. and Liedbergh, D.H., Eds., ASTM Symposium, Tech. Publ. 666, American Society for Testing and Materials, West Conshohocken, PA, 1978.

Macy, W.W., Jr., Two dimensional fringe pattern analysis, *Appl. Opt.*, 22, 3898–3901, 1983.

Mahajan, V.N., Zernike annular polynomials for imaging systems with annular pupils, *J. Opt. Soc. Am.*, 71, 75–85, 1981 (*errata*, 71, 1408–1408, 1981).

Mahajan, V.N., Zernike annular polynomials for imaging systems with annular pupils, *J. Opt. Soc. Am. A*, 1, 685, 1984.

Malacara, D., Set of orthogonal aberration coefficients, *Appl. Opt.*, 22, 1273–1274, 1983.

Malacara, D. and DeVore, S.L., Optical interferogram evaluation and wavefront fitting, in *Optical Shop Testing*, 2nd ed., Malacara, D., Ed., John Wiley & Sons, New York, 1992.

Malacara, D., Cornejo, A., and Morales, A., Computation of Zernike polynomials in optical testing, *Bol. Inst. Tonantzintla*, 2, 121–126, 1976.

Malacara, D., Carpio-Valadéz, J.M., and Sánchez-Mondragón, J.J., Interferometric data fitting on Zernike-like orthogonal basis, *Proc. SPIE*, 813, 35–36, 1987.

Malacara, D., Carpio, J.M., and Sánchez, J.J., Wavefront fitting with discrete orthogonal polynomials in a unit radius circle, *Opt. Eng.*, 29, 672–675, 1990.

Mantravadi, M.V., Newton, Fizeau, and Haidinger interferometers, in *Optical Shop Testing*, 2nd ed., Malacara, D., Ed., John Wiley & Sons, New York, 1992.

Mastin, G.A. and Ghiglia, D.C., Digital extraction of interference fringe contours, *Appl. Opt.*, 24, 1727–1728, 1985.

Matczac, M.J. and Budzinski, J., A software system for skeletonization of interference fringes, *Proc. SPIE*, 1121, 136–141, 1990.

Mieth, U. and Osten, W., Three methods for the interpolation of phase values between fringe pattern skeletons, *Proc. SPIE*, 1121, 151–153, 1990.

Montoya-Hernández, M., Servin, M., Malacara-Hernández, D., and Paez, G., Wavefront Fitting Using Gaussian Functions, *Opt. Comm.*, 163, 259–269, 1999.

Moore, R.C., Automatic method of real-time wavefront analysis, *Opt. Eng.*, 18, 461–463, 1979.

Nakadate, S., Yatagai, T., and Saito, H., Computer-aided speckle pattern interferometry, *Appl. Opt.*, 22, 237–243, 1983.

Osten, W., Höfling, R., and Saedler, J., Two computer methods for data reduction from interferograms, *Proc. SPIE*, 863, 105–113, 1987.

Parthiban, V. and Sirohi, R.J., Use of gray-scale coding in labeling closed fringe patterns, *Proc. SPIE*, 1163, 77–82, 1989.

Platt, B.C., Reynolds, S.G., and Holt, T.R., Determining image quality and wavefront profiles from Interferograms, in *Optical Interferograms: Reduction and Interpretation*, Guenther, A.H. and Liedbergh, D.H., Eds., ASTM Symposium, Tech. Publ. 666, American Society for Testing and Materials, West Conshohocken, PA, 1978.

Plight, A.M., The calculation of the wavefront aberration polynomial, *Opt. Acta*, 27, 717–721, 1980.

Prata, A., Jr., and Rusch, W.V.T., Algorithm for computation of Zernike polynomial expansion coefficients, *Appl. Opt.*, 28, 749–754, 1989.

Press, W.H., Flannery, B.P., Teukolsky, S.A., and Vetterling, W.T., *Numerical Recipes in C*, Cambridge University Press, Cambridge, U.K., 1988.

Reid, G.T., Automatic fringe pattern analysis: a review, *Opt. Lasers Eng.*, 7, 37–68, 1986/87.

Reid, G.T., Image processing techniques for fringe pattern analysis, *Proc. SPIE*, 954, 468–477, 1988.

Robinson, D.W., Automatic fringe analysis with a computer image-processing system, *Appl. Opt.*, 22, 2169–2176, 1983a.

Robinson, D.W., Role for automatic fringe analysis in optical metrology, *Proc. SPIE*, 376, 20–25, 1983b.

Roblin, G. and Prévost, M., A method to interpolate between two-beam interference fringes, *Proc ICO-11 (Madrid)*, 667–670, 1978.

Schemm, J.B. and Vest, C.M., Fringe pattern recognition and interpolation using nonlinear regression analysis, *Appl. Opt.*, 22, 2850–2853, 1983.

Schluter, M., Analysis of holographic interferogram with a TV picture system, *Opt. Laser Technol.*, 12, 93–95, 1980.

Servin, M., Rodríguez-Vera, R., Carpio, M., and Morales, A., Automatic fringe detection algorithm used for moiré deflectometry, *Appl. Opt.*, 29, 3266–3270, 1990.

Snyder, J.J., Algorithm for fast digital analysis of interference fringes, *Appl. Opt.*, 19, 1223–1225, 1980.

Swantner, W.H. and Lowrey, W.H., Zernike–Tatian polynomials for interferogram reduction, *Appl. Opt.*, 19, 161–163, 1980.

Tichenor, D.A. and Madsen, V.P., Computer analysis of holographic interferograms for nondestructive testing, *Proc. SPIE*, 155, 222–227, 1978.

Trolinger, J.D., Automated data reduction in holographic interferometry, *Opt. Eng.*, 24, 840–842, 1985.

Truax, B.E., Programmable interferometry, *Proc. SPIE*, 680, 10–18, 1986.

Truax, B.E. and Selberg, L.A., Programmable interferometry, *Opt. Lasers Eng.*, 7, 195–220, 1986/87.

Varman, C. and Wykes, C., Smoothing of speckle and moiré fringes by computer processing, *Opt. Lasers Eng.*, 3, 87–100, 1982.

Vrooman, H.A. and Maas, A., Interferogram analysis using image processing techniques, *Proc. SPIE*, 1121, 655–659, 1989.

Wang, G.-Y. and Ling, X.-P., Accuracy of fringe pattern analysis, *Proc. SPIE*, 1163, 251–257, 1989.

Wang, J.Y. and Silva, D.E., Wave-front interpretation with Zernike polynomials, *Appl. Opt.*, 19, 1510–1518, 1980.

Womack, K.H., Jonas, J.A., Koliopoulos, C.L., Underwood, K.L., Wyant, J.C., Loomis, J.S., and Hayslett, C.R., Microprocessor-based instrument for analysis of video interferograms, *Proc. SPIE*, 192, 134–139, 1979.

Wyant, J.C. and Creath, K., Basic wavefront aberration theory for optical metrology, in *Applied Optics and Optical Engineering*, Vol. XI, Shannon, R.R. and Wyant, J.C., Eds., Academic Press, New York, 1992.

Yan, D.-P., He, A., and Miao, P.C., Method of rapid fringe thinning for flow-field interferograms, *Proc. SPIE*, 1755, 190–193, 1992.

Yatagai, T., Intensity-based analysis methods, in *Interferogram Analysis, Digital Fringe Pattern Measurement Techniques*, Robinson, D.W. and Reid, G.T., Eds., Institute of Physics, Philadelphia, Pa, 1993.

Yatagai, T., Idesawa, M., Yamaashi, Y., and Suzuki, M., Interactive fringe analysis system: applications to moiré contourgram and interferogram, *Opt. Eng.*, 21, 901–906, 1982a.

Yatagai, T., Nakadate, S., Idesawa, M., and Saito, H., Automatic fringe analysis using digital image processing techniques, *Opt. Eng.*, 21, 432–435, 1982b.

Yatagai, T., Inabu, S., Nakano, H., and Susuki, M., Automatic flatness tester for very large scale integrated circuit wafers, *Opt. Eng.*, 23, 401–405, 1984.

Yi, J.H., Kim, S.H., Kwak, Y.K., and Lee, Y.W., Peak movement detection method of an equally spaced fringe for precise position measurement, *Opt. Eng.*, 41, 428–434, 2002.

Yu, Q., Andersen, K., Osten, W., and Juptner, W.P.O., Analysis and removal of the systematic phase error in interferograms, *Opt. Eng.*, 33, 1630–1637, 1994.

Zanoni, C.A., A new, semiautomatic interferogram evaluation technique, in *Optical Interferograms: Reduction and Interpretation*, Guenther, A.H. and Liedbergh, D.H., Eds., ASTM Symposium, Tech. Publ. 666, American Society for Testing and Materials, West Conshohocken, PA, 1978.

Zernike, F., Begunstheorie des Schneidenver-Fahrens und Seiner Verbesserten Form der Phasenkontrastmethode, *Physica*, 1, 689, 1934.

Zernike, F., The diffraction theory of aberrations, in *Optical Image Evaluation*, Circular 526, National Bureau of Standards, Washington, D.C., 1954.

Zhi, H. and Johansson, R.B., Adaptive filter for enhancement of fringe patterns, *Opt. Lasers Eng.*, 15, 241–251, 1991.

5

Periodic Signal Phase Detection and Algorithm Analysis

5.1 LEAST-SQUARES PHASE DETECTION OF A SINUSOIDAL SIGNAL

An important problem to solve is detection (or measurement) by means of a sampling procedure of the real phase of a real sinusoidal signal for which the frequency is known. Let us begin by studying the least-squares method. From Equation 1.4, the $s(x)$ may be written in a very general manner as:

$$s(x) = a + b\cos(\omega x + \phi) \qquad (5.1)$$

where x is the coordinate (spatial or temporal) at which the irradiance is to be measured, ω is the angular spatial (or temporal) frequency, and ϕ is the phase at the origin ($x = 0$). If we want to make a least-squares fit of these irradiance data to a sinusoidal function, as in Equation 5.1 (see Figure 5.1), we must determine four unknown constants: a, b, ϕ, and ω; however, the analysis is simpler if we assume that the frequency of sinusoidal function ω is known, as is normally the case.

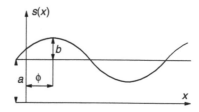

Figure 5.1 Unknown variables when sampling a sinusoidal function. The frequency ω is assumed to be known.

For least-squares analysis following Greivenkamp (1984), it is better to write this expression in an equivalent manner, as follows:

$$s(x) = D_1 + D_2 \cos \omega x + D_3 \sin \omega x \qquad (5.2)$$

where:

$$D_1 = a$$

$$D_2 = b \cos \phi \qquad (5.3)$$

$$D_3 = -b \sin \phi$$

Now, the following N measurements of the signal are taken:

$$s_n = D_1 + D_2 \cos \omega x_n + D_3 \sin \omega x_n, \quad n = 1, \ldots, N \qquad (5.4)$$

where $N \geq 3$, as three constants are to be determined. The best fit of these measurements to the sinusoidal analytical function is obtained if the coefficients D_1, D_2, and D_3 are chosen so that variance ε, defined by:

$$\varepsilon = \frac{1}{N} \sum_{n=1}^{N} (D_1 + D_2 \cos \omega x_n + D_3 \sin \omega x_n - s_n)^2 \qquad (5.5)$$

is minimized. Thus, taking the partial derivatives of variance ε with respect to the three unknown constants (D_1, D_2, and D_3), we find a set of simultaneous equations, which in matrix form may be written as:

$$
\begin{pmatrix}
N & \sum \cos \omega x_n & \sum \sin \omega x_n \\
\sum \cos \omega x_n & \sum \cos^2 \omega x_n & \sum \cos \omega x_n \sin \omega x_n \\
\sum \sin \omega x_n & \sum \cos \omega x_n \sin \omega x_n & \sum \sin^2 \omega x_n
\end{pmatrix}
\begin{pmatrix} D_1 \\ D_2 \\ D_3 \end{pmatrix} =
$$

$$
= \begin{pmatrix}
\sum s_n \\
\sum s_n \cos \omega x_n \\
\sum s_n \sin \omega x_n
\end{pmatrix}
\tag{5.6}
$$

This matrix is evaluated with the values of the phases at which the signal is measured, but it does not depend on the values of the signal. Thus, if necessary, the signal may be measured as many times as desired, without having to calculate the matrix elements every time; it is only necessary to use the same phase values. This is the case for phase-shifting interferometry, for example, as is discussed in Chapter 6. As shown by Greivenkamp (1984), this is a general least-squares procedure for any separation between the measurements, assuming only that frequency ω is known. The system expressed by Equation 5.6 can also be written as:

$$
\begin{pmatrix}
a_{11} & a_{12} & a_{13} \\
a_{12} & a_{22} & a_{23} \\
a_{13} & a_{23} & a_{33}
\end{pmatrix}
\begin{pmatrix} D_1 \\ D_2 \\ D_3 \end{pmatrix} =
\begin{pmatrix}
\sum s_n \\
\sum s_n \cos \omega x_n \\
\sum s_n \sin \omega x_n
\end{pmatrix}
\tag{5.7}
$$

Then, from Equation 5.3, the phase can be found by:

$$
\tan \phi = -\left(\frac{D_3}{D_2} \right)
$$

$$
= - \frac{\displaystyle\sum_{n=1}^{N} s_n \left[A_{11} + A_{12} \cos\left(\frac{2\pi n}{N} \right) + A_{13} \sin\left(\frac{2\pi n}{N} \right) \right]}{\displaystyle\sum_{n=1}^{N} s_n \left[A_{21} + A_{22} \cos\left(\frac{2\pi n}{N} \right) + A_{23} \sin\left(\frac{2\pi n}{N} \right) \right]}
\tag{5.8}
$$

where:

$$A_{11} = (a_{12}a_{23} - a_{13}a_{22})$$

$$A_{12} = (a_{12}a_{13} - a_{11}a_{23})$$

$$A_{13} = (a_{11}a_{22} - a_{12}^2)$$

$$A_{21} = (a_{12}a_{33} - a_{13}a_{23})$$

$$A_{22} = (a_{13}^2 - a_{11}a_{33})$$

$$A_{23} = (a_{11}a_{23} - a_{12}a_{13})$$

(5.9)

A particular least-squares sampling procedure was analyzed by Morgan (1982), who assumed that the measurements were taken at equally spaced intervals, uniformly spaced in k signal periods and defined by:

$$\omega x_n = \frac{2\pi(n-1)}{N} + \omega x_1$$

(5.10)

where x_1 is the location of the first sampling point and $n = 1$, $2, ..., kN$. In the most frequent case, the sampling points are distributed in only one signal period ($k = 1$). To understand this angular distribution, we can plot these sampling points with unit vectors from the origin, each vector having an angle $2\pi(n - 1)/N$ with respect to the x-axis (Figure 5.2). Then, we can see that the sampling distribution for $N \geq 3$ requires that the vector sum of all the vectors from the origin to each point is equal to zero. This condition is expressed by:

$$\sum_{n=1}^{N} \sin \omega x_n = 0, \quad \sum_{n=1}^{N} \cos \omega x_n = 0$$

(5.11)

This condition is necessary but not sufficient to guarantee the equally spaced and uniform distribution in Equation 5.10. As shown in the lower row in Figure 5.2, we also need the following conditions for twice the phase angle:

$$\sum_{n=1}^{N} \sin 2\omega x_n = 0, \quad \sum_{n=1}^{N} \cos 2\omega x_n = 0$$

(5.12)

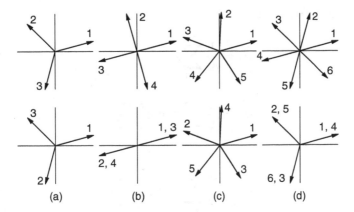

Figure 5.2 Polar representation of the sampling points, uniformly spaced in a signal period: (a) three points, (b) four points, (c) five points, and (d) six points. The upper row plots the phase for Equation 5.11, and the lower row plots twice the phase angle for Equation 5.12.

From the first expression in Equation 5.12 we can see that

$$\sum_{n=1}^{N} \cos \omega x_n \sin \omega x_n = \frac{1}{2}\sum_{n=1}^{N} \sin 2\omega x_n = 0 \qquad (5.13)$$

and, from the second expression and a well-known trigonometric relation, we find:

$$\sum_{n=1}^{N} \cos^2 \omega x_n = \sum_{n=1}^{N} \sin^2 \omega x_n = \frac{N}{2} \qquad (5.14)$$

With these relations, the system matrix becomes diagonal:

$$\begin{pmatrix} N & 0 & 0 \\ 0 & \dfrac{N}{2} & 0 \\ 0 & 0 & \dfrac{N}{2} \end{pmatrix} \begin{pmatrix} \alpha_1 \\ \alpha_2 \\ \alpha_3 \end{pmatrix} = \begin{pmatrix} \sum s_n \\ \sum s_n \cos\left(\dfrac{2\pi n}{N}\right) \\ \sum s_n \sin\left(\dfrac{2\pi n}{N}\right) \end{pmatrix} \qquad (5.15)$$

with the solutions:

$$\alpha_1 = \frac{1}{N}\sum_{n=1}^{N} s_n \tag{5.16}$$

$$\alpha_2 = \frac{2}{N}\sum_{n=1}^{N} s_n \cos\left(\frac{2\pi n}{N}\right) \tag{5.17}$$

and

$$\alpha_3 = \frac{2}{N}\sum_{n=1}^{N} s_n \sin\left(\frac{2\pi n}{N}\right) \tag{5.18}$$

Substituting Equations 5.17 and 5.18 into Equation 5.8, the phase at the origin (ϕ) may be obtained from:

$$\tan\phi = -\left(\frac{\alpha_3}{\alpha_2}\right) = -\left(\frac{\sum_{n=1}^{N} s_n \sin\left(\frac{2\pi n}{N}\right)}{\sum_{n=1}^{N} s_n \cos\left(\frac{2\pi n}{N}\right)}\right) \tag{5.19}$$

Because of its relevance, this algorithm deserves a name. Many different names had been given to it in the past, such as *synchronous detection algorithm*, but here we will call it the *diagonal least-squares algorithm*. The minimum acceptable number of sampling points is $N = 3$, in which case we obtain the sampling spacing as given by Equation 5.10:

$$\Delta x = \frac{2\pi}{3\omega} = \frac{1}{3f} \tag{5.20}$$

and, if $\omega x_1 = 60°$, then the phase ϕ becomes:

$$\tan\phi = \frac{-\sqrt{3}(s_1 - s_3)}{s_1 - 2s_2 + s_3} \tag{5.21}$$

If the sampling points are not properly spaced, as required by Equation 5.20, then the phase value obtained with Equation 5.19 or 5.21 will not be correct, as will be shown later.

5.2 QUADRATURE PHASE DETECTION OF A SINUSOIDAL SIGNAL

Let us consider the sinusoidal signal, $s(x)$, as in Equation 5.1, now written as:

$$s(x) = a + b\cos(2\pi fx + \phi) \tag{5.22}$$

where f is the frequency of this signal. Let us now take the Fourier transform, $S(f)$, of this signal at a reference frequency (f_r):

$$S(f_r) = \int_{-\infty}^{\infty} s(x)\exp(-i2\pi f_r x)\,dx \tag{5.23}$$

to obtain:

$$S(f_r) = a\delta(f_r) + \frac{b}{2}\delta(f_r - f)\exp(i\phi) + \frac{b}{2}\delta(f_r + f)\exp(-i\phi) \tag{5.24}$$

If the reference frequency (f_r) is equal to the frequency of the signal $(f = f_r)$, then this function has the value:

$$S(f_r) = \frac{b}{2}\exp(i\phi)$$
$$= \frac{b}{2}(\cos\phi + i\sin\phi) \tag{5.25}$$

Then, as pointed out in Chapter 2, the phase (ϕ) of the real periodic signal in Equation 5.1, evaluated at the origin $(x = 0)$, is equal to the phase of its Fourier transform at the frequency of the signal $(f = f_r)$. Thus, using Equation 5.23, we obtain:

$$\tan\phi = \left(\frac{\mathrm{Im}\{S(f_r)\}}{\mathrm{Re}\{S(f_r)\}}\right)$$
$$= -\left(\frac{\displaystyle\int_{-\infty}^{\infty} s(x)\sin(2\pi f_r x)\,dx}{\displaystyle\int_{-\infty}^{\infty} s(x)\cos(2\pi f_r x)\,dx}\right) \tag{5.26}$$

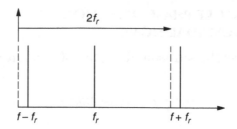

Figure 5.3 Spectrum of functions resulting from the multiplication of the sinusoidal signal by two reference sinusoidal functions, sine and cosine.

To gain some insight into the nature of these integrals, we can multiply the signal with frequency f by sine and cosine functions with frequency f_r:

$$z_S(x) = s(x)\sin\omega_r x$$

$$= -\frac{b}{2}\sin(\omega x - \omega_r x + \phi) + a\sin(\omega_r x) + \frac{b}{2}\sin(\omega x + \omega_r x + \phi) \qquad (5.27)$$

and

$$z_C(x) = s(x)\cos(\omega_r x)$$

$$= \frac{b}{2}\cos(\omega x - \omega_r x + \phi) + a\cos(\omega_r x) + \frac{b}{2}\cos(\omega x + \omega_r x + \phi) \qquad (5.28)$$

where $\omega = 2\pi f$ and $\omega_r = 2\pi f_r$. The functions $z_S(x)$ and $z_C(x)$ are periodical, but they contain three harmonic components: (1) the first term, with a very low frequency, equal to the difference between the signal and the reference frequencies; (2) the second term, with the reference frequency; and (3) the last term, with a frequency equal to the sum of the signal and the reference frequencies. The spectrum of these functions is illustrated in Figure 5.3.

If the terms with frequencies ω_r and $\omega + \omega_r$ are properly eliminated by a suitable low-pass filter that also preserves the ratio of the amplitudes of the low frequency terms, then we obtain the filtered versions of these functions:

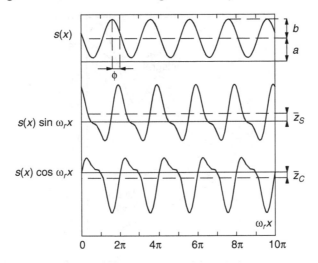

Figure 5.4 Functions resulting from the multiplication of the sinusoidal signal by two reference sinusoidal functions, sine and cosine, with the same frequency as the signal.

$$\bar{z}_S(x) = -\frac{b}{2}\sin(\omega x - \omega_r x + \phi) \tag{5.29}$$

and

$$\bar{z}_C(x) = \frac{b}{2}\cos(\omega x - \omega_r x + \phi) \tag{5.30}$$

Thus, we obtain:

$$\tan(\omega x - \omega_r x + \phi) = -\frac{\bar{z}_S(x)}{\bar{z}_C(x)} \tag{5.31}$$

When the signal and the reference frequencies are equal, functions 5.29 and 5.30 are constants. Figure 5.4 plots Equations 5.27 and 5.28 for this case, where, because the signal is not phase modulated, the filtered functions $\bar{z}_S(x)$ and $\bar{z}_C(x)$ become constants. The phase at the origin (ϕ) ($x = 0$) is calculated by:

$$\tan\phi = -\frac{\bar{z}_S(0)}{\bar{z}_C(0)} \tag{5.32}$$

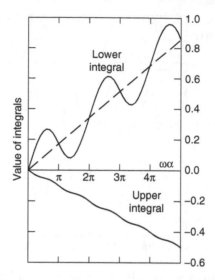

Figure 5.5 Plots of the values of the integrals in Equation 5.23 for a signal phase equal to 30° and signal constants $a = 1.3$ and $b = 1$.

The conditions necessary for this method to produce accurate results and the effects of several possible sources of error have been studied by Nakadate (1988a,b). The next section discusses how the low-pass filtering must be performed in order to obtain the phase at the origin (ϕ) or the phase at any point x ($\omega x - \omega_r x + \phi$).

5.2.1 Low-Pass Filtering in Phase Detection

The simplest case for phase detection is when no detuning is present — that is, when the signal frequency and the reference frequency are equal. In this case, when we evaluate the integrals in Equation 5.26 we obtain the graphs in Figure 5.5. The values of both integrals tend to infinity, although, the ratio of the two integrals has a finite value equal to the ratio of their average slopes.

This finite ratio of the integrals can be found in many ways. For example, because the signal is periodical we can perform the integration only in the finite interval $-1/2f < x < 1/2f$, or integer multiples of this value, as shown in Figure 5.6.

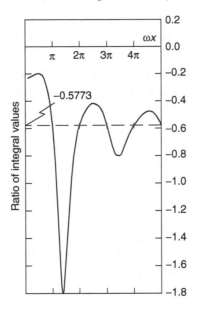

Figure 5.6 Plot of the values of the ratio of the integrals in Equation 5.26 for a signal phase equal to 30° and signal constants $a = 1.3$ and $b = 1$.

Two disadvantages of this method are that a large number of sampling points is needed to emulate a continuous measurement and that the signal frequency must be accurately determined in order to correctly fix the sampling interval.

Another method is a discrete sampling low-pass filtering process that can be performed by means of a convolution, as described in Chapter 2, with a pair of suitable filtering functions: $h_S(x)$ and $h_C(x)$. Let us now consider this method but remove the restriction for no detuning. The entire process of multiplication by the sinusoidal reference and low-pass filtering to obtain the filtered functions $\bar{z}_S(x)$ and $\bar{z}_C(x)$ is expressed by:

$$\bar{z}_S(x) = \int_{-\infty}^{\infty} z_S(\alpha) h_S(x - \alpha) \, d\alpha \qquad (5.33)$$

and, in an analogous manner, with the filtering function $h_C(x)$ we have:

$$\bar{z}_C(x) = \int_{-\infty}^{\infty} z_C(\alpha)\, h_C(x - \alpha)\, d\alpha \qquad (5.34)$$

To use Equation 5.31 to obtain the correct value of the phase $(\omega x - \omega_r x + \phi)$ at any point x in the presence of detuning, we need to satisfy three conditions:

1. The low-pass filtering must be performed using the convolution operation, as expressed by Equations 5.33 and 5.34.
2. The terms with frequencies ω_r and $(\omega + \omega_r)$ must be completely eliminated so this function is zero for any value of x.
3. The ratio of the amplitudes of the low-frequency terms, with frequency $(\omega - \omega_r)$, must be preserved by the filtering process.

In general, the filtering functions for $z_S(x)$ and $z_C(x)$ can be different, although sometimes they are the same, as we will see later. If the filtering function is the same for both functions, the third condition is automatically satisfied, but not if they are different.

Let us now consider the case when we are interested not in the phase at any value of x but only in the phase at the origin (ϕ). In this case, we need to satisfy slightly different conditions. In order to obtain the correct phase using Equation 5.32, the contribution of the high-frequency components of $\bar{z}_S(x)$ or $\bar{z}_C(x)$ to the value of the filtered signals $\bar{z}_S(0)$ or $\bar{z}_C(0)$, respectively, must be zero. In other words, we do not require that the high-frequency components are completely eliminated, only that their value at $x = 0$ is zero. The conditions to be satisfied in this case are:

1. The low-pass filtering must be complete only for the point at the origin, using the convolution with $x = 0$.
2. The contributions to $\bar{z}_S(0)$ and $\bar{z}_C(0)$ of the terms with frequencies ω_r and $(\omega + \omega_r)$, evaluated at the origin, must be zero.
3. The ratio of the amplitudes of the low-frequency terms, with frequency $(\omega - \omega_r)$, must be preserved by the filtering process.

To better understand the second condition, let us assume that we need to avoid any effect on the phase in Equation 5.32 of a certain high-frequency component present in $z_S(x)$ or $z_C(x)$ which is sinusoidal and real. The value of this sinusoidal component must be zero at the origin. The value at the origin of this sinusoidal component is zero not only if its amplitude is zero but also if it is antisymmetrical (a sine function). Then, its Fourier transform at this frequency must be imaginary and antisymmetrical, as shown in Table 2.3.

We have seen in Chapter 2 that the convolution of two functions is equal to the inverse Fourier transform of the product of the Fourier transforms of those two functions. Hence, we may write:

$$F\{\bar{z}_S(x)\} = Z_S(f)\,H_S(f) \tag{5.35}$$

and similarly for $z_C(x)$. Thus, the right-hand side of this expression at the frequency to be filtered, as for the left-hand side, must also be imaginary and antisymmetrical.

On the other hand, the sinusoidal component of $z_S(x)$ that we want to filter out is real; thus, according to Table 2.3, its Fourier transform, $Z_S(f)$, can be (1) real and symmetrical, (2) imaginary and antisymmetrical, or (3) complex and Hermitian. For these cases we can see that $H(f)$ must be (1) imaginary and antisymmetrical, (2) real and symmetrical, or (3) complex and Hermitian, respectively. These results are summarized in Table 5.1.

The second term in Equation 5.27 is real and antisymmetrical; thus, we need a filter function such that its Fourier transform is real and symmetrical at this frequency, satisfying the condition:

$$H_S(f_r) = H_S(-f_r) \tag{5.36}$$

Similarly, the second term in Equation 5.28 is real and symmetrical; thus, we need a filter function such that its Fourier transform is imaginary and antisymmetrical at this frequency, satisfying the condition:

$$H_C(f_r) = -H_C(-f_r) \tag{5.37}$$

TABLE 5.1   Necessary Properties of the Fourier Transform
of the Filtering Function To Make the Right-Hand Side of
Equation 5.35 Imaginary and Antisymmetrical

Sinusoidal Component of $z(x)$	Fourier Transform $Z_S(f_r)$ or $Z_C(f_r)$	Function $H(f_r)$
Real and symmetrical	Real and symmetrical	Imaginary and antisymmetrical
Real and antisymmetrical	Imaginary and antisymmetrical	Real and symmetrical
Real and asymmetrical	Complex and Hermitian	Complex and Hermitian

The terms with frequency $2f_r$ (assuming $f = f_r$) are asymmetrical; that is, they are neither symmetrical nor antisymmetrical. Even more, the degree of asymmetry is not predictable, as it depends on the phase of the signal. So, the only solution is that the Fourier transforms of the filtering functions must have zeros at this frequency, as follows:

$$H_S(2f_r) = H_S(-2f_r) = 0$$
$$H_C(2f_r) = H_C(-2f_r) = 0 \tag{5.38}$$

Besides these conditions, the filtering function $h(x)$ must not modify the ratio between the constant (zero frequency) terms in the functions in Equations 5.27 and 5.28, thus also requiring that

$$H_S(0) = H_C(0) \tag{5.39}$$

These conditions in Equations 5.36 to 5.39 are quite general. The number of possible filter functions, continuous and discrete, that satisfy these conditions is infinite. Each pair of possible filter functions leads to a different algorithm with different properties.

A particular case of the conditions in Equations 5.36 and 5.37 is the stronger condition:

$$H_S(f_r) = H_S(-f_r) = H_C(f_r) = H_C(-f_r) = 0 \qquad (5.40)$$

which occurs when the sampling points distribution satisfies Equation 5.10. In this case, the two filter functions become identical at all frequencies.

A continuous filtering function with continuous sampling points, satisfying Equation 5.10, is the square function:

$$h(\alpha) = 1 \quad \text{for} \; -\frac{1}{2f_r} \le \alpha \le \frac{1}{2f_r}$$

$$= 0 \quad \text{for} \; |\alpha| > \frac{1}{2f_r} \qquad (5.41)$$

for which the Fourier transform has zeros at nf_r, where n is any nonzero integer. We then see that this filtering process is equivalent to performing the integration in a finite limited interval, as suggested before.

5.3 DISCRETE LOW-PASS FILTERING FUNCTIONS

This section describe some discrete sampling low-pass filtering functions. We write the filtering functions $h_S(x)$ and $h_C(x)$ for the sampled signal process as:

$$h_S(x) = \sum_{n=1}^{N} w_{S_n} \delta(x - \alpha_n) \qquad (5.42)$$

and

$$h_C(x) = \sum_{n=1}^{N} w_{C_n} \delta(x - \alpha_n) \qquad (5.43)$$

where α_n are the positions of the sampling points. The Fourier transforms of these functions are given by:

$$H_S(f) = \sum_{n=1}^{N} w_{S_n} \exp(-i2\pi f\alpha_n) \qquad (5.44)$$

and

$$H_C(f) = \sum_{n=1}^{N} w_{C_n} \exp(-i2\pi f\alpha_n) \qquad (5.45)$$

where w_{S_n} and w_{C_n} are the filtering weights.

Filtering functions of special interest are the discrete functions with equally spaced and uniformly distributed sampling points in a signal interval, as stated by Equation 5.10. The filtering functions $h_S(x)$ and $h_C(x)$ satisfy Equation 5.39, thus they are identical and equal to $h(x)$ with all the filtering weights equal to one. With this filtering function, the synchronous detection method (as expressed by Equation 5.26) may become identical to the *diagonal least-squares algorithm*, as expressed by Equation 5.15.

To consider this case, we impose the condition that the sampling points have a constant separation ($\Delta\alpha$) and that the first point is at the position $\alpha = 0$, as in Equation 5.10. This expression then becomes:

$$H(f) = \frac{1 - \exp(-i2\pi fN\Delta\alpha)}{1 - \exp(-i2\pi f\Delta\alpha)}$$

$$= \frac{\sin(\pi fN\Delta\alpha)}{\sin(\pi f\Delta\alpha)} \exp(-i2\pi(N-1)f\Delta\alpha) \qquad (5.46)$$

Hence, the power spectrum of this filtering function is:

$$\left| H(f)^2 \right| = \frac{\sin^2(\pi fN\Delta\alpha)}{\sin^2(\pi f\Delta\alpha)} \qquad (5.47)$$

It is illustrated in Figure 5.7a for the case of an infinite number of points and in Figure 5.7b for the discrete case of five sampling points.

We see that the zeros and peaks of this function occur at frequencies $n/(N\Delta\alpha)$, where n is any integer, and at the zeros when n/N is not an integer; thus, we have $N - 1$ minima (zeros) between two consecutive lobes. A lobe exists at zero

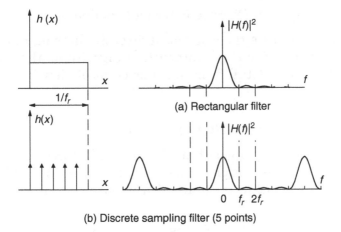

(a) Rectangular filter

(b) Discrete sampling filter (5 points)

Figure 5.7 Spectrum of the filtering function when five points are used to sample a sinusoidal function.

frequency ($n = 0$). Because we want zeros at the signal frequency (f_s) and at twice this frequency, we need at least three sampling points ($N \geq 3$). In order to locate the first two zeros at these frequencies, we require equally and uniformly spaced sampling points on the signal period:

$$\Delta\alpha = \frac{1}{Nf_s} \qquad (5.48)$$

This condition is the same as that in Equation 5.10 and is used in order to make the least-squares matrix diagonal; thus, if we use the filtering function $h(x)$ for equally spaced sampling points, we obtain Equation 5.19.

We may see that the zeros of this function occur at frequencies nf, with the exception of Nf and integer multiples of Nf, where n is any integer and N is the number of sampling points. Because we must filter out frequencies f and $2f$, we must have at least three sampling points ($N = 3$) to have at least two minima (zeros) between two consecutive peaks of the filtering function. Filtering functions and data sampling windows have been studied by de Groot (1995).

5.3.1   Examples of Discrete Filtering Functions

To better illustrate the concept of discrete filtering functions, let us now describe three interesting algorithms that will be studied in more detail from another point of view in the next chapter.

5.3.1.1   Wyant's Three-Step Algorithm

Wyant's three-step algorithm (Wyant et al., 1984; see Section 6.2.3) uses three sampling points, located at −45°, 45°, and 135°. This algorithm is obtained if we use the filtering functions:

$$h_S(x) = \delta\left(x + \frac{X_r}{8}\right) + \delta\left(x - \frac{X_r}{8}\right) \tag{5.49}$$

and

$$h_C(x) = \delta\left(x - \frac{X_r}{8}\right) + \delta\left(x - \frac{3X_r}{8}\right) \tag{5.50}$$

where $X_r = 1/f_r$. These two filtering functions are different. The Fourier transforms of these functions are:

$$H_S(f) = 2\cos\left(\frac{\pi}{4}\frac{f}{f_r}\right) \tag{5.51}$$

and

$$
\begin{aligned}
H_C(f) &= 2\cos\left(\frac{\pi}{4}\frac{f}{f_r}\right)\exp\left(-i\frac{\pi}{2}\frac{f}{f_r}\right) \\
&= 2\cos\left(\frac{\pi}{4}\frac{f}{f_r}\right)\cos\left(\frac{\pi}{2}\frac{f}{f_r}\right) + i2\cos\left(\frac{\pi}{4}\frac{f}{f_r}\right)\sin\left(\frac{\pi}{2}\frac{f}{f_r}\right)
\end{aligned}
\tag{5.52}
$$

We can see that, although the two filtering functions are different, the amplitudes of the two Fourier transforms are equal, as shown in Figure 5.8. A zero of this amplitude occurs at $2f_r$, as required by Equation 5.38. The conditions in Equations 5.36 and 5.39 are also satisfied.

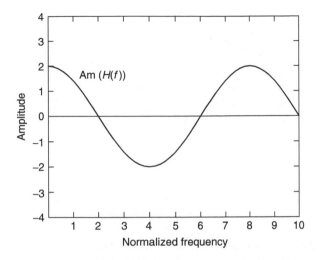

Figure 5.8 Amplitudes of the Fourier transforms of the filtering function for Wyant's algorithm.

5.3.1.2 Four-Steps-in-Cross Algorithm

The four-steps-in-cross algorithm (see Section 6.3.1) uses four sampling points, located at 0°, 90°, 180°, and 270°. This is a diagonal least-squares algorithm. It can be obtained if we use the filtering function:

$$h_S(x) = h_C(x)$$

$$= \delta(x) + \delta\left(x - \frac{X_r}{4}\right) + \delta\left(x - \frac{X_r}{2}\right) + \delta\left(x - \frac{3X_r}{4}\right) \quad (5.53)$$

The Fourier transform of this function is:

$$H_S(f) = 2\left[\cos\left(\frac{3\pi}{4}\frac{f}{f_r}\right) + \cos\left(\frac{\pi}{4}\frac{f}{f_r}\right)\right]\exp\left(-i\frac{3\pi}{4}\frac{f}{f_r}\right) \quad (5.54)$$

and its amplitude is shown in Figure 5.9. We can see that the amplitude has zeros at the reference frequency (f_r) and at twice this frequency. Conditions in Equations 5.38 to 5.40 are thus satisfied.

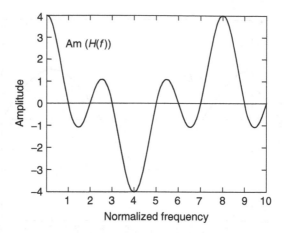

Figure 5.9 Amplitude of the Fourier transform of the filtering function for the four steps in the cross algorithm.

5.3.1.3 Schwider–Hariharan Five-Step (4 + 1) Algorithm

The Schwider–Hariharan five-step (4 + 1) algorithm (Schwider et al., 1983; Hariharan et al., 1987; see Section 6.5.2) uses five sampling points, located at 0°, 90°, 180°, 270°, and 360°. This algorithm is obtained when we use the filtering function:

$$h_S(x) = h_C(x)$$

$$= \frac{1}{2}\delta(x) + \delta\left(x - \frac{X_r}{4}\right) + \delta\left(x - \frac{X_r}{2}\right) + \delta\left(x - \frac{3X_r}{4}\right) + \frac{1}{2}\delta(x - X_r) \quad (5.55)$$

The Fourier transform of this function is:

$$H_S(f) = H_C(f)$$

$$= \left[\cos\left(\pi \frac{f}{f_r}\right) + 2\cos\left(\frac{\pi}{2}\frac{f}{f_r}\right) + 1\right]\exp\left(-i\pi \frac{f}{f_r}\right) \quad (5.56)$$

and its amplitude is shown in Figure 5.10. We can see that the amplitude of this Fourier transform of the filtering functions has zeros at the reference frequency and at twice the reference frequency, thus satisfying Equations 5.38, 5.39, and 5.40.

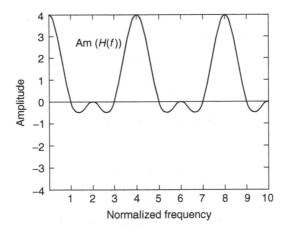

Figure 5.10 Amplitude of the Fourier transform of the filtering function for the Schwider–Hariharan algorithm.

It is interesting to notice in Equations 5.27 and 5.28, as well as in Figure 5.3, that the term with frequency f_r is fixed, and its position is independent of any possible difference between the reference frequency (f_r) and the signal frequency (f) (detuning). On the other hand, the Fourier components with the lowest frequency and with frequency $f + f_r$ may have slight frequency variations with this frequency deviation. The slope of the amplitude in these two regions is nearly zero, making this algorithm insensitive to small detuning.

5.4 FOURIER DESCRIPTION OF SYNCHRONOUS PHASE DETECTION

In this section we will study the synchronous detection in a more general manner, from a Fourier domain point of view, as developed by Freischlad and Koliopoulos (1990) and Parker (1991) and later reviewed by Larkin and Oreb (1992). If we want to remove the restriction of equally and uniformly spaced sampling points, the product of the sine function and the low-pass filtering function $h(-x)$ must be more generally considered, as the function $g_1(x)$. This function does not necessarily have to be the product of a sine function by a filtering function. In

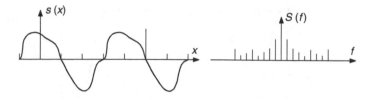

Figure 5.11   A periodic distorted signal and its spectrum.

an analogous manner, the function $g_2(x)$ replaces the product of the cosine function by the filtering function. These two functions will be referred to as the *sampling reference functions*.

The treatment here considers synchronous detection with the following two general assumptions:

1. The signal to be detected is periodic but not necessarily sinusoidal; in other words, it may contain harmonics.
2. The two reference functions, $g_1(x)$ and $g_2(x)$, are used instead of the products of the sine and cosine functions by the low-pass filtering function.

This approach will allow us to analyze many possible sources of errors. It will also permit the study of the detection of a sinusoidal signal with a frequency other than that of the reference functions.

A real periodic distorted signal, $s(x)$, as shown in Figure 5.11, has several harmonic frequencies — that is, frequencies that are integer multiples of the fundamental frequency f — and may be written as:

$$s(x) = S_0 + 2\sum_{m=1}^{\infty} S_m \cos(2\pi mfx + \phi_m) \tag{5.57}$$

or, equivalently,

$$s(x) = \sum_{m=-\infty}^{\infty} S_m \exp i(2\pi mfx + \phi_m) \tag{5.58}$$

where we have defined $S_{-m} = -S_m$, $\phi_{-m} = -\phi_m$, and $\phi_0 = 0$.

Thus, the Fourier transform of this signal may be represented by:

$$S(f) = \sum_{m=-\infty}^{\infty} S_m \delta(f - mf) \exp(i\phi_m) \qquad (5.59)$$

In this expression, m is the harmonic component number; S_m and ϕ_m are the amplitude and phase at the origin, respectively, of the harmonic component m; and f is the fundamental frequency of the signal.

The two sampling reference functions, $g_i(x)$, are real and not necessarily periodical but they do have a continuous Fourier transform with many sinusoidal components with different frequencies. Also, the sinusoidal elements of the two functions do not necessarily have the same amplitude nor are they necessarily orthogonal at any frequency, only at certain selected frequencies. In order to use these sampling functions as references, their Fourier elements at the desired *reference frequency* of these functions must be orthogonal, must have the same amplitude, and must not have any DC bias. Ideally, the reference frequency is the fundamental frequency of the signal to be detected. Because in general this is not known with a high degree of accuracy, we define the reference frequency as the assumed fundamental frequency of the signal. In other words, the elemental reference components $\delta g_i(x)$ at the reference frequency ideally should be the typical sine and cosine functions:

$$\delta g_1(x) = \pm A \sin(2\pi f_r x - \psi(f_r))\delta f$$
$$= A \cos(2\pi f_r x - \psi(f_r) \mp \pi/2)\delta f \qquad (5.60)$$

and

$$\delta g_2(x) = A \cos(2\pi f_r x - \psi(f_r))\delta f \qquad (5.61)$$

where $\psi(f_r)$ is the displacement in the positive direction of the Fourier element $\delta g_i(x)$ with frequency f_r of the reference function $g_i(x)$, with respect to the origin of the phase. The frequency interval, δf, is formed by two symmetrical intervals placed to cover positive as well as negative frequencies with value $|f_r|$. The first maxima of the Fourier transform $G_i(f)$ is frequently located near the reference frequency (f_r) but not necessarily.

We have seen before that the phase is the ratio of the two convolutions in Equations 5.33 and 5.34, using the proper filtering function. On the other hand, we also have seen that if the goal is to find the phase at the origin (ϕ), we need to evaluate the convolution only at this origin. So, it is reasonable to expect that the phase will be given by the ratio $r(f)$ of the correlations:

$$r(f) = \frac{C_1}{C_2} = \frac{\displaystyle\int_{-\infty}^{+\infty} s(x)g_1(x)\mathrm{d}x}{\displaystyle\int_{-\infty}^{+\infty} s(x)g_2(x)\mathrm{d}x} \qquad (5.62)$$

if the functions $g_1(x)$ and $g_2(x)$ are properly selected. This correlation ratio is a function of signal frequency f, as well as of the signal phase (ϕ). If the two reference functions, $g_1(x)$ and $g_2(x)$, satisfy the intuitive conditions stated earlier, by analogy with Equation 5.28 we can expect the phase (ϕ) of the signal harmonic with frequency f being detected to be given by:

$$\tan\big(\phi - \psi(f_r)\big) = \mp r(f_r) \qquad (5.63)$$

We will prove this expression to be correct if these conditions are satisfied; otherwise, the phase ϕ cannot be found with this expression. Let us now study with some detail when these conditions are satisfied. The quantity C_j has been defined as:

$$C_j = \int_{-\infty}^{\infty} s(x)g_j(x)\mathrm{d}x, \quad j = 1,2 \qquad (5.64)$$

which is the cross-correlation of the two functions evaluated at the origin, $s(x)$ and $g_i(x)$. For simplicity, we will simply call these quantities *correlations*.

We can see that the ratio of the correlations $r(f)$ is a function of the reference and signal frequencies and that it is directly related to the phase of the real signal only if the proper conditions for the functions $g_j(x)$ are met. From the central ordinate theorem expressed by Equation 2.14 we find:

$$C_j = \left(\mathrm{F}\{s(x)g_j(x)\}\right)_{f=0}, \quad j = 1,2 \tag{5.65}$$

evaluated at the origin ($f = 0$), because the quantity to be determined is the phase of the fundamental frequency of the signal with respect to the phase of the reference functions. Now, using the convolution theorem in Equation 2.18, we find:

$$C_j = \left(S(f) * G_j(f)\right)_{f=0}, \quad j = 1,2 \tag{5.66}$$

where $S(f)$ and $G_j(f)$ are the Fourier transforms of $s(x)$ and $g_j(x)$, respectively. Hence, writing the convolution at $f = 0$, we obtain:

$$C_j = \int_{-\infty}^{\infty} \left(S(v)G_j(-v)\right)dv, \quad j = 1,2 \tag{5.67}$$

where v is the dummy variable used in the convolution. Because $s(x)$ and $g_j(x)$ are real, $S(f)$ and $G_j(f)$ are Hermitian and we obtain:

$$C_j = 2\,\mathrm{Re}\int_0^{\infty} S(f)G_j^*(f)df, \quad j = 1,2 \tag{5.68}$$

where Re stands for the real part, and the symbol * denotes the complex conjugate. For clarity, the dummy variable v has been changed to the frequency variable f.

If we substitute here the value of $S(f)$ from Equation 5.59 we obtain:

$$C_j = 2\,\mathrm{Re} \sum_{m=-\infty}^{\infty} S_m G_j^*(mf)\exp(i\phi_m), \quad j = 1,2 \tag{5.69}$$

The reference functions $g_1(x)$ and $g_2(x)$ are real; hence, their Fourier transforms are complex and Hermitian. Quite generally, using Equation 2.5 we may express these functions $G_j(f)$ as:

$$G_j(f) = \text{Am}(G_j(f))\exp(i\gamma_j(f)), \quad j = 1, 2 \qquad (5.70)$$

where $\gamma_j(f)$ is the phase of the Fourier element with frequency f of the reference functions $g_j(x)$. Also, $\gamma_j(-mf) = -\gamma_j(-mf)$ because $G_j(f)$ is Hermitian. Hence,

$$C_j = 2\,\text{Re}\sum_{m=-\infty}^{\infty} S_m\text{Am}(G_j(mf))\exp(i(\phi_m - \gamma_j(mf))), \quad j = 1, 2 \quad (5.71)$$

Because the argument of the exponential function is antisymmetric with respect to m, this equation may also be written as:

$$C_j = 2S_0\text{Am}(G_j(0)) +$$
$$+ 4\sum_{m=1}^{\infty} S_m\text{Am}(G_j(mf))\cos(\phi_m - \gamma_j(mf)), \quad j = 1, 2 \qquad (5.72)$$

This expression is valid for C_1 as well as for C_2 and for any harmonic component of the signal with frequency mf. The correlation ratio, $r(f)$, is then given by:

$$r(f) = \frac{S_0\text{Am}(G_1(0)) + 2\sum_{m=1}^{\infty} S_m\text{Am}(G_1(mf))\cos(\phi_m - \gamma_1(mf))}{S_0\text{Am}(G_2(0)) + 2\sum_{m=1}^{\infty} S_m\text{Am}(G_2(mf))\cos(\phi_m - \gamma_2(mf))} \qquad (5.73)$$

This is a completely general expression for the value of $r(f)$, but, as pointed out before, it does not produce correct results for the signal phase unless certain conditions are met, as will be seen next. The elemental Fourier components of these functions at the frequency of the signal being selected must satisfy the following conditions, briefly mentioned previously:

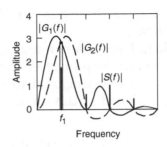

Frequency

Figure 5.12   Fourier spectra of the two reference functions and a signal.

1. The Fourier elements of the reference functions $g_1(x)$ and $g_2(x)$ must have a zero DC term. Also, the Fourier transforms $G_1(f)$ and $G_2(f)$ of the two reference functions at zero frequency must be equal to zero.
2. All interference (cross-talk) between undesired harmonics in the signal and in the reference functions must be avoided.
3. The Fourier elements of the reference functions $g_1(x)$ and $g_2(x)$ at frequency f_r must be orthogonal to each other. This means that the Fourier transforms $G_1(f)$ and $G_2(f)$ of the two reference functions at frequency f_r must have a phase difference equal to $\pm\pi/2$. The plus sign corresponds to the upper sign in Equation 5.57, and the phase of $G_2(f)$ is $\pi/2$ greater than the phase of $G_1(f)$.
4. The Fourier transforms $G_1(f)$ and $G_2(f)$ of the two reference functions at frequency f_r must have the same amplitude.

Given a reference frequency, these four conditions can in general be satisfied only at certain signal frequencies. To illustrate these conditions, Figure 5.12 illustrates the Fourier spectra of two reference functions plotted together with the Fourier spectra of a periodical signal. Here, we notice the following for the functions $G_1(f)$ and $G_2(f)$:

1. They pass through the origin, indicating that their DC bias is zero.

2. The harmonics of the signal are located at zeros of these functions.
3. The functions have the same amplitude and sign at the fundamental frequency of the signal, f. If these functions are also orthogonal to each other, all conditions are satisfied at the fundamental frequency of the signal.

Let us now consider the four conditions listed above and apply them to Equation 5.71. The first condition of a zero DC term may be easily satisfied if, from the central theorem studied in Chapter 2, we write:

$$G_1(0) = G_2(0) = 0 \qquad (5.74)$$

Then Equation 5.73 becomes:

$$r(f) = \frac{\displaystyle\sum_{m=1}^{\infty} S_m \mathrm{Am}(G_1(mf))\cos(\phi_m - \gamma_1(mf))}{\displaystyle\sum_{m=1}^{\infty} S_m \mathrm{Am}(G_2(mf))\cos(\phi_m - \gamma_2(mf))} \qquad (5.75)$$

The second condition (no interference from undesired harmonics) is satisfied if, for all harmonics m, with the exception of the fundamental frequency, which is being measured, we have:

$$S_m G_i(mf) = 0; \quad \text{for } m > 1 \qquad (5.76)$$

This means that the harmonic components $m > 1$ should not be present, either in the signal or in the reference functions. Obviously, if the signal is perfectly sinusoidal, this condition is always satisfied.

Applying these two conditions to a sinusoidal signal with frequency f, Equation 5.73 becomes:

$$r(f) = \frac{\mathrm{Am}(G_1(f))\cos(\phi - \gamma_1(f))}{\mathrm{Am}(G_2(f))\cos(\phi - \gamma_2(f))}$$

$$= \frac{\mathrm{Re}\{G_1(f)\exp(-i\phi)\}}{\mathrm{Re}\{G_2(f)\exp(-i\phi)\}} \qquad (5.77)$$

During the phase-detection process, the frequency of the signal has to be estimated so the reference frequency (f_r) is as close as possible to this value. We say that a detuning error has occurred if the reference frequency (f_r) is different from the signal frequency (f).

Now, we need to satisfy only two more conditions. For the two elements of the two reference functions to be orthogonal to each other at the reference frequency (f_r), we need:

$$G_1(f_r) = \mp iz(f_r)G_2(f_r) = z(f_r)G_2(f_r)\exp\left(\mp i\frac{\pi}{2}\right) \qquad (5.78)$$

at the harmonic m being considered. The sign of the reference sampling functions is chosen so that the Fourier transforms of the reference sampling functions at the reference frequency are both positive (or both negative). Then, the upper (minus) sign is taken when the phase of $G_2(f_r)$ is $\pi/2$ greater than the phase of $G_1(f_r)$. This case corresponds to the upper sign in Equation 5.60. Thus, the phases $\gamma_1(f_r)$ and $\gamma_2(f_r)$ at the reference frequency in Equation 5.70 are related by:

$$\gamma_1(f_r) = \gamma_2(f_r) \mp \frac{\pi}{2} \qquad (5.79)$$

The values of these angles depend on the location of the point selected as the origin of the coordinates $(x = 0)$.

The condition that the amplitudes of the Fourier components at the frequency being detected are equal requires that

$$\mathrm{Am}\big(G_1(f_r)\big) = \mathrm{Am}\big(G_2(f_r)\big) \qquad (5.80)$$

Thus, applying these last two conditions, we finally obtain:

$$r(f_r) = \mp\left(\frac{\sin\big(\phi - \gamma_2(f_r)\big)}{\cos\big(\phi - \gamma_2(f_r)\big)}\right) = \mp\tan\big(\phi - \gamma_2(f_r)\big) \qquad (5.81)$$

where, as noted previously, the upper sign is taken when the phase of $G_2(f_r)$ is $\pi/2$ greater than the phase of $G_1(f_r)$ (i.e., $\gamma_2(f_r) > \gamma_1(f_r)$), and vice versa.

We have defined $\psi(f_r)$ as the phase displacement in the positive direction of the zero phase point of the Fourier elements of the reference functions with frequency f_r, with respect to the origin of coordinates, which now we can identify with $\gamma_2(f_r)$. Thus, we can write:

$$\psi(f_r) = \gamma_2(f_r) \tag{5.82}$$

We see that when $\psi(f_r)$ is equal to zero, the function $G_2(f)$ becomes real at the reference frequency. In this case, the function element $\delta g_1(x)$ is antisymmetrical. In other words, the origin of coordinates is located at the zero phase point of this sine function.

To conclude, the signal phase is given by:

$$\tan(\phi - \psi(f_r)) = \tan(\phi - \gamma_2(f_r)) = \mp r(f_r) \tag{5.83}$$

as was intuitively expected.

5.5 SYNCHRONOUS DETECTION USING A FEW SAMPLING POINTS

Let us now apply the general theory of synchronous detection just developed to the particular case of a discrete sampling procedure using only a few sampling points. As illustrated in Figure 5.13, let us take $N \geq 3$ points with their relative phases α_n, referred to the origin $\mathbf{O}_\alpha$. The phases of the sampling points are measured with respect to the origin of the reference function, which may be located at any arbitrary position, not necessarily the origin of coordinates or any sampling point in particular. Thus, we obtain N equations from which the signal phase (ϕ) at the origin of the reference function may be calculated.

The location of the phase origin, $\mathbf{O}_\alpha$, for the sampling points is the same as the zero phase point for the sampling reference functions at the reference frequency, but not necessarily at any other frequency. According to the translation property in Fourier theory, because the two reference functions are orthogonal to each other at the reference frequency

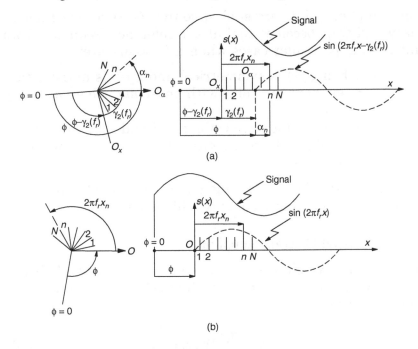

Figure 5.13   Sampling a signal with equally spaced points.

(f_r), the location of the zero phase point with respect to the sampling points may be selected so that the Fourier transform $G_1(f_r)$ is real and the Fourier transform $G_2(f_r)$ is imaginary, or vice versa.

Given a phase-detecting sampling algorithm for which we have defined the positions of the sampling points with respect to the origin of coordinates ($x = 0$) and their associated sampling weights, the value of $\gamma_2(f_r)$ is already determined and its value can be found after the Fourier transform $G_2(f)$ has been calculated. Thus, we have:

$$\alpha_n(x) = 2\pi f_r x_n - \gamma_2(f_r) \tag{5.84}$$

A common approach in most sampling algorithms is to place the zero phase origin, $\mathbf{O}_\alpha$ (i.e., the origin of the reference functions $\cos(2\pi f_r x)$ and $\sin(2\pi f_r x)$), at the coordinate origin, $\mathbf{O}_x$, thus making $\gamma_2(f_r) = 0$, as shown in Figure 5.13b. Then, the sampling points are shifted so that $G_1(f)$ becomes imaginary and $G_2(f)$ becomes real at the reference frequency. Two interesting particular cases when this occurs are:

1. When $g_1(x)$ is symmetrical and $g_2(x)$ is antisymmetrical about the point with phase $m\pi$, where m is any integer
2. When $g_1(x)$ is antisymmetrical and $g_2(x)$ is symmetrical about the point with phase $(m + 1/2)\pi$, where m is any integer

If desired, the first sampling point may be placed at the coordinate origin, but frequently this is not the case.

5.5.1   General Discrete Sampling

If we sample N points, with an arbitrary separation between them, we can see that the sampling reference functions are then given by:

$$g_1(x) = \sum_{n=1}^{N} W_{1n}\delta(x - x_n) \qquad (5.85)$$

and

$$g_2(x) = \sum_{n=1}^{N} W_{2n}\delta(x - x_n) \qquad (5.86)$$

where the W_{in} are the sampling weights for each sampling point, and N is the number of sampling points with coordinates $x = x_n$. The Fourier transforms of these sampling reference functions are:

$$G_1(f) = \sum_{n=1}^{N} W_{1n}\exp(-i2\pi f x_n) \qquad (5.87)$$

and

$$G_2(f) = \sum_{n=1}^{N} W_{2n}\exp(-i2\pi f x_n) \qquad (5.88)$$

but from Equation 5.84 we can write:

$$2\pi f x_n = \left(\alpha_n + \gamma_2(f_r)\right)\frac{f}{f_r} \tag{5.89}$$

Hence, these Fourier transforms become:

$$G_1(f) = \exp\left(-i\gamma_2(f_r)\frac{f}{f_r}\right)\sum_{n=1}^{N} W_{1n}\exp\left(-i\alpha_n\frac{f}{f_r}\right) \tag{5.90}$$

and

$$G_2(f) = \exp\left(-i\gamma_2(f_r)\frac{f}{f_r}\right)\sum_{n=1}^{N} W_{2n}\exp\left(-i\alpha_n\frac{f}{f_r}\right) \tag{5.91}$$

Now, because the reference functions are to be orthogonal to each other and have the same amplitude at the frequency $f = f_r$, we need, as in Equation 5.74,

$$G_1(f_r) = \mp i G_2(f_r) \tag{5.92}$$

where as usual the upper (minus) sign indicates that the phase of $G_2(f_r)$ is $\pi/2$ greater than the phase of $G_1(f_r)$; that is, $\gamma_1(f_r) < \gamma_2(f_r)$. Using this expression with Equations 5.87 and 5.88, we find:

$$\sum_{n=1}^{N}\left(W_{2n} \mp i W_{1n}\right)\exp(-i2\pi f_r x_n) = 0 \tag{5.93}$$

Thus, we have:

$$\sum_{n=1}^{N}\left(W_{2n} \mp i W_{1n}\right)\cos(2\pi f_r x_n) - i\sum_{n=1}^{N}\left(W_{2n} \mp i W_{1n}\right)\sin(2\pi f_r x_n) = 0 \tag{5.94}$$

or

$$\sum_{n=1}^{N}\left[W_{2n}\cos(2\pi f_r x_n) \mp W_{1n}\sin(2\pi f_r x_n)\right] - $$

$$- i\sum_{n=1}^{N}\left[W_{2n}\sin(2\pi f_r x_n) \pm W_{1n}\cos(2\pi f_r x_n)\right] = 0 \tag{5.95}$$

which can be true only if:

$$\sum_{n=1}^{N}\left[W_{2n}\cos(2\pi f_r x_n)\mp W_{1n}\sin(2\pi f_r x_n)\right]=0 \qquad (5.96)$$

and

$$\sum_{n=1}^{N}\left[W_{2n}\sin(2\pi f_r x_n)\pm W_{1n}\cos(2\pi f_r x_n)\right]=0 \qquad (5.97)$$

We can now define the Fourier transform vectors $\mathbf{G}_1$ and $\mathbf{G}_2$ as:

$$\mathbf{G}_1 = \left(\sum_{n=1}^{N}W_{1n}\cos(2\pi f_r x_n),\ \sum_{n=1}^{N}W_{1n}\sin(2\pi f_r x_n)\right) \qquad (5.98)$$

and

$$\mathbf{G}_2 = \left(\sum_{n=1}^{N}W_{2n}\cos(2\pi f_r x_n),\ \sum_{n=1}^{N}W_{2n}\sin(2\pi f_r x_n)\right) \qquad (5.99)$$

where, from Equations 5.87 and 5.88, we see that the x and y components of the vector are the real and imaginary parts of the Fourier transforms of the reference functions. These Fourier transform vectors can also be written as:

$$\mathbf{G}_1 = \mathbf{G}_{11}+\mathbf{G}_{12}+\mathbf{G}_{13}+...+\mathbf{G}_{1N} \qquad (5.100)$$

and

$$\mathbf{G}_2 = \mathbf{G}_{21}+\mathbf{G}_{22}+\mathbf{G}_{23}+...+\mathbf{G}_{2N} \qquad (5.101)$$

where this is a vector sum of the vectors $\mathbf{G}_{in}$ defined by:

$$\mathbf{G}_{in} = \left(W_{in}\cos(2\pi f_r x_n),\ W_{in}\sin(2\pi f_r x_n)\right) \qquad (5.102)$$

If we use these vectors in Equations 5.96 and 5.97, we will see that the vectors are orthonormal; that is, they are mutually perpendicular and have the same magnitude at the frequency f_r. Thus, we may say that the two reference sampling functions are orthogonal and have the same amplitude

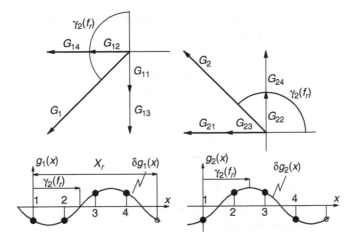

Figure 5.14   Sampling reference vectors for a sampling algorithm.

if the two Fourier transform vectors are mutually perpendicular and have the same magnitude, as illustrated in Figure 5.14. The angle of $\mathbf{G}_1$ is $\pi/2$ greater than that of $\mathbf{G}_2$ for the upper sign. The angle of $\mathbf{G}_1$ with respect to the positive horizontal axis is equal to $\gamma_1(f_r)$. In the same manner, the angle of $\mathbf{G}_2$ with the positive horizontal axis is equal to $\gamma_2(f_r)$.

Quite frequently, the phase origin in algorithms is located at a point such that $G_1(f)$ is imaginary and $G_2(f)$ is real at the reference frequency. Under these conditions, vector $\mathbf{G}_1$ is vertical, vector $\mathbf{G}_2$ is horizontal, and Equations 5.96 and 5.97 may be written as:

$$\sum_{n=1}^{N}\left[W_{1n}\cos(2\pi f_r x_n)\right]=0 \tag{5.103}$$

$$\sum_{n=1}^{N}\left[W_{2n}\sin(2\pi f_r x_n)\right]=0 \tag{5.104}$$

and

$$\sum_{n=1}^{N}\left[W_{1n}\sin(2\pi f_r x_n)\right]=\sum_{n=1}^{N}\left[W_{2n}\cos(2\pi f_r x_n)\right] \tag{5.105}$$

Additionally, we must have no bias in the reference functions, which is true if:

$$\sum_{n=1}^{N} W_{1n} = 0 \tag{5.106}$$

and

$$\sum_{n=1}^{N} W_{2n} = 0 \tag{5.107}$$

The value of the phase ϕ may be calculated by using Equations 5.75 and 5.86 in Equation 5.62 and then using Equation 5.82 to obtain:

$$\tan\left(\phi - \gamma_2(f_r)\right) = \mp \frac{\displaystyle\sum_{n=1}^{N} s(x_n) W_{1n}}{\displaystyle\sum_{n=1}^{N} s(x_n) W_{2n}} \tag{5.108}$$

The upper sign corresponds to the cases when $\gamma_1(f_r) - \gamma_2(f_r) < 0$, and the lower sign otherwise. As pointed out before, the constant phase $\gamma_2(f_r)$ in most algorithms is equal to zero.

5.5.2 Equally Spaced and Uniform Sampling

A frequent, particular case is when the sampling points are equally separated and uniformly distributed in the signal period X_r, with the positions defined as in Equation 5.10 by:

$$x_n = \frac{(n-1)X_r}{N} + x_1 = \frac{(n-1)}{Nf_r} + x_1; \quad n = 1,...,N \tag{5.109}$$

In this expression, the origin $(\mathbf{O}_\alpha)$ for the reference function and the first sampling point was taken at the origin of coordinates $(\mathbf{O}_x)$, as shown in Figure 5.13b. The reference frequency (f_r) is defined as $1/X_r$ and is usually equal to the signal frequency but may differ.

As described in Section 5.1, with this sampling distribution we have:

$$\sum_{n=1}^{N} \sin\left(2\pi f_r x_n - \gamma_2(f_r)\right) = \sum_{n=1}^{N} \sin(2\pi f_r x_n) = 0 \quad (5.110)$$

$$\sum_{n=1}^{N} \cos\left(2\pi f_r x_n - \gamma_2(f_r)\right) = \sum_{n=1}^{N} \cos(2\pi f_r x_n) = 0 \quad (5.111)$$

$$\sum_{n=1}^{N} \cos\left(4\pi f_r x_n - \gamma_2(f_r)\right) = \sum_{n=1}^{N} \cos(4\pi f_r x_n) = 0 \quad (5.112)$$

and

$$\sum_{n=1}^{N} \sin\left(4\pi f_r x_n - \gamma_2(f_r)\right) = \sum_{n=1}^{N} \sin(4\pi f_r x_n) = 0 \quad (5.113)$$

These results are independent of the location of the origin for the phases — that is, for any value of $\gamma_2(f_r)$. The reason for this becomes clear if we notice that the vector diagram in Figure 5.2 remains in equilibrium when all vectors are rotated by an angle $\gamma_2(f_r)$.

The condition of no DC term (bias) on the reference functions is expressed by Equations 5.106 and 5.107. From Equation 5.112, we can see that:

$$\sum_{n=1}^{N} \cos\left(2\pi f_r x_n - \gamma_2(f_r)\right) \cos(2\pi f_r x_n) -$$

$$-\sum_{n=1}^{N} \sin\left(2\pi f_r x_n - \gamma_2(f_r)\right) \sin(2\pi f_r x_n) = 0 \quad (5.114)$$

and from Equation 5.113:

$$\sum_{n=1}^{N} \cos\left(2\pi f_r x_n - \gamma_2(f_r)\right) \sin(2\pi f_r x_n) +$$

$$+\sum_{n=1}^{N} \sin\left(2\pi f_r x_n - \gamma_2(f_r)\right) \cos(2\pi f_r x_n) = 0 \quad (5.115)$$

Now, we can see that these two last expressions become identical to Equations 5.96 and 5.97 if the sampling weights are defined by:

$$W_{1n} = \pm\sin(2\pi f_r x_n - \gamma_2(f_r)) \qquad (5.116)$$

and

$$W_{2n} = \cos(2\pi f_r x_n - \gamma_2(f_r)) \qquad (5.117)$$

When $\gamma_2(f_r) = 0$, Equations 5.110, 5.111, 5.114, and 5.115 are the same as those used in Section 5.1 in order to make the least-squares matrix diagonal.

Now, we can obtain the phase value with the ratio of the correlations by using the sampling weights in Equation 5.108, assuming that $\gamma_2(f_r) = 0$:

$$\tan\phi = \mp\left(\frac{\displaystyle\sum_{n=1}^{N} s(x_n)\sin(2\pi f_r x_n)}{\displaystyle\sum_{n=1}^{N} s(x_n)\cos(2\pi f_r x_n)} \right) \qquad (5.118)$$

and the signal may be calculated with Equation 5.83. As pointed out before, the upper sign is used when $\gamma_1(f_r) < 0$. This result is the diagonal least-squares algorithm.

We have pointed out before that the location of the origin of coordinates is important because it affects the algebraic appearance (phase) of the result; however, for any selected origin location, the relative phase for all points is the same. The two typical locations for the origin are (1) the first sampling point or (2) the zero phase point for the Fourier elements.

5.5.3 Applications of Graphical Vector Representation

Graphical vector representation has three quite interesting properties:

1. By examining the vectors of any two algorithms that satisfy the conditions for orthogonality and equal

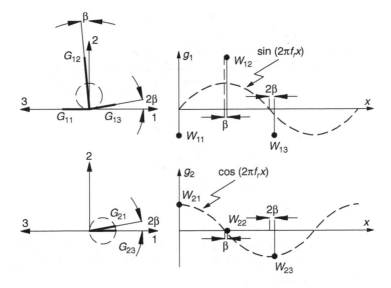

Figure 5.15 Effect of detuning in a three-point algorithm (inverted T); the upper part shows the effects on g_1 and $\mathbf{G}_{1n}$, and the lower part shows the effects on g_2 and $\mathbf{G}_{2n}$.

amplitudes of $G_1(f)$ and $G_2(f)$, we can see that a super-position of both algorithms also satisfies the required conditions.

2. Any vector system with zero bias and in equilibrium may be added to the system without changing the conditions of either orthogonality or equal amplitudes at the reference frequency.

3. A detuning shifts the angular orientations of the vectors W_{ij} a small angle (β) directly proportional to their phase (α_n).

To illustrate, let us consider the effect of detuning using vector representation in two algorithms with three sampling points. The first one to be considered is shown in Figure 5.15. The three points have phases 0°, 90°, and 180°; however, in the presence of detuning, as shown in this figure, the sampling points have phases 0°, 90° + β, and 180° + 2β. Examining the vector plots on the left side of this figure, we see that the

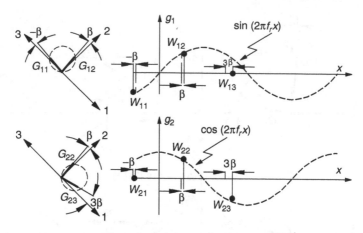

Figure 5.16 Effect of detuning in a three-point algorithm (Wyant's); the upper part shows the effects on g_1 and $\mathbf{G}_{1n}$, and the lower part shows the effects on g_2 and $\mathbf{G}_{2n}$.

vector sums $\mathbf{G}_1$ and $\mathbf{G}_2$ are both rotated by angle β, thus preserving their orthogonality. Because β is arbitrary, the orthogonality condition is preserved at all frequencies, but the amplitudes are not.

Figure 5.16 shows another algorithm, where the sampling points are located at $-45°$, $45°$, and $135°$. In the presence of detuning, the three phases will be $-(45° + \beta)$, $(45° + \beta)$, and $(135° + 3\beta)$, and the vectors on the left side of the figure are angularly displaced. We may easily observe that the angle between vectors $\mathbf{G}_{1n}$ is preserved, as is the angle between vectors $\mathbf{G}_{2n}$. Thus, the amplitudes of $G_1(f)$ and $G_2(f)$ are preserved, but their orthogonality is not.

5.5.4 Graphic Method To Design Phase-Shifting Algorithms

Using this theory of phase-shifting algorithms, Malacara-Doblado et al. (2000) proposed a method to design such algorithms with particular desired properties. The reference functions $g_1(x)$ and $g_2(x)$ are assumed to be formed by a linear combination of symmetric and antisymmetric components, respectively. Thus, we can write Equation 5.85 and 5.86 as:

$$g_1(x) = \sum_{n=1}^{N} W_{1n}\delta(x - x_n) = \sum_{k=1}^{K} w_{1k}h_{1k}(x)$$

$$g_2(x) = \sum_{n=1}^{N} W_{2n}\delta(x - x_n) = \sum_{k=1}^{K} w_{2k}h_{2k}(x)$$

(5.119)

where $h_{1k}(x)$ and $h_{2k}(x)$ are the symmetric and antisymmetric harmonic components, respectively. The number of sampling points is N, and the number of harmonic components is K. In this case, the reference functions $g_1(x)$ and $g_2(x)$ will always be orthogonal at all frequencies. The zero bias condition is guaranteed if the weight of the central sampling points for the symmetrical harmonic components is set such that the sum of all weights is zero, thus obtaining:

$$h_{1k}(x) = \delta(x - x_k) - \delta(x + x_k)$$

$$h_{2k}(x) = -\delta(x - x_k) + 2\delta(x) - \delta(x - x_k)$$

(5.120)

where the coordinate x_k is given by:

$$x_k = \frac{\alpha_k}{2\pi f_r}$$

(5.121)

and $\alpha_k = \beta$, where α is the angle of separation between two consecutive sampling points.

The Fourier transform amplitudes of these harmonic components, $H_{1k}(f)$ and $H_{2k}(f)$, are shown in Figure 5.17 for a phase separation between the sampling points equal to $\beta = \pi/2$. The Fourier transforms of the sampling functions, $G_1(f)$ and $G_2(f)$, are given by:

$$G_1(f) = \sum_{k=1}^{K} w_{1k}H_{1k}(f)$$

$$G_2(f) = \sum_{k=1}^{K} w_{2k}H_{2k}(f)$$

(5.122)

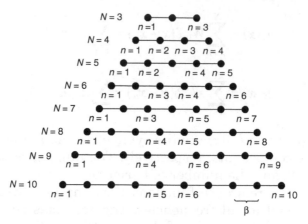

Figure 5.17 Symmetrical location of sampling points.

These Fourier transforms of the harmonic components of the sampling functions can be used to design a sampling algorithm with the desired properties. For example, let us consider those shown in Figure 5.18:

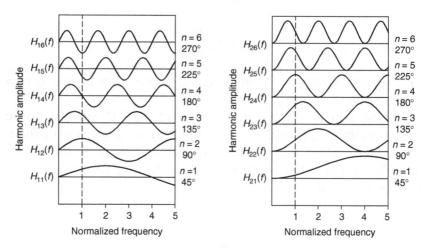

Figure 5.18 Fourier transforms of harmonic components produced by a pair of symmetrically located sampling points.

1. The component $H_{14}(f)$ has a zero at the normalized frequency equal to one $(f = f_r)$; thus, this component can be added with any multiplying weight (w_{14}) without modifying the final value of $G_1(f)$ at the frequency $f = f_r$. Its only effect would be to change the slope of this function at this frequency.

2. The components $H_{12}(f)$ and $H_{16}(f)$ have zero slope at the normalized frequency equal to one; thus, they can be added with any desired weight without modifying the slope of $G_1(f)$ at this frequency. Only the amplitude will be changed.

In general, by examining the zeros and slopes of these harmonic components at the fundamental frequency of the signal $(f = f_r)$ and its harmonics $(f = kf_r)$, the desired properties for the algorithm can be obtained.

5.6 SIGNAL AMPLITUDE MEASUREMENT

Not only can the phase of the signal be obtained with phase-shifting algorithms but also its amplitude. Assuming for simplicity that $\gamma_2(f_r) = 0$, as in most phase-shifting algorithms, then from Equations 5.73 and 5.108 we can write:

$$\frac{\displaystyle\sum_{n=1}^{N} W_{1n}s(x_n)}{\displaystyle\sum_{n=1}^{N} W_{2n}s(x_n)} = \frac{S_1\text{Am}(G_1(f_r))\sin\phi}{S_1\text{Am}(G_2(f_r))\cos\phi} \qquad (5.123)$$

where S_1 is the signal amplitude (fundamental component). We know that at the reference frequency the amplitudes of the Fourier transforms of $G_1(f_r)$ and $G_2(f_r)$ are equal and we assume that $\gamma_2(f_r) = 0$, so from Equations 5.102 and 5.103 we obtain:

$$\text{Am}(G_1(f_r)) = \frac{\sqrt{\left(\sum_{n=1}^{N} W_{1n} \sin(2\pi f_r x_n)\right)^2 + \left(\sum_{n=1}^{N} W_{1n} \cos(2\pi f_r x_n)\right)^2}}{\sqrt{\left(\sum_{n=1}^{N} W_{1n} \sin(\alpha_n)\right)^2 + \left(\sum_{n=1}^{N} W_{1n} \cos(\alpha_n)\right)^2}} \qquad (5.124)$$

If we equate the numerators and the denominators in Equation 5.108, we obtain:

$$S_1 \sin\phi = \frac{\sum_{n=1}^{N} W_{1n} s(x_n)}{\sqrt{\left(\sum_{n=1}^{N} W_{1n} \sin(\alpha_n)\right)^2 + \left(\sum_{n=1}^{N} W_{1n} \cos(\alpha_n)\right)^2}} \qquad (5.125)$$

and

$$S_1 \cos\phi = \frac{\sum_{n=1}^{N} W_{2n} s(x_n)}{\sqrt{\left(\sum_{n=1}^{N} W_{1n} \sin(\alpha_n)\right)^2 + \left(\sum_{n=1}^{N} W_{1n} \cos(\alpha_n)\right)^2}} \qquad (5.126)$$

Squaring these two last expressions we finally obtain:

$$S_1^2 = \frac{\left(\sum_{n=1}^{N} W_{2n} s(x_n)\right)^2 + \left(\sum_{n=1}^{N} W_{2n} s(x_n)\right)^2}{\left(\sum_{n=1}^{N} W_{1n} \sin(\alpha_n)\right)^2 + \left(\sum_{n=1}^{N} W_{1n} \cos(\alpha_n)\right)^2} \qquad (5.127)$$

Thus, any phase-shifting algorithm can be used to measure the signal amplitude. The second term in the denominator becomes zero if $\gamma_2(f_r) = 0$.

5.7 CHARACTERISTIC POLYNOMIAL OF A SAMPLING ALGORITHM

A characteristic polynomial that can be used with a discrete sampling algorithm was proposed by Surrel (1996). This polynomial can be used to derive all the main properties of the algorithm in a manner closely resembling the Fourier theory just described. To define this polynomial, let us use Equation 5.108, considering that the phase ϕ is given by the phase of the complex function, $V(\phi)$, defined by:

$$V(\phi) = \sum_{n=1}^{N} (W_{1n} + iW_{2n})s(x_n) \qquad (5.128)$$

where $\psi(f_r) = 0$. Then, using the Fourier expansion for the signal given by Equation 5.58 in this expression, we find:

$$V(\phi) = \sum_{m=-\infty}^{\infty} S_m \exp(i\phi_m) \sum_{n=1}^{N} (W_{1n} + iW_{2n}) \exp(i2\pi mfx_n) \qquad (5.129)$$

where $\phi = \phi_1$ is the phase of the signal at the fundamental frequency. Different harmonic components have different phases. Now, from Equation 5.89 we have:

$$V(\phi) = \sum_{m=-\infty}^{\infty} S_m \exp(i\phi_m) \sum_{n=1}^{N} (W_{1n} + iW_{2n}) \exp\left(im\alpha_n \frac{f}{f_r} \right) \qquad (5.130)$$

where α_n is the phase for the sampling point n. This phase may be assumed to be equal to $\alpha_n = (n - 1)\Delta\alpha$, where $\Delta\alpha$ is the phase interval separation between the sampling points, transforming this expression into:

$$V(\phi) = \sum_{m=-\infty}^{\infty} S_m \exp(i\phi_m) \sum_{n=1}^{N} (W_{1n} + iW_{2n}) \exp\left(im(n - 1)\frac{f}{f_r}\Delta\alpha \right) \qquad (5.131)$$

In the absence of detuning, such that $f = f_r$, then this expression can be written as:

$$V(\phi) = \sum_{m=-\infty}^{\infty} S_m \exp(i\phi_m)P[\exp(im\Delta\alpha)] \qquad (5.132)$$

where the polynomial $P(z)$ is defined by:

$$P(z) = \sum_{n=1}^{N} (W_{1n} + iW_{2n})[\exp(im\Delta\alpha)]^{(n-1)} = \sum_{n=1}^{N} \sigma_n z^{(n-1)} \quad (5.133)$$

This is the characteristic polynomial proposed by Surrel (1996) that is associated with any sampling algorithm. It is quite simple to derive this polynomial from the sampling weights W_{in}. From this characteristic polynomial we can determine many interesting properties of the sampling algorithm with which it is associated.

Let us first consider the case of no detuning ($f = f_r$). We assume, however, that the signal has harmonic distortion. The signal harmonic component m ($m \neq 1$) will not influence the value of the complex function $V(\phi)$ if the polynomial $P(z)$ has a root (zero value) for the value of z that corresponds to that harmonic.

Each complex value of z is associated with a harmonic number (m) by:

$$\exp(im\Delta\alpha) = z \qquad (5.134)$$

These values of z may be represented in a unit circle in the complex plane. Given a sampling algorithm, the value of the phase interval $\Delta\alpha$ between sampling points is fixed; that is, each possible value of the harmonic number (positive and negative) has a point, as illustrated in Figure 5.19, which is a characteristic diagram of the sampling algorithm.

In the presence of detuning ($f \neq f_r$) we can expand a Taylor series to obtain:

$$\sum_{n=1}^{N} \sigma_n \left[\exp\left(im\frac{f}{f_r}\Delta\alpha\right)\right]^{(n-1)} = P(z) + im\left(\frac{f}{f_r} - 1\right)\exp(im\Delta\alpha)P'(z) \quad (5.135)$$

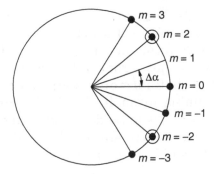

Figure 5.19 Points for each harmonic number for a sampling algorithm. If a polynomial root exists at any sampling point, the point is plotted with a large dot. If a double root exists, it is plotted with a circle around the dot.

In this case, we observe insensitivity to the harmonic component (m) as well as to detuning of that harmonic only if both $P(z)$ and its derivative have roots at the corresponding value of z. In other words, a double root must lie at that value of z.

Following are some of the important properties of this characteristic diagram:

1. An algorithm is insensitive to the harmonic component (m) if the characteristic polynomial has zeros for the values of z corresponding to $\pm m$. To state it in a different manner, the algorithm is insensitive to harmonic m when $m \neq 1$ if both $\exp(-im\Delta\alpha)$ and $\exp(im\Delta\alpha)$ are roots of the characteristic polynomial.
2. If only $\exp(-im\Delta\alpha)$ with $m > 0$ is a root and $\exp(im\Delta\alpha)$ is not a root of the characteristic polynomial, then that harmonic component can be detected. If the fundamental frequency ($m = 1$) is to be detected, as is normally the case, $\exp(-i\Delta\alpha)$ should be a root and $\exp(i\Delta\alpha)$ should not be.
3. In an analogous manner, it is possible to prove insensitivity, as well as detuning insensitivity, to harmonic m ($m = 1$) if a double zero occurs at the values of z corresponding to the αm harmonic components. In

other words, both exp $(im\Delta\alpha)$ and exp $(-im\Delta\alpha)$ are double roots of the characteristic polynomial.

4. If only exp$(-im\Delta\alpha)$ with $m > 0$ is a double root and exp$(im\Delta\alpha)$ is not a root of the characteristic polynomial, then that harmonic component can be detected with detuning insensitivity. If the fundamental frequency ($m = 1$) is to be detected with detuning insensitivity, exp$(-i\Delta\alpha)$ should be a double root and exp$(i\Delta\alpha)$ should not be a root.

As an example, let us consider the Schwider–Hariharan algorithm with $\Delta\alpha = 90°$ (studied in greater detail in Chapter 6). The phase equation is:

$$\tan\phi = -\frac{2(s_2 - s_4)}{s_1 - 2s_3 + s_5} \tag{5.136}$$

thus, the corresponding characteristic polynomial is:

$$V(\phi) = 1 - 2iz - 2z^2 + 2iz^3 + z^4$$
$$= (z - 1)(z + 1)(z + 1)^2 \tag{5.137}$$

We can observe that the signal may be detected with detuning insensitivity at the fundamental frequency and also at the fifth harmonic. The characteristic diagram for this algorithm is shown in Figure 5.20.

Many other properties can be derived from a detailed analysis of the characteristic diagram of a sampling algorithm. A close connection exists between such a characteristic diagram and the Fourier theory studied earlier. The characteristic diagrams for many sampling algorithms have been described by Surrel (1997).

5.8　GENERAL ERROR ANALYSIS OF SYNCHRONOUS PHASE-DETECTION ALGORITHMS

The theory developed in this chapter permits error analysis of sampling algorithms used for the synchronous detection of periodical signals. Some possible sources of error are discussed

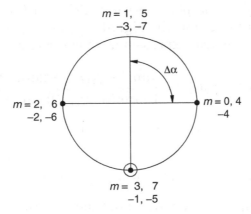

Figure 5.20 Characteristic diagram for a detuning-insensitive algorithm (Schwider–Hariharan).

in this section. In the treatment by Freischlad and Koliopoulos (1990), we have seen that, if the four conditions required in Section 5.4 are satisfied, the phase can be determined without any error. With proper algorithm design, these conditions are satisfied when the reference frequency (f_r) is equal to the frequency of a harmonic component of the signal to be detected. If one or more of the four conditions is not satisfied, an error may occur regarding the calculated phase.

5.8.1 Exact Phase-Error Analysis

We will now perform an exact phase-error analysis for the case of no harmonic components — that is, when the signal is sinusoidal and the phase shifts are linear. In the absence of any phase error, when the four conditions are satisfied, the phase is calculated with:

$$\tan\big(\phi - \gamma_2(f_r)\big) = \mp r(f_r) \tag{5.138}$$

but, in the presence of an error, the calculated phase with the phase error introduced becomes:

$$\tan\big(\phi_{err} - \gamma_2(f_r)\big) = \tan\big(\phi - \gamma_2(f_r) + \delta\phi(\phi, f)\big) = \mp r(f) \tag{5.139}$$

where $\delta\phi(\phi, f)$ is the phase error, which is a function both of the signal phase ϕ and the signal frequency f. Using a well-known trigonometric expression, we can write:

$$\tan(\phi + \delta\phi(\phi, f) - \gamma_2(f_r)) = \frac{\tan(\phi - \gamma_2(f_r)) + \tan\delta\phi(\phi, f)}{1 - \tan(\phi - \gamma_2(f_r))\tan\delta\phi(\phi, f)} \quad (5.140)$$

and from this expression we can find:

$$\tan\delta(\phi, f) = \frac{\sin(\phi - \gamma_2(f_r)) - \tan(\phi - \gamma_2(f_r) + \delta(\phi, f))\cos(\phi - \gamma_2(f_r))}{-\tan(\phi - \gamma_2(f_r) + \delta(\phi, f))\sin(\phi - \gamma_2(f_r)) - \cos(\phi - \gamma_2(f_r))} \quad (5.141)$$

This is a completely general expression for the phase error if one or more of the four required conditions is not fulfilled. Depending on which condition is not met, the ratio of the two correlations $r(f)$ defined by Equation 5.62 can be calculated as follows:

1. In the general case, Equation 5.73 can be used when one or more of the four conditions fails.
2. If the zero bias condition is the only one being satisfied, Equation 5.75 can be used.
3. If, besides satisfying the zero bias condition, the signal is perfectly sinusoidal or no cross-talk between harmonic components is present in the signal and in the reference functions, then only the orthogonality condition or the condition for equal amplitudes may be not satisfied. In this case, Equation 5.77 can be used.

We define the ratio, $\Delta(f)$, of the amplitudes of the Fourier transforms of the sampling functions as:

$$\rho(f) = \frac{\text{Am}(G_1(f))}{\text{Am}(G_2(f))} \quad (5.142)$$

By using this definition in Equation 5.77 (valid only if the signal is sinusoidal) and substituting in Equation 5.141, we obtain:

$$\tan(\phi - \gamma_2(f_r) + \delta\phi(\phi, f)) = \mp r(f) = \mp\rho(f)\frac{\cos(\phi - \gamma_1(f))}{\cos(\phi - \gamma_2(f))} \quad (5.143)$$

Now, using this expression in Equation 5.41 we find:

$$\tan \delta\phi(\phi, f) =$$

$$= -\frac{\cos(\phi - \gamma_2(f))\sin(\phi - \gamma_2(f_r)) \mp \rho(f)\cos(\phi - \gamma_1(f))\cos(\phi - \gamma_2(f_r))}{\mp \rho(f)\cos(\phi - \gamma_1(f))\sin(\phi - \gamma_2(f_r)) - \cos(\phi - \gamma_2(f))\cos(\phi - \gamma_2(f_r))} \quad (5.144)$$

which can also be written as:

$$\tan \delta\phi(\phi, f) = \frac{H_{01} + H_{11}\cos 2\phi - H_{12}\sin 2\phi}{H_{02} + H_{12}\cos 2\phi - H_{11}\sin 2\phi} \quad (5.145)$$

where:

$$H_{01} = \sin(\gamma_2(f) - \gamma_2(f_r)) + \rho(f)\sin(\gamma_2(f) - \gamma_1(f_r))$$

$$H_{02} = -\cos(\gamma_2(f) - \gamma_2(f_r)) - \rho(f)\cos(\gamma_1(f) - \gamma_1(f_r)) \quad (5.146)$$

$$H_{11} = -\sin(\gamma_2(f) - \gamma_2(f_r)) - \rho(f)\sin(\gamma_1(f) - \gamma_1(f_r))$$

$$H_{12} = -\cos(\gamma_2(f) - \gamma_2(f_r)) - \rho(f)\cos(\gamma_1(f) - \gamma_1(f_r))$$

This is a general and exact expression for phase error due to a lack of orthogonality of the sampling reference functions or failure of the condition that their Fourier transform amplitudes must be equal. This phase error is a function of the signal phase ϕ and signal frequency f, but it can be decomposed into two additive components, one that depends only on the frequency and another that depends on both variables, as follows:

$$\delta\phi(\phi, f) = \delta\phi_0(f) + \delta\phi_1(\phi, f) \quad (5.147)$$

For a given frequency of the signal, the first term is a constant (assuming the signal frequency is constant), thus acting as a piston term when an interferogram is being evaluated. We can easily see that the phase error is a periodic function with the phase ϕ. So, the first or piston term can be evaluated with:

$$\delta\phi_0(f) = \frac{1}{2\pi} \int_0^{2\pi} \delta\phi(\phi, f) d\phi \quad (5.148)$$

5.8.2 Phase-Error Approximation in Two Particular Cases

The preceding analysis is exact if the two sampling functions are not orthogonal or if their Fourier transforms do not have the same amplitude, which may happen when the signal frequency is different from the reference frequency. Let us assume that the signal frequency is different but relatively close to the reference frequency, so we can write:

$$\delta\gamma_1 = \gamma_1(f) - \gamma_1(f_r)$$
$$\delta\gamma_2 = \gamma_2(f) - \gamma_2(f_r) \tag{5.149}$$

We also assume that $\psi(f_r) = \gamma_2(f_r) = 0$, which, as we said before, is true in most phase-detecting algorithms. Then, we can approximate the functions H_{ij} by:

$$H_{01} = \rho(f)\delta\gamma_1 + \delta\gamma_2$$

$$H_{02} = -(\rho(f) + 1)$$

$$H_{11} = \rho(f)\delta\gamma_1 - \delta\gamma_2 \tag{5.150}$$

$$H_{12} = -(\rho(f) - 1)$$

hence obtaining:

$$\delta\phi(f) = \frac{[\rho(f)\delta\gamma_1(f) + \delta\gamma_2(f)] + [\rho(f)\delta\gamma_1(f) - \delta\gamma_2(f)]\cos(2\phi) - [p(f) - 1]\sin(2\phi)}{-[\rho(f) + 1] - [p(f) - 1]\cos(2\phi) + [p(f)\delta\gamma_1(f) - \delta\gamma_2(f)]\sin(2\phi)} \tag{5.151}$$

which can further be approximated by:

$$\delta\phi(f) = \frac{1}{2}[\rho(f) - 1]\sin(2\phi) +$$

$$+ \frac{1}{2}[p(f)\delta\gamma_1(f) - \delta\gamma_2(f)]\cos(2\phi) - \tag{5.152}$$

$$- \frac{1}{2}[p(f)\delta\gamma_1(f) + \delta\gamma_2(f)]$$

where we should keep in mind that the signal is assumed to be sinusoidal and that the phase shifts are linear.

Given a detuning magnitude, when measuring an interferogram the signal frequency is a constant in most cases, with a few rare exceptions to be described later. The last term in this expression is a constant phase shift for all points in the wavefront, thus it acts like a piston term. In general, this term does not have any practical importance and can be ignored, so we obtain:

$$\delta\phi(f) = \frac{1}{2}[\rho(f) - 1]\sin(2\phi) +$$

$$+ \frac{1}{2}[\rho(f)\delta\gamma_1(f) - \delta\gamma_2(f)]\cos(2\phi) \qquad (5.153)$$

The phase error $\delta\phi(f)$ has a sinusoidal variation with the signal phase at twice the frequency of the signal. This result is valid for any kind of error where the conditions of orthogonality and equal amplitudes fail; however, when cross-talk between harmonics is present (for example, when the signal has harmonic distortion), this conclusion might not be true. As pointed out by Cheng and Wyant (1985), the phase error may be eliminated by averaging the results of two measurements with opposite errors (see Chapter 6). The two measurements must only have an offset of 90° with respect to each other.

When only the condition of equal amplitudes fails, $\Delta(f)$ is not equal to one and $\delta\gamma_1(f) = \delta\gamma_2(f)$. Then, the $\cos(2\phi)$ term is sufficiently small so that we can neglect it and write:

$$\delta\phi(f) = \frac{1}{2}[\rho(f) - 1]\sin(2\phi) \qquad (5.154)$$

As shown in Figure 5.21, in this case the phase error becomes zero when the phase to be measured (ϕ) is an integer multiple of $\pi/2$. This error has a peak value equal to $(\rho(f) - 1)/2$.

Finally, if only the orthogonality condition fails, $\rho(f)$ is equal to one, and the phase error is:

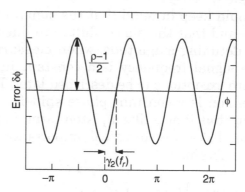

Figure 5.21   Phase error as a function of the measured phase for an algorithm where the Fourier transforms $G_1(f)$ and $G_2(f)$ are orthogonal at all frequencies.

$$\delta\phi(f) = \frac{1}{2}[\delta\gamma_1(f) - \delta\gamma_2(f)]\cos(2\phi)$$

$$= \frac{1}{2}\left[\frac{d(\gamma_1(f) - \gamma_2(f))}{df}\right]\delta f\cos(2\phi) \qquad (5.155)$$

We can see that, in this case, the phase error again oscillates sinusoidally with the signal phase, between zero and a peak value equal to the derivative of the phase difference $\gamma_2(f) - \gamma_1(f)$ with respect to the signal frequency (Figure 5.22). This phase error becomes zero even in the presence of some detuning, when the phase to be measured (ϕ) is equal to $\pi/4$ plus an integer multiple of $\pi/2$. These expressions are the basis for analysis of errors in phase-shifting interferometry, as is described further in the next few sections.

5.9 SOME SOURCES OF PHASE ERROR

The sources of error in phase-shifting interferometry are many. These errors have been studied by several researchers (e.g., Schwider et al., 1983; Cheng and Wyant, 1985; Creath, 1986, 1991; Ohyama et al., 1988; Brophy, 1990). Wingerden et al. (1991) made a general study of many phase errors in phase-detecting algorithms. They classified these errors as follows:

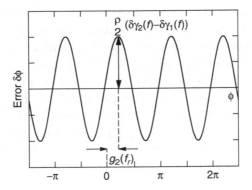

Figure 5.22 Phase error as a function of the measured phase for an algorithm where the Fourier transforms $G_1(f)$ and $G_2(f)$ have equal amplitudes at all frequencies.

1. *Systematic errors.* The value of these errors varies sinusoidally with respect to the signal phase with a frequency equal to twice the signal frequency. These errors have a constant amplitude and phase. By averaging the measurements made with two algorithms for which the sampling points in one algorithm are displaced 90° with respect to those on the other algorithm, the error can be canceled out.

2. *Random errors with sinusoidal phase dependence.* Random additive noise affects the signal measurements in such a manner that the noise errors corresponding to any two different signal measurements are statistically independent. Also, the noise is independent of the signal frequency. Thus, we can consider the noise amplitude and phase to be random, not constant. As for systematic errors, these have a sinusoidal phase dependence. The effect of the presence of additive noise on sampling algorithms has been studied in detail by Surrel (1997). Mechanical vibrations introduce this kind of noise if the frequency is not too high, as is discussed later. Hariharan (2000) has proposed using an average of many measurements with different phase differences to reduce these systematic phase errors. Hibino (1997)

has proved that a phase-detection algorithm designed to compensate for systematic phase errors may become more susceptible to random noise and give larger random errors in the phase.

3. *Random errors without phase dependence.* The value of these errors is independent of the phase of the measured signal. The case of additive random errors with a Gaussian distribution has been studied in depth by Rathjen (1995) and is described here in some detail.

We have seen that the phase error when any of four conditions are not fulfilled can be calculated by means of Equation 5.145, and several particular cases were considered. Expressions for the analysis of phase errors were given that can be applied to the calculation of errors in phase-shifting interferometry, as described in the next few sections.

5.9.1　Phase-Shifter Miscalibration and Nonlinearities

If the phase-shifter device is not well calibrated or its response is not linear, the target phase shift (α) is not the real phase shift (α'). This effect can be represented by the expression:

$$\alpha'_n = \alpha_n \left(1 - \varepsilon_1 - \varepsilon_2\alpha - \varepsilon_3\alpha + \ldots\right)$$

$$= \alpha_n + \left(\varepsilon_1 - \varepsilon_2\alpha - \varepsilon_3\alpha^2 + \ldots\right)\alpha_n \qquad (5.156)$$

$$= \alpha_n + \Delta\alpha_n$$

where α is the target or reference value of the phase shift and α' is the real obtained value. The linear and quadratic error coefficients are γ_1 and γ_2, respectively.

When we have only linear and quadratic errors and we require the total error to be zero at the beginning $(\alpha = \alpha_1 = 0)$ and at the end $(\alpha = \alpha_N)$ of the reference period, we need to add an extra linear term so the total linear error coefficient becomes:

$$\varepsilon_1 = -\varepsilon_2\alpha_n \qquad (5.157)$$

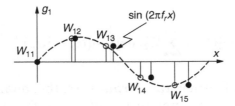

Figure 5.23 Displaced sampling points due to linear phase error.

which can be done only after measuring the phase errors. The phase errors may be interpreted in two different ways.

5.9.1.1 Error in the Sampling Reference Functions

An error is in the actual phase shift or, equivalently, on the interferometer optical path difference, so the sampling points are displaced from their correct positions, as shown in Figure 5.23, but the signal to be detected remains unmodified. The phase (α_n) for each sampling point with the error being introduced is used in the sampling reference functions in Equations 5.85 and 5.86, thus giving us a modified set of functions $g_1'(x)$ and $g_2'(x)$:

$$g_1'(x) = \sum_{n=1}^{N} W_{1n}\delta(x - x_n - \Delta x_n) \qquad (5.158)$$

and

$$g_2'(x) = \sum_{n=1}^{N} W_{2n}\delta(x - x_n - \Delta x_n) \qquad (5.159)$$

where $\Delta x_n = \Delta\alpha_n/(2\pi f_r)$. Thus, from Equations 5.87 and 5.88, the Fourier transforms of these sampling reference functions are:

$$G_1'(f) = \sum_{n=1}^{N} W_{1n}\exp\left[-i(\alpha_n + \Delta\alpha_n)\frac{f}{f_r}\right] \qquad (5.160)$$

and

$$G_2'(f) = \sum_{n=1}^{N} W_{2n} \exp\left[-i(\alpha_n + \Delta\alpha_n)\frac{f}{f_r}\right] \qquad (5.161)$$

The error-free Fourier transforms are orthogonal to each other and have the same magnitude at the reference frequency; nevertheless, with the phase error added, either of the two conditions or both will fail. These modified Fourier transforms then allow us to compute the phase error, as will be described later in some detail.

5.9.1.2 Error in the Measured Signal

In this model, we consider that the signal is phase modulated by the error and that the sampling point positions are correct. If we consider a phase-modulated signal, we see that the phase modulation is a nonperiodic function of α; thus, the signal is not periodic and the Fourier transform of the signal is no longer discrete but continuous. Figure 5.24a shows the

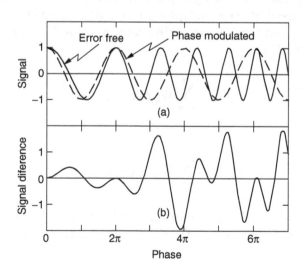

Figure 5.24 (a) Plots of the error-free signal (dotted curve) and the signal with error (continuous curve); (b) difference between these two signals. The value $\varepsilon_2 = 0.05$ was used.

error-free signal and the signal phase modulated with the error. The difference between these two signals is shown in Figure 5.24b. Because the Fourier transform is not discrete, in order to find the correct phase the correlations between the reference sampling functions and the signal must be found using the integrals in Equation 5.62. The phase errors would have no importance at all if their values were independent of the signal phase. In that case, the error would be just a constant piston term on the measured wavefront. Unfortunately, this is not the case. We have seen before that the phase errors have a value that varies sinusoidally with the signal phase.

5.9.2 Measurement and Compensation of Phase-Shift Errors

This problem has been studied by several authors (e.g., Ramson and Kokal, 1986). In the case of small detuning and a signal frequency deviating from the reference frequency, the zero bias condition is preserved. If the signal is assumed to be sinusoidal, the condition for no cross-talk between the signal and reference function harmonics is also preserved. The conditions for orthogonality and equal magnitudes of $G_1(f_r)$ and $G_2(f_r)$, however, may not be satisfied; thus, the phase error in this case is given in general by Equations 5.152, 5.154, or 5.155, depending on the case. In the case of no quadratic (nonlinear) error and only linear error, we have $\varepsilon_2 = 0$. To eliminate the linear error it is necessary to calibrate the phase shifter using an asynchronous algorithm, as described, for example, by Cheng and Wyant (1985).

The presence of *linear phase error* may be detected by measuring a flat wavefront when a large linear carrier has been introduced with tilt fringes. If a phase error occurs, a sinusoidally corrugated wavefront will be detected with twice the spatial frequency of the tilt fringes being introduced, as shown in Figure 5.25.

The presence of *phase-shifter error* may also be detected with a procedure suggested by Cheng and Wyant (1985). Tilt fringes are introduced and measurements of the signal are

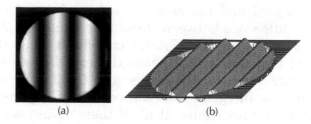

Figure 5.25 Detection of phase error by the presence of a corrugated wavefront: (a) interferogram, and (b) wavefront.

taken across the interferogram in a direction perpendicular to the fringes. These measurements are then plotted to obtain a sinusoidal curve. This plot is repeated $N + 1$ times, with shift increments of $2\pi/N$. The first and the $(N + 1)$th measurements should overlap each other, unless phase error has occurred, as shown in Figure 5.26.

Another interesting method to detect phase errors has been proposed by Kinnstaetter et al. (1988). Two points in quadrature (phase difference equal to 90°) are selected in the fringe pattern, then the signal values at these two points are plotted in a diagram for several values of the phase shift. These diagrams are referred to as *Lissajous displays*, which have the following characteristics (Figure 5.27):

1. For no phase errors and when the points being selected have the same signal amplitude and are exactly in quadrature, the diagram is a circle with equidistant points.

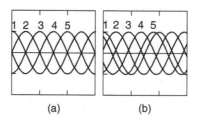

Figure 5.26 Plots to detect phase error.

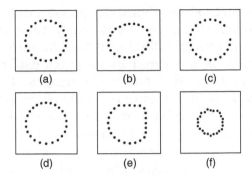

Figure 5.27 Lissajous curves with different types of phase error.

2. For no phase error but when the interferogram points being selected do not have the same signal amplitude or are not in perfect quadrature, the diagram is an ellipse.
3. If linear error is present, the ellipse or circle does not close or leaves a gap open. In other words, the first dot and the last are not at the same place in the diagram.
4. For nonlinear error, the distance between the dots is not constant.
5. For a nonlinear response or saturation in the light detector, the ellipse is deformed, with some parts having a different local curvature.
6. If there is vibrational noise, the curve is smaller and irregular.

Alcalá-Ochoa and Huntley (1998) proposed a calibration method in wh^ich many measurements are taken with a series of equidistant and close phase differences. The Fourier transforms of the measurements are then calculated to obtain not only the frequency of the signal but also its harmonic content.

Sometimes measurement of the phase difference between any two interferograms with different phases is difficult because of a large amount of noise. In this case, direct measurement of the phase difference between two fixed interferograms is possible if many tilt fringes are present, as described by Wang et al. (1996).

Another method to eliminate phase shift errors is to directly measure the phase shift every time the phase is shifted. Lai and Yatagai (1991) proposed an interferometer in which the phase is measured in an extra calibration fringe interference pattern with many tilt fringes. This auxiliary interferogram is projected onto one side of the interferogram to be measured using a high-precision tilted mirror.

A different approach was proposed by Huang and Yatagai (1999), where the measurements are taken at unknown phases with unknown steps. The number of steps is sufficiently large so they can establish a linear system of equations where $\sin\phi$, $\cos\phi$, and the signal bias appear as unknown variables. The system is then solved with an iterative least-squares fitting algorithm to find the optimum value for these unknown variables.

5.9.3   Linear or Detuning Phase-Shift Error

In spite of all efforts to eliminate linear phase-shift errors, they are frequently unavoidable. An ideal algorithm is one for which the Fourier transform amplitudes of the reference sampling functions as well as the orthogonality conditions are preserved for all signal frequencies. In other words, Equation 5.92 should be true for all frequencies. This is not possible in practical algorithms, so, to obtain at least a small frequency range on which the sensitivity to detuning is small, we require that

$$\left(\frac{dG_2(f)}{df}\right)_{f=f_r} = \left(\frac{dG_1(f)}{df}\right)_{f=f_r} \tag{5.162}$$

Thus, the Fourier transform amplitudes should be equal at the reference frequency and should also be tangential to each other at that point; that is,

$$\left(\frac{d\mathrm{Am}(G_2(f))}{df}\right)_{f=f_r} = \left(\frac{d\mathrm{Am}(G_1(f))}{df}\right)_{f=f_r} \tag{5.163}$$

with the same slope requirement for the phases, as follows:

$$\left(\frac{d\gamma_2(f)}{df}\right)_{f=f_r} = \left(\frac{d\gamma_1(f)}{df}\right)_{f=f_r} \qquad (5.164)$$

In some algorithms, the orthogonality condition holds for all frequencies so only the condition in Equation 5.163 is required. In other algorithms, the orthogonality condition fails when f is different from f_r, but the ratio between the two magnitudes of the Fourier transforms is valid at all frequencies. In this case, only the condition in Equation 5.164 is necessary.

When the signal is not sinusoidal, the treatment of detuning is more complicated, because any detuning affects not only the fundamental frequency of the signal but also its harmonic components, as will be described later. We explained before that these phase errors are sinusoidally dependent on the measured phase with twice the signal frequency. This fact was used to design special detuning-insensitive algorithms. As described in this book, special algorithms can be devised to detect or reduce phase errors due to phase-shifter miscalibration and nonlinearity (Joenathan, 1994). Schwider (1989) also used this sinusoidal variation of the phase error to calculate an error function which is then subtracted from the calculated phase values to substantially reduce the linear phase error.

5.9.4 Quadratic Phase-Shift Errors

Even when the linear error has been properly eliminated by calibration of the phase shifter, quadratic error may still be present. The phase error expression allows us to apply either of the two previously described models. We can modify the sampling point positions and calculate the Fourier transforms of the reference sampling functions, or we can modify the measured signal that has been phase modulated by the phase error.

Let us now analyze the case of only linear and quadratic error. To use the first model, it is convenient to express the phase error in such a way that the quadratic error becomes zero at the first sampling point ($n = 1$) and at the last sampling point ($n = N$). Thus, we can write:

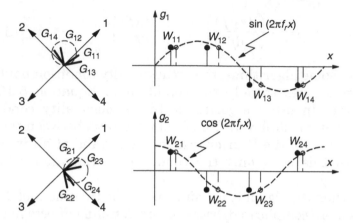

Figure 5.28 Effect of quadratic phase error in an algorithm.

$$\Delta\alpha_n = \varepsilon_1\alpha_n + \varepsilon_2(\alpha_n - \alpha_N)\alpha_r$$

$$= -\varepsilon_2\left(\frac{\alpha_N}{2}\right)^2 + \varepsilon_1\alpha_n + \varepsilon_2\left(\alpha_n - \frac{(\alpha_N - \alpha_1)}{2}\right)^2 \quad (5.165)$$

The first term is a piston or phase-offset term of no practical importance. We see that in this expression the quadratic error is symmetric about the central point between the first and last sampling points. Thus, the significant term for the quadratic error can be written as:

$$\Delta\alpha_n = \varepsilon_2\left(\alpha_n - \frac{(\alpha_N - \alpha_1)}{2}\right)^2 \quad (5.166)$$

which leads us to

$$\Delta\alpha_n = \Delta\alpha_{N-n+1} \quad (5.167)$$

Figure 5.28 illustrates a sample application of these concepts for an algorithm with four sampling points in **X**. We can see that this algorithm is insensitive to quadratic nonlinear phase error. Other algorithms may be analyzed in a similar manner.

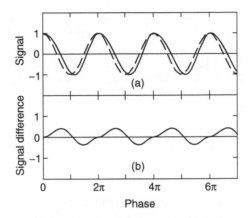

Figure 5.29 Periodic distorted signal due to nonlinear phase error.

To apply the second model to analyzing this error, the signal may be represented by:

$$s(z) = a + b\cos\left(2\pi fz + 4\pi^2\varepsilon_2(fz - 1)fz + \phi\right) \qquad (5.168)$$

where, for notational simplicity, the x,y dependence has been omitted and the optical path difference (OPD) has been replaced by z. Also, because no change in the signal period is introduced by the compensated nonlinear error, no detuning occurs and the reference frequency (f_r) becomes equal to the signal frequency (f).

In our examination of the Fourier theory of algorithms in this chapter, we have assumed that the signal is periodic so its Fourier transform is discrete. If we assume that the phase value α is applied to each period of the signal, taking the beginning of each period as the new origin, we obtain a periodicity of the signal (Figure 5.29), and its Fourier transform is discrete. This approach is valid only when the sampling points are within one signal period, as is true for most phase-detecting algorithms.

The Fourier coefficients in Equation 2.6 may then be found using Equations 2.7 and 2.8. Unfortunately, evaluation of these integrals is not simple and leads to Fresnel integrals,

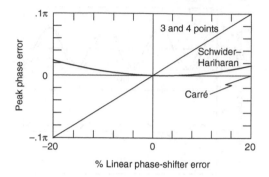

Figure 5.30 Nonlinear phase error and some common phase-detecting algorithms. (From Creath, K., in *Progress in Optics*, Vol. XXVI, Wolf, E., Ed., Elsevier Science, Amsterdam, 1988. With permission.)

as shown by Ai and Wyant (1987). Creath (1988) has performed numerical simulations to gain insight into the nature of this phase error (Figure 5.30).

5.9.5 High-Order, Nonlinear, Phase-Shift Errors with a Sinusoidal Signal

Let now study the most general case of nonlinearities up to order p with a sinusoidal signal. As shown in Section 5.9.1, the effective Fourier transforms, $G'(f)$, of the sampling reference functions in the presence of nonlinear phase steps can be found by substituting Equation 5.156 for the phase shift in Equations 5.160 and 5.161:

$$G_1'(f) = \sum_{n=1}^{N} W_{1n} \exp\left[-i\alpha_n\left(1 + \varepsilon_1 + \varepsilon_2\alpha_n + \varepsilon_3\alpha_n^2 + ...\right)\frac{f}{f_r}\right] \quad (5.169)$$

and

$$G_2'(f) = \sum_{n=1}^{N} W_{2n} \exp\left[-i\alpha_n\left(1 + \varepsilon_1 + \varepsilon_2\alpha_n + \varepsilon_3\alpha_n^2 + ...\right)\frac{f}{f_r}\right] \quad (5.170)$$

where N is the number of sampling points. Equation 5.169 can also be written as:

$$G_1'(f) = \sum_{n=1}^{N} W_{1n} \exp\left[-i\alpha_n \frac{f}{f_r}\right] \exp\left[-i\alpha_n \left(\varepsilon_1 + \varepsilon_2 \alpha_n + \varepsilon_3 \alpha_n^2 + ...\right) \frac{f}{f_r}\right] \quad (5.171)$$

Assuming now that the phase error is much smaller than $\pi/2$ we can approximate it by:

$$G_1'(f) = \sum_{n=1}^{N} W_{1n} \exp\left[-i\alpha_n \frac{f}{f_r}\right] \left[1 - i\alpha_n \left(\varepsilon_1 + \varepsilon_2 \alpha_n + \varepsilon_3 \alpha_n^2 + ...\right) \frac{f}{f_r}\right] (5.172)$$

which is equal to:

$$G_1'(f) = G_1(f) - i\frac{f}{f_r} \sum_{n=1}^{N} \left[\alpha_n W_{1n} \left(\varepsilon_1 + \varepsilon_2 \alpha_n + \varepsilon_3 \alpha_n^2 + ...\right) \exp\left(-i\alpha_n \frac{f}{f_r}\right)\right] \quad (5.173)$$

where $G_1(f)$ is the Fourier transform in the absence of any phase errors. Then, by taking the derivatives of $G_1(f)$ in Equation 5.90 with $\gamma_2(f_r) = 0$, it can be shown that this expression can be transformed into:

$$G_1'(f) = G_1(f) + f \sum_{k=1}^{K} i^{(k-1)} \varepsilon_k f_r^{(k-1)} \frac{d^k G_1(f)}{df^k} \quad (5.174)$$

where K is the maximum order of the nonlinear error. In a similar manner, we can obtain from Equation 5.169:

$$G_2'(f) = G_2(f) + f \sum_{k=1}^{K} i^{(k-1)} \varepsilon_k f_r^{(k-1)} \frac{d^k G_2(f)}{df^k} \quad (5.175)$$

Thus, if we impose the condition:

$$G_1'(f_r) = \pm i G_2'(f_r) \quad (5.176)$$

to eliminate all phase errors, we finally obtain:

$$G_1(f_r) = \pm i G_2(f_r) \quad (5.177)$$

(which includes the conditions of equal magnitudes and orthogonality) and

$$\left(\frac{d^k G_1(f)}{df^k}\right)_{f=f_r} = \left(\frac{d^k G_2(f)}{df^k}\right)_{f=f_r} \tag{5.178}$$

where k is the phase-shift deformation order present in the system.

5.9.6 High-Order, Nonlinear, Phase-Shift Errors with a Distorted Signal

To study the detection of a harmonically distorted signal when there is high-order nonlinear phase-shift error, we can use Equations 5.75, 5.79, and 5.63, assuming an algorithm for which $\gamma_2(f_r) = 0$, as is true in most cases, to obtain:

$$\tan\phi = \mp \frac{S_1 \text{Am}(G_1(f))\sin\phi_1 + \sum_{m=2}^{\infty} S_m \text{Am}(G_1(mf))\sin\phi_m}{S_1 \text{Am}(G_2(f))\cos\phi_1 + \sum_{m=2}^{\infty} S_m \text{Am}(G_2(mf))\cos\phi_m} \tag{5.179}$$

Ideally, all of the terms in the sum in the numerator and all of the terms in the sum in the denominator must be zero; however, if the signal has harmonic components above the fundamental frequency, some of them will be different from zero. Furthermore we will see that the value of these terms depends not only on the amplitudes (S_m) of the harmonic components but also on the phase-shift nonlinearities that might be present.

As shown by Hibino (1997), the analysis is quite similar to that given in Section 5.9.5 for the case of phase-shifting nonlinearities affecting only the first term in the numerator and the denominator of Equation 5.179. The effective Fourier transforms, $G'(mf)$, of the sampling reference functions in the presence of nonlinear phase steps are given by:

$$G_1'(mf) = \sum_{n=1}^{N} W_{1n} \exp\left[-im\alpha_n\left(1+\varepsilon_1+\varepsilon_2\alpha_n+\varepsilon_3\alpha_n^2+\ldots\right)\frac{f}{f_r}\right] \quad (5.180)$$

and

$$G_2'(mf) = \sum_{n=1}^{N} W_{2n} \exp\left[-im\alpha_n\left(1+\varepsilon_1+\varepsilon_2\alpha_n+\varepsilon_3\alpha_n^2+\ldots\right)\frac{f}{f_r}\right] \quad (5.181)$$

where N is the number of sampling points. Equation 5.180 can also be written as:

$$G_1'(mf) = \sum_{n=1}^{N} W_{1n} \exp\left[-im\alpha_n\frac{f}{f_r}\right]\exp\left[-im\alpha_n\left(\varepsilon_1+\varepsilon_2\alpha_n+\varepsilon_3\alpha_n^2+\ldots\right)\frac{f}{f_r}\right] \quad (5.182)$$

Assuming now that the phase error is much smaller than $\pi/2$, we can approximate it by:

$$G_1'(mf) = \sum_{n=1}^{N} W_{1n} \exp\left[-im\alpha_n\frac{f}{f_r}\right]\left[1-im\alpha_n\left(\varepsilon_1+\varepsilon_2\alpha_n+\varepsilon_3\alpha_n^2+\ldots\right)\frac{f}{f_r}\right] \quad (5.183)$$

which is equal to:

$$G_1'(mf) = G_1(mf) -$$

$$-i\frac{f}{f_r}\sum_{n=1}^{N}\left[m\alpha_n W_{1n}\left(\varepsilon_1+\varepsilon_2\alpha_n+\varepsilon_3\alpha_n^2+\ldots\right)\exp\left(-im\alpha_n\frac{f}{f_r}\right)\right] \qquad (5.184)$$

where $G(mf)$ is the Fourier transform for the harmonic component (m) in the absence of any phase-shift errors. This expression can now be transformed into:

$$G_1'(mf) = G_1(mf) + f\sum_{k=1}^{K} i^{(k-1)}\varepsilon_k\alpha_n^{(k-1)}f_r^{(k-1)}\frac{d^k G_1(mf)}{df^k} \quad (5.185)$$

where K is the maximum order of the nonlinear error. In a similar manner, we can obtain from Equation 5.169:

$$G_2'(mf) = G_2(mf) + f \sum_{k=1}^{K} i^{(k-1)} \varepsilon_k \alpha_n^{(k-1)} f_r^{(k-1)} \frac{d^k G_1(mf)}{df^k} \quad (5.186)$$

If the signal is sinusoidal ($m = 1$), we obtain the results in the previous section. If signal harmonic components above the fundamental frequency are present, in order to obtain the sum terms in the numerator and all of the sum terms in the denominator of Equation 5.181, we need to impose the condition:

$$G_1'(mf_r) = G_2'(mf_r) = 0, \quad \text{for } m \geq 2 \quad (5.187)$$

So, to eliminate phase error due to the presence of harmonic components ($m \geq 2$) and their associated nonlinear phase-shifting errors, we finally obtain:

$$G_1(mf_r) = G_2(mf_r) = 0 \quad (5.188)$$

and

$$\left(\frac{d^k G_1(mf)}{df^k} \right)_{f=f_r} = \left(\frac{d^k G_2(mf)}{df^k} \right)_{f=f_r} = 0 \quad (5.189)$$

where k is the phase shift deformation order present in the system, and m is the harmonic component above the fundamental also present.

In conclusion, the nonlinear phase-shift error of order k is corrected in an algorithm only if the following two conditions are satisfied:

1. The kth derivatives of the Fourier transforms of the sampling reference functions at the reference frequency are equal.
2. The kth derivatives of the Fourier transforms of the sampling reference functions at the frequency of the $m \geq 2$ harmonic component present are zero.

We should remember that these Fourier transforms are complex functions. If they are orthogonal to all frequencies, the amplitudes of these functions should be equal to zero. Nonlinear phase-shift errors in the presence of harmonic distortion have been studied by Hibino et al. (1995), who later applied their results to design algorithms corrected for nonuniform phase shifting (Hibino et al., 1997). In response to this work, Surrel (1998) noted that these new algorithms are corrected for nonuniform shifting but they have a large sensitivity to random noise. Random noise is described later in this chapter.

5.9.7 Nonuniform Phase-Shifting Errors

Nonuniform phase shifting appears when a given applied phase step is not the same real phase step at different points in the interferogram. In other words, the applied phase steps are spatially nonuniform. As reported by Hibino et al. (1997) and by Hibino and Yamauchi (2000), this occurs in many practical situations. An example is a liquid-crystal modulator, for which the phase shift is nonlinear as well as nonuniform. Two other examples are illustrated in Figure 5.35. Figure 5.35a shows a Twyman–Green interferometer for which a large mirror is driven with several (two or three) piezoelectric transducers. Each one of them has different linear and nonlinear characteristics. Figure 7.35b shows a Fizeau interferometer for which the phase change is produced in a convergent beam by moving a spherical mirror. The total phase shift on the axis is different from the total phase shift close to the edge of the fringe pattern.

In the presence of nonuniform phase shifting, the signal from different points in the interferogram will be different in two ways:

1. The different linear calibrations of the phase displacements will produce the effect of different signal frequencies from different points.
2. The different nonlinear phase displacements will produce the effect of different phase modulation from different points.

The nonuniform phase error appears when:

1. The nonlinear phase shift error of any order k is not corrected.
2. The nonlinear phase shift error coefficient (ε_k) has different values for each point in the interferogram.

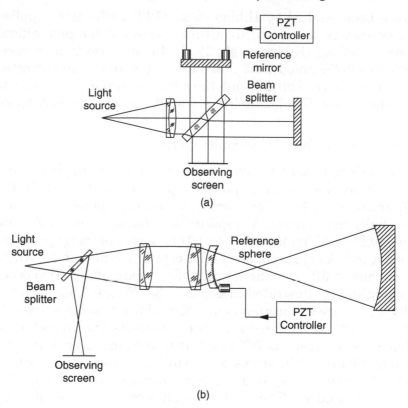

Figure 5.35 Nonlinear phase shift error in (a) a Twyman–Green interferometer, where the displacing mirror is driven by two or three piezoelectric controllers; and (b) a Fizeau interferometer with a moving spherical reference surface and convergent light beam.

Hibino (1999) and Hibino and Yamauchi (2000) designed some algorithms to correct as much as possible for nonuniform phase error and random noise. Some of these algorithms are described in Chapter 6. Hibino et al. have shown that algorithms with fewer than six samples have no error-compensating capability for phase nonlinearity. When the number of samples reaches a value of eleven, a substantial reduction in these errors is achieved.

5.9.8   Phase Detection of a Harmonically Distorted Signal

A distorted periodic signal may be phase detected with a synchronous detection sampling method without any error only if the signal harmonic frequencies are located at places having a zero value for the amplitudes of the Fourier transforms of the reference functions. Many sampling algorithms, such as some described in this chapter, have zeros of the reference functions spectra at some harmonics. As shown in the preceding sections, signal harmonics may appear for many reasons, for example:

1. When the signal is not sinusoidal, such as in the measurement of aspherical wavefronts by means of spatial phase-shifting analysis of interferograms
2. When the signal is sinusoidal but the phase-shifting device has a nonlinear response in the phase scale, such as in the case of temporal phase-shifting interferometry with a nonlinear phase shifter
3. When the signal is sinusoidal but the response of the light detector is not linear with the signal
4. In multiple-beam interferograms, or Ronchigrams (Hariharan, 1987)

We have shown before that, to make the algorithm insensitive to the signal harmonic (m), we must have zeros of the amplitudes of the Fourier transforms of the sampling reference functions for the harmonic (m) to which the algorithm should be insensitive; however, this condition may not be satisfied. Stetson and Brohinsky (1985), Hibino et al. (1995), and Hibino (1997) have shown that to suppress all harmonics up to the mth order in algorithms with equally spaced points the following conditions are necessary:

1. The maximum phase spacing between sampling points should be equal to $2\pi/(m + 2)$.
2. The minimum number of sampling points is $m + 2$ when the phase interval is set to its maximum value. A smaller phase interval would require more sampling points.

TABLE 5.2 Sensitivity to Signal Harmonics of Algorithms with Equally and Uniformly Spaced Points

Number of Sampling Points	Harmonics Being Suppressed									
	2	3	4	5	6	7	8	9	10	11
3	—	y	—	—	y	—	—	y	—	—
4	y	—	y	—	y	—	y	—	y	—
5	y	y	—	y	—	y	y	—	y	—
6	y	y	y	—	y	—	y	y	y	—

Source: From Stetson, K.A. and Brohinsky, W.R., *Appl. Opt.*, 24, 3631–3637, 1985. With permission.

To clarify, let us assume that we have N equally spaced sampling points with a phase separation equal to $2\pi/N$. In this case, *all* harmonic components up to the $m = N - 2$ order will be eliminated. Of course, some other higher harmonics may also be eliminated. Stetson and Brohinsky (1985) have shown that an algorithm with equally and uniformly spaced sampling points, as given in Equation 5.10, is sensitive to the harmonics given by:

$$m = N \pm 1 + pN \tag{5.190}$$

where p is an integer. These results are shown in Table 5.2. If the phase-detecting algorithm is sensitive to undesired harmonics, the response to these harmonics may be reduced by additional filtering provided by bucket integration or with an additional filtering function, as described in Section 5.7.

In order to provide insensitivity to a given harmonic order in the presence of detuning, we must meet the following two requirements regarding the Fourier transforms $G_1(f)$ and $G_2(f)$ of the reference sampling functions:

1. Both Fourier transforms must have zero amplitude at the harmonic frequency.
2. Both Fourier transforms must have a stationary amplitude with respect to the frequency (zero slope) at the harmonic frequency.

Hibino et al. (1995) have shown that, to obtain an algorithm that is insensitive up to the mth harmonic order and is also insensitive to detuning of the fundamental frequency and its harmonics, the following must be true:

1. The maximum phase interval between sampling points must be equal to $2\pi/(m + 2)$.
2. The minimum number of sampling points must be equal to $2m + 3$ when the phase interval is set to its maximum value.

Surrel (1996) later showed, however, that the minimum number of sampling points should be equal to $2m + 2$. A smaller phase interval than its maximum value would require a greater number of sampling points. An exception is when the algorithm requires detuning insensitivity only at the fundamental frequency, in which case the phase interval may be reduced from its maximum value of 120° to any smaller value, without the need for more than five sampling points.

Given an unfiltered signal with harmonics, for which the amplitude and phase are known, the phase error may be calculated by means of the general expression with the ratio of the correlations $r(f)$ given by Equation 5.75, where the only condition being satisfied is the zero bias. If we assume that (1) the conditions for orthogonality and equal amplitudes are fulfilled at the signal frequency, and (2) that the algorithm has the relatively common property that the orthogonality of the reference sampling functions is preserved at all signal frequencies, then we can write this expression as:

$$r(f) = \mp \frac{S_1 \mathrm{Am}(G_1(f))\sin\phi + \displaystyle\sum_{m=2}^{\infty} S_m \mathrm{Am}(G_1(mf))\sin\phi_m}{S_1 \mathrm{Am}(G_1(f))\cos\phi + \displaystyle\sum_{m=2}^{\infty} S_m \mathrm{Am}(G_2(mf))\cos\phi_m} \qquad (5.191)$$

Hence, using Equation 5.138 and 5.141 with $\gamma_2(f_r)$, the phase error may be shown to be given by:

$$\delta\phi = \sum_{m=2}^{\infty} \frac{S_m}{S_1} \left(\begin{array}{c} \dfrac{\mathrm{Am}(G_1(mf))}{\mathrm{Am}(G_1(f))} \sin\phi_m \cos\phi - \\[2ex] \dfrac{\mathrm{Am}(G_2(mf))}{\mathrm{Am}(G_1(f))} \cos\phi_m \sin\phi \end{array} \right) \qquad (5.192)$$

The values of the amplitudes (S_m) and of the phases (ϕ_m) of the harmonic components of the signal depend on the signal characteristics. The phase ϕ_m may be written as $\phi_m = m\phi + \beta_m$. We observe that the phase error does not change in a purely sinusoidal manner with the signal phase as do the other phase errors considered previously. The functional dependence with the signal phase ϕ is more complicated, but in a first approximation it has oscillations with the same frequency of the signal.

5.9.9 Light-Detector Nonlinearities

The light detector may have an electric output with a nonlinear relationship with the signal, even though they are normally adjusted to work in its most linear region. If s' is the detector signal output and s is the input signal, we can write:

$$s' = s + \varepsilon s^2 \qquad (5.193)$$

where ε is the nonlinear error coefficient. Thus, the output from the detector is:

$$s' = a(1+\varepsilon a) + (1+2\varepsilon a)b\cos(\alpha_n + \phi) +$$
$$+\frac{1}{2}\varepsilon b^2\left(1+\cos^2 2(\alpha_n + \phi)\right) \qquad (5.194)$$

We can see that a second harmonic component appears in the signal. If the value of the coefficient ε for this nonlinearity is known, the compensation can be made; otherwise, a phase error appears. As pointed out by Creath (1991), no error of this nature is present for algorithms with four and five samples; however, the three-sample algorithm and Carré's algorithm have noticeable errors with four times the fringe frequency.

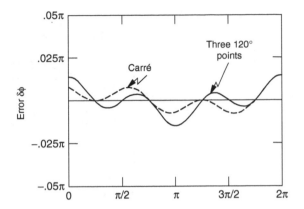

Figure 5.31 Phase error as a function of the phase, due to detector second-order nonlinearities, for two common phase-detecting algorithms. (From Creath, K., in *Progress in Optics*, Vol. XXVI, Wolf, E., Ed., Elsevier Science, Amsterdam, 1988. With permission.)

Some corrections can be made on the video camera after the image has been digitized, but care must be taken to avoid saturating the detector, which increases the harmonic content. Creath made numerical calculations of this phase error, and Figure 5.31 shows the peak phase error as a function of the phase, due to detector second-order nonlinearities, in some common phase-detecting algorithms. The peak phase errors for various amounts of nonlinear error due to detector second-order nonlinearities for some common phase-detecting algorithms are shown in Figure 5.32. Third-order detector nonlinearities may also appear. Figure 5.33 shows the peak phase error as a function of the phase, due to detector third-order nonlinearities, in some common phase-detecting algorithms. Figure 5.34 shows the peak phase errors for various amounts of nonlinear error due to detector third-order nonlinearities for some common phase-detecting algorithms.

5.9.10 Random Phase Error

In a manner similar to that in Equation 5.141, by differentiating tan ϕ and assuming that $\gamma_2(f_r) = 0$ as in most phase-shifting algorithms, we obtain:

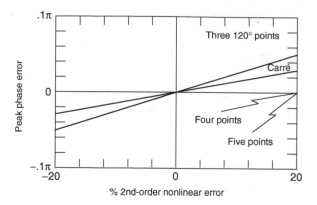

Figure 5.32 Peak phase error as a function of the amount of nonlinear error, due to detector second-order nonlinearities, for some common phase-detecting algorithms. (From Creath, K., in *Progress in Optics*, Vol. XXVI, Wolf, E., Ed., Elsevier Science, Amsterdam, 1988. With permission.)

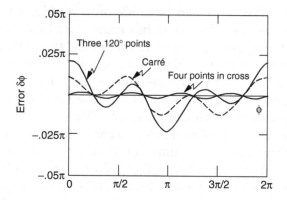

Figure 5.33 Phase error as a function of the phase, due to detector third-order nonlinearities, for some common phase-detecting algorithms. (From Creath, K., in *Progress in Optics*, Vol. XXVI, Wolf, E., Ed., Elsevier Science, Amsterdam, 1988. With permission.)

$$\tan \delta(\phi, f) = \frac{\tan(\phi + \delta(\phi, f)) - \tan \phi}{1 + \tan \phi \tan(\phi + \delta(\phi, f))} \qquad (5.195)$$

which can be approximated by:

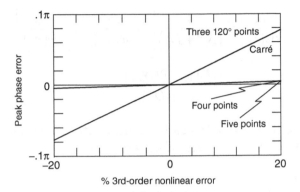

Figure 5.34　Peak phase error as a function of the amount of nonlinear error, due to detector third-order nonlinearities, for some common phase-detecting algorithms. (From Creath, K., in *Progress in Optics*, Vol. XXVI, Wolf, E., Ed., Elsevier Science, Amsterdam, 1988. With permission.)

$$\delta(\phi, f) = \frac{\delta \tan \phi}{1 + \tan^2 \phi} \tag{5.196}$$

If we now assume that this phase error is due to an error in the measurement of the signal $s(x_n)$ we have:

$$\frac{\delta(\phi, f)}{\delta s(x_n)} = \frac{1}{1 + \tan^2 \phi} \frac{\delta \tan \phi}{\delta s(x_n)} \tag{5.197}$$

We can now write Equation 5.108 as:

$$\tan \phi = \frac{\displaystyle\sum_{n=1}^{N} s(x_n) W_{1n}}{\displaystyle\sum_{n=1}^{N} s(x_n) W_{2n}} = \frac{N}{D} \tag{5.198}$$

Hence, from the two expressions we can find:

$$\frac{d(\phi, f)}{ds(x_n)} = \frac{1}{\left(N^2 + D^2\right)}\left[D\frac{\partial N}{\partial s(x_n)} - N\frac{\partial D}{\partial s(x_n)}\right]$$

$$= \frac{\left[DW_{1n} - NW_{2n}\right]}{\left(N^2 + D^2\right)} \tag{5.199}$$

We can identify $(N^2 + D^2)$ as the numerator in Equation 5.127; thus, this equation is transformed into:

$$\frac{d(\phi, f)}{ds(x_n)} = \frac{\left[W_{1n}\cos\phi - W_{2n}\sin\phi\right]}{S_1\sqrt{\left(\displaystyle\sum_{n=1}^{N} W_{1n}\sin\alpha_n\right)^2 + \left(\displaystyle\sum_{n=1}^{N} W_{1n}\cos\alpha_n\right)^2}} \tag{5.200}$$

and then into:

$$\delta(\phi, f) = \frac{\sqrt{W_{1n}^2 + W_{2n}^2}\cos(\phi + \beta)}{S_1\sqrt{\left(\displaystyle\sum_{n=1}^{N} W_{1n}\sin\alpha_n\right)^2 + \left(\displaystyle\sum_{n=1}^{N} W_{1n}\cos\alpha_n\right)^2}}\,\delta s(x_n) \tag{5.201}$$

where β_n is given by:

$$\tan\beta_n = \frac{W_{2n}}{W_{1n}} \tag{5.202}$$

This is the phase error due to an error in the signal sample $s(x_n)$ being measured. We now assume that the signal errors are uncorrelated between the samples and that the standard deviation of all measurements is the same. Then, the statistical phase error variance $\langle \Delta\phi^2 \rangle$ can be expressed by:

$$\langle \Delta\phi^2 \rangle = \frac{1}{S_1^2}\left[\frac{\displaystyle\sum_{n=1}^{N}\left(W_{1n}^2 + W_{2n}^2\right)\cos^2(\phi + \beta_n)\langle \Delta s(x_n)^2 \rangle}{\left(\sqrt{\left(\displaystyle\sum_{n=1}^{N} W_{1n}\sin\alpha_n\right)^2 + \left(\displaystyle\sum_{n=1}^{N} W_{1n}\cos\alpha_n\right)^2}\right)}\right] \tag{5.203}$$

where $\langle \Delta s(x_n)^2 \rangle$ is the statistical error variance of the signal. The second term in the denominator becomes zero if $\gamma_2(f_r) = 0$. If we neglect the phase dependence and average over all possible values of ϕ, the *rms* average $\delta\phi$ is given approximately by:

$$\delta\phi = \frac{1}{\sqrt{2}S_1} \sqrt{\frac{\displaystyle\sum_{n=1}^{N}\left(W_{1n}^2 + W_{2n}^2\right)}{\left(\displaystyle\sum_{n=1}^{N} W_{1n}\sin\alpha_n\right)^2 + \left(\displaystyle\sum_{n=1}^{N} W_{1n}\cos\alpha_n\right)^2}}\,\delta s(x_n) \qquad (5.204)$$

$$= R\delta s(x_n)$$

This result has been obtained by Hibino and Yamauchi (2000), and an equivalent result was derived by Hibino (1997) and Brophy (1990). The conclusion is that the susceptibility (R) of a phase-shifting algorithm to random uncorrelated noise is directly proportional to the root mean square of all of the sampling weights. Hibino (1997) showed that the minimum possible value of this *rms* value is given by:

$$\left[\sqrt{W_{1n}^2 + W_{2n}^2}\right]_{\min} = \frac{2}{\sqrt{m}} \qquad (5.205)$$

This is the case for the diagonal least-squares algorithms represented by Equation 5.19. Hibino (1997) also proved that when an algorithm is designed to reduce systematic errors, it becomes more susceptible to random errors.

5.10 SHIFTING ALGORITHMS WITH RESPECT TO THE PHASE ORIGIN

The sampling weights of an algorithm change if the sampling points of an algorithm are shifted with respect to the origin by the phase distance ε. This section studies how the sampling weights change, thus modifying the algorithm structure. Shifting an algorithm in this manner does not change its basic properties with respect to immunity to harmonic components, insensitivity to detuning, etc.; however, shifting an algorithm

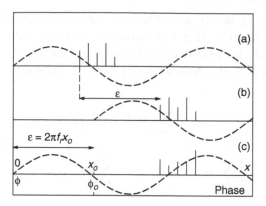

Figure 5.36   Shifting an algorithm.

can change the symmetry properties of the sampling reference functions. Thus, an algorithm that has equal magnitudes of the Fourier transforms of the sampling reference functions at all frequencies can be transformed by shifting it into one that is orthogonal at all frequencies and vice versa.

To learn how to shift an algorithm, let us first consider one in which the x origin ($\mathbf{O}_x$) and the phase origin ($\mathbf{O}_\alpha$) are at the same point, as in Figure 5.36a. Using Equations 5.62 and 5.63, the phase of the signal at the origin is then given by:

$$\tan\phi = \mp \frac{\int_{-\infty}^{\infty} s(x)g_1(x)dx}{\int_{-\infty}^{\infty} s(x)g_2(x)dx} \qquad (5.206)$$

If the sampling points are shifted together with the sinusoidal reference functions in the positive direction of x (Figure 5.36b), the reference sampling functions values are preserved but their positions are shifted. Thus, the new shifted phase, $\phi_0 = \phi + \varepsilon$, at position x_0 where $\varepsilon = 2\pi f_r x_0$, is now given by:

$$\tan\phi_0 = \mp \frac{\int_{-\infty}^{\infty} s(x)g_1(x-x_0)dx}{\int_{-\infty}^{\infty} s(x)g_2(x-x_0)dx} \qquad (5.207)$$

where $\varepsilon > 0$ and $x_0 > 0$ if the sampling reference functions are shifted in the positive direction.

The phase with respect to the nonshifted sinusoidal reference functions with these shifted sampling points (Figure 5.36c) can be obtained only if the values of the reference sampling functions are properly modified by using the phase equation:

$$\tan \phi = \mp \frac{\int_{-\infty}^{\infty} s(x) g_1'(x) dx}{\int_{-\infty}^{\infty} s(x) g_2'(x) dx} \qquad (5.208)$$

Applying a well-known trigonometric relation, we see that

$$\tan \phi = \tan(\phi_0 - \varepsilon) = \frac{\tan \phi_0 - \tan \varepsilon}{1 + \tan \varepsilon \tan \phi_0} \qquad (5.209)$$

From Equations 5.207 to 5.209 we find:

$$\frac{g_1'(x)}{g_2'(x)} = \frac{\cos \varepsilon \, g_1(x - x_0) \pm \sin \varepsilon \, g_2(x - x_0)}{\cos \varepsilon \, g_2(x - x_0) \pm \sin \varepsilon \, g_1(x - x_0)} \qquad (5.210)$$

Thus, we may write:

$$g_1'(x) = \cos \varepsilon \, g_1(x - x_0) \pm \sin \varepsilon \, g_2(x - x_0) \qquad (5.211)$$

and

$$g_2'(x) = \cos \varepsilon \, g_2(x - x_0) \mp \sin \varepsilon \, g_1(x - x_0) \qquad (5.212)$$

Hence, we may also write for the Fourier transforms of these reference sampling functions:

$$G_1'(f) = (\cos \varepsilon \, G_1(f) \pm \sin \varepsilon \, G_2(f)) \exp\left(-i\varepsilon \frac{f}{f_r}\right) \qquad (5.213)$$

and

$$G_2'(f) = (\cos \varepsilon \, G_2(f) \mp \sin \varepsilon \, G_1(f)) \exp\left(-i\varepsilon \frac{f}{f_r}\right) \qquad (5.214)$$

or, in terms of the amplitudes and phases:

$$G_1'(f) = \begin{pmatrix} \cos\varepsilon \, \text{Am}(G_1(f))\exp(i\gamma_1(f)) \\ \pm\sin\varepsilon \, \text{Am}(G_2(f))\exp(i\gamma_2(f)) \end{pmatrix} \exp\left(-i\varepsilon\frac{f}{f_r}\right) \quad (5.215)$$

and

$$G_2'(f) = \begin{pmatrix} \cos\varepsilon \, \text{Am}(G_2(f))\exp(i\gamma_2(f)) \\ \mp\sin\varepsilon \, \text{Am}(G_1(f))\exp(i\gamma_1(f)) \end{pmatrix} \exp\left(-i\varepsilon\frac{f}{f_r}\right) \quad (5.216)$$

The upper sign is used when $\gamma_1(f_r) - \gamma_2(f_r) < 0$. It is easy to show that in the original algorithm $\gamma_2(f_r) = 0$ and $\gamma_1(f_r) = \mp\pi$, and in the shifted algorithm we also have $\gamma_2'(f_r) = 0$ and $\gamma_1'(f_r) = \mp\pi$.

5.10.1 Shifting the Algorithm by $\pm\pi/2$

Of special interest is the case when the sampling points are shifted a phase ε equal to $\pm\pi/2$. In this case, we may see from Equation 5.211 that

$$g_1'(x) = \pm g_2(x - x_0) = \pm g_2\left(x - \frac{X_r}{4}\right) \quad (5.217)$$

and from Equation 5.212:

$$g_2'(x) = \mp g_1(x - x_0) = \mp g_1\left(x - \frac{X_r}{4}\right) \quad (5.218)$$

where $X_r = 1/f_r$. The plus or minus sign is used according to Table 5.3.

In other words, we can say that, after shifting, the sampling reference functions are just exchanged (with a change in sign) for one and only one of these functions. We can also write:

$$W_{1n}' = \pm W_{2n} \quad (5.219)$$

and

$$W_{2n}' = \mp W_{1n} \quad (5.220)$$

TABLE 5.3 Sign To Be Used in the Transformation Equations When Shifting an Algorithm

Relation between Phases $\gamma_1(f_r)$ and $\gamma_1(f_r)$	Sign of Shift	Sign To Be Used
$\gamma_1(f_r) - \gamma_2(f_r) < 0$	$\varepsilon > 0$	Upper
	$\varepsilon < 0$	Lower
$\gamma_1(f_r) - \gamma_2(f_r) > 0$	$\varepsilon > 0$	Upper
	$\varepsilon < 0$	Lower

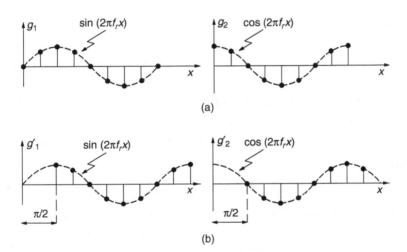

(a)

(b)

Figure 5.37 Sampling point movement when shifting an algorithm by $\pi/2$.

with the new sampling points located at phases displaced $\pm\pi/2$ with respect to those in the original algorithm. Figure 5.37 illustrates how the sampling points move for a shift of the algorithm equal to $\pi/2$.

5.10.2 Shifting the Algorithm by $\pm\pi/4$

This is another particular case of special interest. In this case, from Equation 5.211 we can see that:

$$g_1'(x) = \frac{1}{\sqrt{2}}\left(g_1(x - x_0) \pm g_2(x - x_0)\right) \qquad (5.221)$$

and from Equation 5.212:

$$g_2'(x) = \frac{1}{\sqrt{2}}\left(g_2(x - x_0) \mp g_1(x - x_0)\right) \qquad (5.222)$$

Thus, if we ignore the unimportant constant factor, we have:

$$g_1'(x) = g_1\left(x - \frac{X_r}{8}\right) \pm g_2\left(x - \frac{X_r}{8}\right) \qquad (5.223)$$

and

$$g_2'(x) = g_2\left(x - \frac{X_r}{8}\right) \mp g_1\left(x - \frac{X_r}{8}\right) \qquad (5.224)$$

where the signs are selected according to Table 5.3. We can also write:

$$W_{1n}' = W_{1n} \pm W_{2n} \qquad (5.225)$$

and

$$W_{2n}' = W_{2n} \mp W_{1n} \qquad (5.226)$$

with the new sampling points located at phases displaced $\pm\pi/4$ with respect to those in the original algorithm. Figure 5.38 illustrates how the sampling points move for a shift of the algorithm equal to $\pi/4$.

Let us now compare the sensitivity to detuning of the original and shifted algorithms. The Fourier transforms of these sampling reference functions from Equations 5.215 and 5.216 are:

$$G_1'(f) = \frac{1}{\sqrt{2}}\begin{pmatrix} \mathrm{Am}(G_1(f))\exp(i\gamma_1(f)) \\ \pm\,\mathrm{Am}(G_2(f))\exp(i\gamma_2(f)) \end{pmatrix}\exp\left(-i\frac{\pi}{4}\frac{f}{f_r}\right) \qquad (5.227)$$

and

$$G_2'(f) = \frac{1}{\sqrt{2}}\begin{pmatrix} \mathrm{Am}(G_2(f))\exp(i\gamma_2(f)) \\ \mp\,\mathrm{Am}(G_1(f))\exp(i\gamma_1(f)) \end{pmatrix}\exp\left(-i\frac{\pi}{4}\frac{f}{f_r}\right) \qquad (5.228)$$

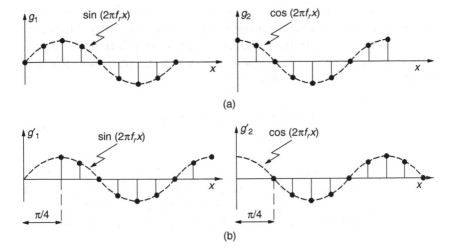

Figure 5.38 Sampling point movement when shifting an algorithm by $\pi/4$.

Let us now study two different particular cases of this algorithm shifted by $\pi/4$. The first case is when the original reference functions have the same amplitudes but are not orthogonal. In this case, from Equations 5.227 and 5.228 we have:

$$G_1'(f) = \frac{1}{\sqrt{2}} \text{Am}(G_1(f))\left(\exp(i\gamma_1(f)) \pm \exp(i\gamma_2(f))\right)\exp\left(-i\frac{\pi}{4}\frac{f}{f_r}\right) \quad (5.229)$$

and

$$G_2'(f) = \frac{1}{\sqrt{2}} \text{Am}(G_2(f))\left(\exp(i\gamma_2(f)) \mp \exp(i\gamma_1(f))\right)\exp\left(-i\frac{\pi}{4}\frac{f}{f_r}\right) \quad (5.230)$$

which may be transformed into:

$$G_1'(f) = \sqrt{2}\text{Am}(G_1(f))\cos\left(\frac{\gamma_1(f)-\gamma_2(f)}{2}\right)\exp i\left(\frac{\gamma_1(f)+\gamma_2(f)}{2} - \frac{\pi}{4}\frac{f}{f_r}\right) \quad (5.231)$$

and

$$G_2'(f) = \sqrt{2}\,i\text{Am}(G_1(f))\sin\left(\frac{\gamma_1(f)-\gamma_2(f)}{2}\right)\exp i\left(\frac{\gamma_1(f)+\gamma_2(f)}{2} - \frac{\pi}{4}\frac{f}{f_r}\right) \quad (5.232)$$

These values are for the upper signs. For the lower signs, these values are interchanged. The important conclusion is that these Fourier transforms are orthogonal, but their amplitudes are not the same. The ratio of the amplitudes of these Fourier transforms is given by:

$$\frac{\text{Am}(G_1'(f))}{\text{Am}(G_2'(f))} = \cot\left(\frac{\gamma_1(f) - \gamma_2(f)}{2}\right) \qquad (5.233)$$

The second case to study is when the original reference sampling functions are orthogonal but their amplitudes are not the same. From Equations 5.227 and 5.228 and by using the orthogonality condition in Equation 5.79, we have:

$$G_1'(f) = \frac{1}{\sqrt{2}}\left(\text{Am}(G_1(f)) + i\,\text{Am}(G_2(f))\right)\exp i\left(\gamma_1(f) - \frac{\pi}{4}\frac{f}{f_r}\right) \qquad (5.234)$$

and

$$G_2'(f) = \frac{1}{\sqrt{2}}\left(\text{Am}(G_2(f)) + i\,\text{Am}(G_1(f))\right)\exp i\left(\gamma_2(f) - \frac{\pi}{4}\frac{f}{f_r}\right) \qquad (5.235)$$

Thus, the shifted algorithm in this case has the same amplitudes, but it is not orthogonal.

A consequence of these last two results is that an algorithm for which the reference sampling functions are orthogonal to all frequencies but their amplitudes are not equal at all frequencies will convert, after shifting by $\pi/4$, to an algorithm for which the sampling reference functions have equal amplitudes at all frequencies but are orthogonal only at some frequencies.

Let us now consider the detuning properties of the shifted algorithm. Assuming detuning from the reference frequency (f_r) that shifts the phases γ_1 and γ_2, we can use Equation 5.232 to find:

$$\frac{\text{Am}(G_1'(f))}{\text{Am}(G_2'(f))} = \cot\left(\frac{\delta\gamma_1(f) - \delta\gamma_2(f) - \dfrac{\pi}{2}}{2}\right) \qquad (5.236)$$

Then, if the detuning is relatively small, we can obtain:

$$\frac{1}{2}\left(\frac{Am(G_1'(f))}{Am(G_2'(f))} - 1\right) = \frac{(\delta\gamma_1(f) - \delta\gamma_2(f))}{2} \qquad (5.237)$$

If we examine Equations 5.152 we can see that the amplitude of the detuning effect is the same for the original and the shifted algorithms, so shifting the algorithm will not modify its detuning sensitivity.

5.11 OPTIMIZATION OF PHASE-DETECTION ALGORITHMS

Given a number of sampling points and their phase positions, an infinite number of sampling weight sets can define the algorithm. In this chapter, we have developed some methods to find algorithms with the desired properties but this was done primarily to evaluate them. Another approach is to use optimization techniques to find the optimum sampling weights for some desired algorithm properties (Servín et al., 1997). To simplify the analysis we assume that the sampling reference functions $g_1(x)$ and $g_2(x)$ are antisymmetrical and symmetrical, respectively. No loss in generality has occurred, because, as described before, any algorithm can be shifted without losing its properties until the symmetry conditions are satisfied. Then, it is possible to show that the Fourier transforms of the reference functions are given by:

$$G_1(f) = -2i \sum_{n=1}^{N/2} W_{1n} \sin\left(\alpha_n \frac{f}{f_r}\right) \qquad (5.238)$$

and

$$G_2(f) = 2 \sum_{n=1}^{N/2} W_{2n} \cos\left(\alpha_n \frac{f}{f_r}\right) + \sigma_1 W_{2\left(\frac{N+1}{2}\right)} \qquad (5.239)$$

with:

$$\alpha_n = \frac{2\pi}{N}\left(n - \sigma_2 \frac{1}{2}\right) \qquad (5.240)$$

where:

$$\sigma_1 = 0; \quad \sigma_2 = 1; \quad \text{for } N \text{ even}$$
$$\sigma_1 = 1; \quad \sigma_2 = 0; \quad \text{for } N \text{ odd} \tag{5.241}$$

These symmetries ensure that the two sampling functions are orthogonal at all signal frequencies. The sampling weight values can now be found by minimizing the merit function $U(W_1, W_2, ..., W_N)$, defined by:

$$U(W_1, W_2, ..., W_N) = \rho_0 G_2(0)^2 +$$

$$+ \rho_1 \int_{f=f_r-\Delta_1}^{f_r+\Delta_1} [G_1(f) - G_2(f)]^2 df + \tag{5.242}$$

$$+ \rho_2 \int_{f=2f_r-\Delta_2}^{2f_r+\Delta_2} [G_1(f)^2 - G_2(f)^2] df + ...$$

The first term minimizes the bias (DC) component of the second sampling function. The bias of the second reference function is zero due to its antisymmetry. The second term minimizes the differences between the magnitudes of the sampling reference functions at the reference frequency. The third term minimizes the sensitivity of the algorithm to the second signal harmonic. More terms may be added if insensitivity to other signal harmonics is desired. The constants ρ_m are the weights assigned to each term. The constants Δ_m are the half-widths of the frequency intervals on which the optimizations for each signal harmonic are desired.

The optimum values of the sampling weights (W_n) may now be obtained by minimizing the merit function $U(W_1, W, ..., W_N)$ for the parameters W_n by solving the linear system of equations:

$$\frac{\partial U(W_1, W_2, ..., W_N)}{\partial W_n} = 0 \tag{5.243}$$

where the maximum value of n is $N/2$ if N is even or $(N + 1)/2$ if N is odd.

When solving the linear system, analytical or numerical integration may be used in the expression for the merit function. For practical convenience, numerical integration has been preferred.

To optimize the algorithm, a minimum of four sampling points is required. Servín et al. (1997) obtained optimized algorithms with four, five, and seven sampling points. An example of an algorithm designed using this method is provided in the next chapter.

5.12 INFLUENCE OF WINDOW FUNCTION OF SAMPLING ALGORITHMS

A signal that has harmonics that the signal algorithms cannot eliminate can be reduced by a suitable additional filtering function, sometimes called a *window function*, as described by de Groot (1995) and Schmit and Creath (1996). Any algorithm with reference sampling functions $g_1(x)$ and $g_2(x)$ may be modified by means of the window function $h(x)$. Then, the new reference sampling functions $g_1'(x)$ and $g_2'(x)$ would be given by:

$$g_1'(x) = h(x)g_1(x) \tag{5.244}$$

and

$$g_2'(x) = h(x)g_2(x) \tag{5.245}$$

With the convolution theorem, the Fourier transforms of these functions are:

$$G_1'(f) = H(f) * G_1(f) \tag{5.246}$$

and

$$G_2'(f) = H(f) * G_2(f) \tag{5.247}$$

These new reference sampling functions must satisfy the conditions of orthogonality and equal magnitudes at the reference frequency; hence, we require:

$$G_1'(f_r) \pm iG_2'(f_r) = \left(H(f) * \left[G_1(f) \pm iG_2(f)\right]\right)_{f=f_r} = 0 \tag{5.248}$$

The zero bias condition must also be satisfied. Thus, from Equations 5.106 and 5.107, we can write:

$$\sum_{n=1}^{N} W'_{1n} = \sum_{n=1}^{N} h(x_n)W_{1n} = 0 \tag{5.249}$$

and

$$\sum_{n=1}^{N} W'_{2n} = \sum_{n=1}^{N} h(x_n)W_{2n} = 0 \tag{5.250}$$

Any window function satisfying these conditions transforms an algorithm into another with different properties. A formal mathematical derivation of the general conditions required by the window function is possible using these relations; nevertheless, we will restrict ourselves to the simple particular case of an algorithm with sampling points in two periods of the reference function, with an identical distribution on each of the two periods, so if the sampling function for the basic one-period algorithm is $g_{bi}(x)$ then the sampling function $g_i(x)$ for the two periods is:

$$g_1(x) = g_{bi}(x) + g_{bi}(x + 2\pi) \tag{5.251}$$

A particular case of this kind of algorithm is when the points are equally spaced in the two periods and the number of points is even. Thus, its Fourier transform is:

$$G_i(f) = G_{bi}(f)\left(1 + \exp\left(i2\pi\frac{f}{f_r}\right)\right) \tag{5.252}$$

It is relatively simple to prove either mathematically or graphically that any window function that satisfies the condition:

$$h(x) = 2 - h\left(x + \frac{1}{2f_r}\right) \tag{5.253}$$

preserves the magnitude and phase of the Fourier transforms of the reference sampling functions at the reference frequency

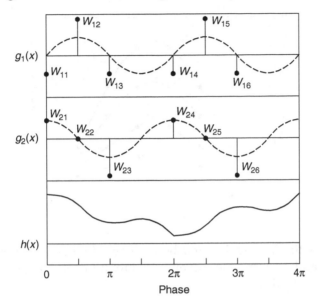

Figure 5.39 Reference sampling functions and window function when two periods of the signal are sampled.

as well as the zero bias. Figure 5.39 illustrates a particular case of these functions. This window function, then, can be expressed by a Fourier series as:

$$h(x) = 2 + \sum_{m=1}^{\infty} A_m \cos(m\pi f_r x) \qquad (5.254)$$

where m is an odd integer. The Fourier transform of this filter function thus becomes:

$$H(f) = 2\delta(f) + \frac{1}{2} \sum_{m=-\infty}^{\infty} A_m \delta\left(f - \frac{m f_r}{2}\right) \qquad (5.255)$$

Using the merit function defined in the preceding section, the best value for these A_m coefficients can be calculated.

Schmit and Creath (1996) described in some detail triangular and bell functions, which can be considered particular

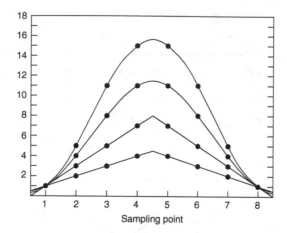

Figure 5.40 Triangular and bell window functions (described by Schmit and Creath) for an eight-sampling-point, diagonal least-squares algorithm.

cases of the one described here. Improved algorithms are obtained if these window functions are applied to the eight-sampling-point diagonal least-squares algorithms, with an even number of points. These window functions, shown in Figure 5.40, improve the characteristics of the algorithm.

Schmit and Creath proved that the triangular window produces the same effect as the multiple sequential technique, while the bell window produces the same effect as the multiple averaging technique. de Groot (1995) also studied the effect of a window function, using an approach more similar to the filtering function studied earlier.

5.13 CONCLUSIONS

In this chapter, we have established the foundations for the analysis of phase-detection algorithms. This theory permits us to analyze the properties of any algorithm and even allows us to design better ones.

APPENDIX. DERIVATIVE OF THE AMPLITUDE OF THE FOURIER TRANSFORM OF THE REFERENCE SAMPLING FUNCTIONS

The derivative of the Fourier transform of the sampling functions is frequently needed. In this appendix, we derive the expression for this derivative. Equation 5.54 may be written as:

$$Am(G_j(f))\exp(i\gamma(f)) = X(f) + iY(f) \tag{A.1}$$

where $X(f)$ is the real part and $Y(f)$ is the imaginary part. Taking the derivative of this expression with respect to f we find:

$$iAm(G_j(f))\frac{d\gamma_j(f)}{df}\exp(i\gamma(f)) +$$
$$+ \frac{dAm(G_j(f))}{df}\exp(i\gamma_j(f)) = \frac{dX(f)}{df} + i\frac{dY(f)}{df} \tag{A.2}$$

which can be transformed into:

$$\frac{dAm(G_j(f))}{df} = \left(\frac{dX(f)}{df} + i\frac{dY(f)}{df}\right)\exp(-i\gamma_j(f)) -$$
$$- iAm(G_j(f))\frac{d\gamma_j(f)}{df} \tag{A.3}$$

Because the left-hand side of this expression is real, the right-hand side must also be real. Thus, we obtain:

$$\frac{dAm(G_j(f))}{df} = \left(\frac{dX(f)}{df}\cos\gamma_j(f) + \frac{dY(f)}{df}\sin\gamma_j(f)\right) \tag{A.4}$$

To apply this expression to an algorithm with N sampling points, we now use Equations 5.74 and 5.75 in this expression, with $\psi(f_r) = 0$:

$$\frac{\mathrm{dAm}\left(G_j(f)\right)}{\mathrm{d}f} = -\frac{1}{f_r}\cos\gamma_j(f)\sum_{n=1}^{N} W_{jn}\alpha_n \sin\left(\alpha_n \frac{f}{f_r}\right) -$$

$$-\frac{1}{f_r}\sin\gamma_j(f)\sum_{n=1}^{N} W_{jn}\alpha_n \cos\left(\alpha_n \frac{f}{f_r}\right) \tag{A.5}$$

Thus, this derivative at the signal harmonic k (including the signal frequency, f_r, with $k = 1$) becomes:

$$\left(\frac{\mathrm{dAm}\left(G_j(f)\right)}{\mathrm{d}f}\right)_{f=kf_r} = -\frac{1}{f_r}\cos\gamma_j(kf_r)\sum_{n=1}^{N} W_{jn}\alpha_n \sin(k\alpha_n) -$$

$$-\frac{1}{f_r}\sin\gamma_j(kf_r)\sum_{n=1}^{N} W_{jn}\alpha_n \cos(k\alpha_n) \tag{A.6}$$

REFERENCES

Ai, C. and Wyant, J.C., Effect of piezoelectric transducer nonlinearity on phase shift interferometry, *Appl. Opt.*, 26, 1112–1116, 1987.

Alcalá-Ochoa, N. and Huntley, J.M., Convenient method for calibrating nonlinear phase modulators for use in phase-shifting interferometry, *Opt. Eng.*, 37, 2501–2505, 1998.

Brophy, C.P., Effect of intensity error correlation on the computed phase of phase-shifting interferometry, *J. Opt. Soc. Am. A*, 7, 537–540, 1990.

Cheng, Y.-Y. and Wyant, J.C., Phase shifter calibration in phase-shifting interferometry, *Appl. Opt.*, 24, 3049–3052, 1985.

Creath, K., Comparison of phase measuring algorithms, *Proc. SPIE*, 680, 19–28, 1986.

Creath, K., Phase-measurement interferometry techniques, in *Progress in Optics*, Vol. XXVI, Wolf, E., Ed., Elsevier Science, Amsterdam, 1988.

Creath, K., Phase measurement interferometry: beware these errors, *Proc. SPIE*, 1553, 213–220, 1991.

de Groot, P., Derivation of algorithms for phase shifting interferometry using the concept of a data-sampling window, *Appl. Opt.*, 34, 4723–4730, 1995.

Freischlad, K. and Koliopoulos, C.L., Fourier description of digital phase measuring interferometry, *J. Opt. Soc. Am. A*, 7, 542–551, 1990.

Greivenkamp, J.E., Generalized data reduction for heterodine interferometry, *Opt. Eng.*, 23, 350–352, 1984.

Hariharan, P., Phase-shifting interferometry: minimization of systematic errors, *Opt. Eng.*, 39, 967–969, 2000.

Hariharan, P., Oreb, B.F., and Eiju, T., Digital phase-shifting interferometry: a simple error-compensating phase calculation algorithm, *Appl. Opt.* 26, 2504–2505, 1987.

Hibino, K., Susceptibility of systematic error-compensating algorithms to random noise in phase shifting interferometry, *Appl. Opt.*, 36, 2084–2092, 1997.

Hibino, K. and Yamauchi, M., Phase-measuring algorithms to suppress spatially nonuniform phase modulation in a two beam interferometer, *Opt. Rev.*, 7, 543–549, 2000.

Hibino, K., Oreb, B.F., Farrant, D.I., and Larkin, K.G., Phase shifting for non-sinusoidal waveforms with phase shift errors, *J. Opt. Soc. Am. A*, 12, 761–768, 1995.

Hibino, K., Oreb, B.F., Farrant, D.I., and Larkin, K.G., Phase shifting algorithms for nonlinear and spatially nonuniform phase shifts, *J. Opt. Soc. Am. A*, 12, 918–930, 1997.

Huang, H., Itoh, M., and Yatagai, T., Phase retrieval of phase-shifting interferometry with iterative least squares fitting algorithm: experiments, *Opt. Rev.*, 6, 196–203, 1999.

Joenathan, C., Phase-measurement interferometry: new methods and error analysis, *Appl. Opt.*, 33, 4147–4155, 1994.

Kinnstaetter, K., Lohmann, A., Schwider, W., and Streibl, J.N., Accuracy of phase shifting interferometry, *Appl. Opt.*, 27, 5082–5089, 1988.

Lai, G. and Yatagai, T., Generalized phase shifting interferometry, *J. Opt. Soc. Am. A*, 8, 822–827, 1991.

Larkin, K.G. and Oreb, B.F., Design and assessment of symmetrical phase-shifting algorithm, *J. Opt. Soc. Am.*, 9, 1740–1748, 1992.

Malacara-Doblado, D, Dorrío B.V., and Malacara-Hernández, D., Graphic tool to produce tailored symmetrical phase shifting algorithms, *Opt. Lett.*, 25, 64–66, 2000.

Morgan, C.J., Least squares estimation in phase-measurement interferometry, *Opt. Lett.*, 7, 368–370, 1982.

Nakadate, S., Phase detection of equidistant fringes for highly sensitive optical sensing. I. Principle and error analysis, *J. Opt. Soc. Am. A*, 5, 1258–1264, 1988a.

Nakadate, S., Phase detection of equidistant fringes for highly sensitive optical sensing. II. Experiments, *J. Opt. Soc. Am. A*, 5, 1265–1269, 1988b.

Ohyama, N., Kinoshita, S., Cornejo-Rodríguez, A., Honda, T., and Tsujiuchi, J., Accuracy of determination with unequal reference phase shift, *J. Opt. Soc. Am. A*, 5, 2019–2025, 1988.

Parker, D.H., Moiré patterns in three-dimensional Fourier space, *Opt. Eng.*, 30, 1534–1541, 1991.

Ransom, P.L. and Kokal, J.B., Interferogram analysis by a modified sinusoid fitting technique, *Appl. Opt.*, 25, 4199–4204, 1986.

Rathjen, C., Statistical properties of phase-shift algorithms, *J. Opt. Soc. Am. A*, 12, 1997–2008, 1995.

Schmit, J. and Creath, K., Window function influence on phase error in phase-shifting algorithms, *Appl. Opt.*, 35, 5642–5649, 1996.

Schwider, J., Phase shifting interferometry: reference phase error reduction, *Appl. Opt.*, 28, 3889–3892, 1989.

Schwider, J., Burow, R., Elssner, K.-E., Grzanna, J., Spolaczyk, R., and Mertel, K., Digital wave-front measuring interferometry: some systematic error sources, *Appl. Opt*, 22, 3421–3432, 1983.

Servín, M., Malacara, D., Marroquin, J.L., and Cuevas, F.J., Complex linear filters for phase shifting with low detuning sensitivity, *J. Mod. Opt.*, 44, 1269–1278, 1997.

Stetson, K.A. and Brohinsky, W.R., Electrooptic holography and its applications to hologram interferometry, *Appl. Opt.*, 24, 3631–3637, 1985.

Surrel, I., Design of algorithms for phase measurements by the use of phase stepping, *Appl. Opt.*, 35, 51–60, 1996.

Surrel, I., Additive noise effect in digital phase detection, *Appl. Opt.*, 36, 271–276, 1997.

Surrel, I., Phase-shifting algorithms for nonlinear and spatially nonuniform phase shifts, *Opt. Soc. Am. A*, 15, 1227–1233, 1998.

Wang, Z., Graça, M.S., Bryanston-Cross, P.J., and Whitehouse, D.J., Phase-shifted image matching algorithm for displacement measurement, *Opt. Eng.*, 35, 2327–2332, 1996.

Wingerden, J. van, Frankena, H.J., and Smorenburg, C., Linear approximation for measurement errors in phase-shifting interferometry, *Appl. Opt.*, 30, 2718–2729, 1991.

Wyant, J.C., Koliopoulos, C.L., Bushan, B., and George, O.E., An optical profilometer for surface characterization of magnetic media, *ASLE Trans.*, 27, 101, 1984.

6

Phase-Detection Algorithms

6.1 GENERAL PROPERTIES OF SYNCHRONOUS PHASE-DETECTION ALGORITHMS

Various phase-measuring algorithms have been reviewed by many authors (e.g., Schwider et al., 1983; Creath, 1986, 1991). In this chapter, we describe several of the phase-detection algorithms, each of which has different properties, and we apply the Fourier theory developed in Chapter 5 to the analysis of some of these phase-detection schemes.

Because we have three unknowns in Equation 1.4 (i.e., a, b, and $\phi\Delta$), we need a minimum of three signal measurements to determine the phase ϕ. The measurements can have any phase, as long as they are known. We can assume that the first measurement is at phase α_1, the second at α_2, the third at α_3, and so on. Here, the zero-value position for these phases (α_n) will be considered to be at the origin of coordinates, thus making $\psi(f_r) = 0$. In this case, the Fourier transforms of the sampling functions (from Equations 5.90 and 5.91) are:

$$G_1(f) = \sum_{n=1}^{N} W_{1n} \exp\left(-i\alpha_n \frac{f}{f_r}\right) \qquad (6.1)$$

and

$$G_2(f) = \sum_{n=1}^{N} W_{2n} \exp\left(-i\alpha_n \frac{f}{f_r}\right) \qquad (6.2)$$

where the phase shift (α_n) is measured with respect to the reference frequency.

A sampling phase-detecting algorithm is defined by the number of sampling points, their phase positions, and their associated sampling weights. The minimum number of sampling points is three. In this case, their positions automatically define the values of the sampling weights. When the number of sampling points is greater than three, the phase positions of the sampling points do not completely define the algorithm, as an infinite number of sampling weight sets satisfies the conditions studied in Chapter 5; however, only one of these possible solutions is a least-squares fit.

In Chapter 5 we found that, in the presence of detuning, the conditions requiring equal magnitudes or orthogonality of the Fourier transforms of the sampling points, or both, are lost. Given a number of sampling points, these properties are defined by the phase locations of the sampling points.

If we consider only nonzero sampling weights, we can show that:

1. If $g_1(f)$ is symmetric and $g_2(f)$ is antisymmetric, or vice versa, about the same phase point, then the two functions are orthogonal at all frequencies.
2. If $g_1(f)$ and $g_2(f)$ are equal and only one is shifted with respect to the other (for example, if both are symmetric or antisymmetric about different points separated by 90°), then they will have the same magnitudes at all frequencies.

6.2 THREE-STEP ALGORITHMS TO MEASURE THE PHASE

We have seen before that, to determine the phase without any ambiguity, a minimum of three sampling points is necessary. Let us now consider the case of three sampling points with any phases α_1, α_2, and α_3. Hence, we can write:

$$s_1 = a + b\cos(\phi + \alpha_1)$$

$$s_2 = a + b\cos(\phi + \alpha_2) \qquad (6.3)$$

$$s_3 = a + b\cos(\phi + \alpha_3)$$

where the x,y dependence is implicit. These expressions can also be written as:

$$s_1 = a + b\cos\alpha_1\cos\phi - b\sin\alpha_1\sin\phi$$

$$s_2 = a + b\cos\alpha_2\cos\phi - b\sin\alpha_2\sin\phi \qquad (6.4)$$

$$s_3 = a + b\cos\alpha_3\cos\phi - b\sin\alpha_3\sin\phi$$

Hence, we can find:

$$\frac{s_2 - s_3}{2s_1 - s_2 - s_3} =$$

$$\frac{(\cos\alpha_2 - \cos\alpha_3) - (\sin\alpha_2 - \sin\alpha_3)\tan\phi}{(2\cos\alpha_1 - \cos\alpha_2 - \cos\alpha_3) - (2\sin\alpha_1 - \sin\alpha_2 - \sin\alpha_3)\tan\phi} \qquad (6.5)$$

This is a general expression for three-point sampling algorithms. Let us now consider some particular cases.

6.2.1   120° Three-Step Algorithm

A particular case of the three-step method is to take $\alpha_1 = 60°$, $\alpha_2 = 180°$, and $\alpha_3 = 300°$, as shown in Figure 6.1. Thus, we obtain the following result for the phase:

$$\tan\phi = -\sqrt{3}\,\frac{s_1 - s_3}{s_1 - 2s_2 + s_3} \qquad (6.6)$$

From this expression (by comparing with Equation 5.108), we can see that the reference sampling weights have the values $W_{11} = \sqrt{3}/2$, $W_{12} = 0$, $W_{13} = -\sqrt{3}/2$, $W_{21} = 1/2$, $W_{22} = -1$, and $W_{23} = 1/2$. Thus, the reference sampling functions (Figure 6.1) are:

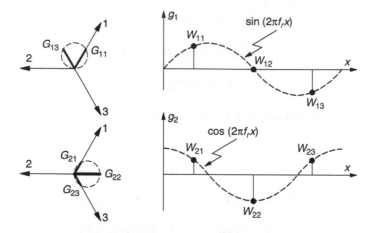

Figure 6.1   A 120° three-step algorithm to measure the phase.

$$g_1(x) = \frac{\sqrt{3}}{2}\delta\!\left(x - \frac{X_r}{6}\right) - \frac{\sqrt{3}}{2}\delta\!\left(x - \frac{5X_r}{6}\right) \qquad (6.7)$$

and

$$g_2(x) = \frac{1}{2}\delta\!\left(x - \frac{X_r}{6}\right) - \delta\!\left(x - \frac{3X_r}{6}\right) + \frac{1}{2}\delta\!\left(x - \frac{5X_r}{6}\right) \qquad (6.8)$$

Because these three sampling points are equally spaced and uniformly distributed along the reference function period, as described by Equation 5.19, the values of W_{1n} are equal to $\sin(2\pi f_r x_n)$ and the values of W_{2n} are equal to $\cos(2\pi f_r x_n)$. Thus, this is a diagonal least-squares algorithm, and Equation 5.19 for the phase is valid. It can easily be shown that Equation 5.19 reduces to Equation 6.4 for these sampling points.

The sampling weights represented in a polar diagram are shown on the left side of Figure 6.1. We can see that the sampling vectors $\mathbf{G}_1$ and $\mathbf{G}_2$ are perpendicular to each other. We can also see on the right side of this figure that the sum of all sampling weights W_{1n} and similarly the sum of all sampling weights W_{2n} are equal to zero, as the functions $g_i(x)$ have no DC term.

The Fourier transforms of the sampling functions, using Equations 5.90 and 5.91, are:

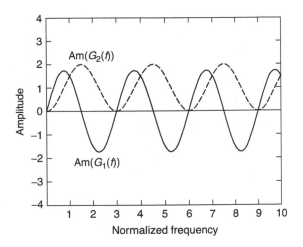

Figure 6.2 Amplitudes of the Fourier transforms of sampling functions for the 120° three-step algorithm.

$$G_1(f) = \sqrt{3}\sin\left(\frac{2\pi}{3}\frac{f}{f_r}\right)\exp\left[-i\pi\left(\frac{f}{f_r}-\frac{1}{2}\right)\right] \qquad (6.9)$$

and

$$G_2(f) = \left[1-\cos\left(\frac{2\pi}{3}\frac{f}{f_r}\right)\right]\exp\left[-i\pi\left(\frac{f}{f_r}-1\right)\right] \qquad (6.10)$$

The amplitudes of these functions are plotted in Figure 6.2. Observing Equations 6.9 and 6.10, we see that these two functions are orthogonal at all frequencies. The normalized frequency is defined as the ratio of the frequency f to the reference frequency f_r. With a detuning, the condition for equal magnitudes is lost. It must be pointed out here that a phase π has been added, if necessary, to all expressions for the Fourier transforms $G_1(f)$ and $G_2(f)$ in this chapter, in order to change their sign and make their amplitudes positive at the reference frequency f_r. The phases as functions of the normalized frequency are linear and are orthogonal for all frequencies as illustrated in Figure 6.3.

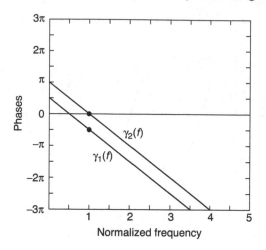

Figure 6.3 Sampling function phases for the 120° three-step algorithm.

Given a reference frequency (f_r), the value of $r(f)$ is a function of the signal phase and the signal frequency and is expressed by Equation 5.77. The value of $r(f)$ is thus given by:

$$r(f) = -\frac{\sqrt{3}\sin\left(\frac{2\pi}{3}\frac{f}{f_r}\right)\tan\left(\pi\frac{f}{f_r}+\phi\right)}{1-\cos\left(\frac{2\pi}{3}\frac{f}{f_r}\right)} \qquad (6.11)$$

If both the reference and signal frequencies are known, the phase can be obtained when the value of $r(f)$ has been determined. If $f = f_r$, this expression reduces to Equation 5.47. From Figure 6.2 we can see that this algorithm has the following properties:

1. It is sensitive to detuning error, as shown in Figure 6.3, as the magnitudes of the Fourier transforms of the sampling functions are altered by small detunings. The phase error as a function of the normalized frequency is shown in Figure 6.4.

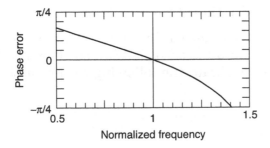

Figure 6.4 Detuning error for the 120° three-step algorithm.

2. Signals with frequencies f_r, $2f_r$, $4f_r$, $5f_r$, $7f_r$, etc. can be detected, as the amplitudes of the Fourier transforms are the same (even if of different sign) at these frequencies.
3. Phase errors can be introduced by the presence in the signal of second, fourth, fifth, seventh, and eight harmonics; however, it is insensitive to third, sixth, and ninth harmonics.

As expected, the phase error is also a function of the signal phase and has an almost sinusoidal shape, as shown in Figure 6.5.

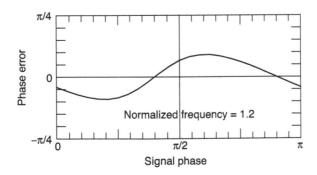

Figure 6.5 Periodic phase error as a function of the signal phase for the 120° three-step algorithm. This is for a normalized frequency equal to 1.2.

6.2.2 Inverted T Three-Step Algorithm

Another particular case of the three-step method is when we use $\alpha_1 = 0°$, $\alpha_2 = 90°$, and $\alpha_3 = 180°$, as shown in Figure 6.6. In this case, we obtain the following result for the phase:

$$\tan\phi = -\frac{-s_1 + 2s_2 - s_3}{s_1 - s_3} \tag{6.12}$$

These three points are equally but not uniformly spaced along the reference sampling function period. As a consequence, the sampling weights W_{1n} and W_{2n} are not equal to the functions $\sin(2\pi f_r \alpha_n)$ and $\cos(2\pi f_r \alpha_n)$, respectively, as in the case of uniformly spaced sampling points.

The sampling weights have the values $W_{11} = -1$, $W_{12} = 2$, $W_{13} = -1$, $W_{21} = 1$, $W_{22} = 0$, and $W_{23} = -1$. Thus, the reference sampling functions are:

$$g_1(x) = -\delta(x) + 2\delta\left(x - \frac{X_r}{4}\right) - \delta\left(x - \frac{2X_r}{4}\right) \tag{6.13}$$

and

$$g_2(x) = \delta(x) - \delta\left(x - \frac{X_r}{2}\right) \tag{6.14}$$

and the Fourier transforms of the sampling functions become:

$$G_1(f) = 4\sin^2\left(\frac{\pi}{4}\frac{f}{f_r}\right)\exp\left(-i\frac{\pi}{2}\frac{f}{f_r}\right) \tag{6.15}$$

and

$$G_2(f) = 4\left[\sin\left(\frac{\pi}{4}\frac{f}{f_r}\right)\cos\left(\frac{\pi}{4}\frac{f}{f_r}\right)\right]\exp\left[-i\frac{\pi}{2}\left(\frac{f}{f_r} - 1\right)\right] \tag{6.16}$$

We can see that these functions are orthogonal at all frequencies and that their magnitudes are equal only at the reference frequency (f_r) and all of its harmonics. Their amplitudes are shown in Figure 6.7. The value of $r(f)$, from Equation 5.77, is:

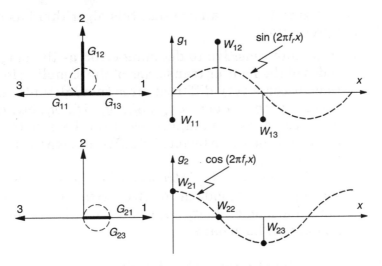

Figure 6.6 A three-step inverted T algorithm to measure the phase.

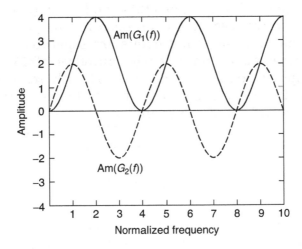

Figure 6.7 Amplitudes of the Fourier transforms of sampling functions for the three-step inverted T algorithm.

$$r(f) = -\tan\left(\frac{\pi}{4}\frac{f}{f_r}\right)\tan\left(\phi + \frac{\pi}{2}\frac{f}{f_r} - \frac{\pi}{2}\right) \quad (6.17)$$

which, as expected, for $f = f_r$, becomes Equation 5.81.

From Figure 6.7 we can see that this algorithm has the following properties:

1. It is quite sensitive to detuning error, as the magnitudes of the Fourier transforms of the sampling functions become very different after small detunings.
2. Signals with frequencies f_r, $3f_r$, $5f_r$, $7f_r$, $9f_r$, etc. can be detected, as the amplitudes of the Fourier transforms are the same (even if of different sign) at these frequencies.
3. Phase errors can be introduced by the presence in the signal of second, third, fifth, sixth, seventh, and ninth harmonics; however, it is insensitive to fourth and eighth harmonics.

6.2.3 Wyant's Tilted T Three-Step Algorithm

A particularly interesting version of a three-step algorithm was proposed by Wyant et al. (1984) and later by Bhushan et al. (1985). In this case, the expression for the phase is quite simple. The three sampling points are separated by 90°, as in the former algorithm, but with an offset of 45° (i.e., the first sampling point is taken at −45° with respect to the origin). It is interesting to note that a change in this offset changes the values of the sampling weights. These authors used $\alpha_1 = -45°$, $\alpha_2 = 45°$, and $\alpha_3 = 135°$, as shown in Figure 6.8. Thus, we obtain the following result for the phase:

$$\tan\phi = -\frac{-s_1 + s_2}{s_2 - s_3} \tag{6.18}$$

The sampling weights have the following values: $W_{11} = -1$, $W_{12} = 1$, $W_{13} = 0$, $W_{21} = 0$, $W_{22} = 1$, and $W_{23} = -1$. The reference sampling functions are:

$$g_1(x) = -\delta\left(x + \frac{X_r}{8}\right) + \delta\left(x - \frac{X_r}{8}\right) \tag{6.19}$$

and

$$g_2(x) = \delta\left(x - \frac{X_r}{8}\right) - \delta\left(x - \frac{3X_r}{8}\right) \tag{6.20}$$

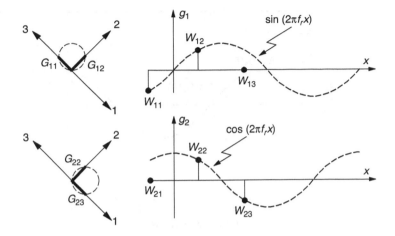

Figure 6.8 Wyant's three-step algorithm.

Thus, the Fourier transform amplitudes of the sampling functions, as illustrated in Figure 6.9, are:

$$G_1(f) = \sqrt{2}\sin\left(\frac{\pi}{4}\frac{f}{f_r}\right)\exp\left(-i\frac{\pi}{2}\right) \tag{6.21}$$

and

$$G_2(f) = \sqrt{2}\sin\left(\frac{\pi}{4}\frac{f}{f_r}\right)\exp\left[-i\frac{\pi}{2}\left(\frac{f}{f_r}-1\right)\right] \tag{6.22}$$

These functions have the same amplitudes at all frequencies so their graphs superimpose one over the other. They are orthogonal only at the reference frequency (f_r) and at its odd harmonics, as shown in Figure 6.10. From Equation 5.77, the coefficient $r(f)$ is given by:

$$r(f) = -\frac{\sin\phi}{\sin\left(\phi + \frac{\pi}{2}\frac{f}{f_r}\right)} \tag{6.23}$$

which can be used to find the phase in the presence of detuning, if the magnitude of this detuning is known.

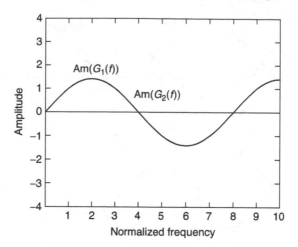

Figure 6.9 Amplitudes of Fourier transforms for reference sampling functions in Wyant's three-step algorithm.

From Figure 6.10 we can see that this algorithm has the following properties:

1. It is quite sensitive to detuning error, as the orthogonality of the Fourier transforms of the sampling functions is lost after small detunings. The phase error is illustrated in Figure 6.11.
2. Just as in the preceding algorithm, signals with frequencies f_r, $2f_r$, $4f_r$, $5f_r$, $7f_r$, etc. can be detected, as the amplitudes of the Fourier transforms are the same (even if of different sign) at these frequencies.
3. Also as in the preceding algorithm, phase errors can be introduced by the presence in the signal of second, third, fifth, sixth, seventh, and ninth harmonics, and it is also insensitive to fourth and eighth harmonics.

6.2.4　Two-Steps-Plus-One Algorithm

If the constant term or bias is removed from the signal measurements, the phase can be determined using only two sampling points having a phase difference of 90°. The tangent of

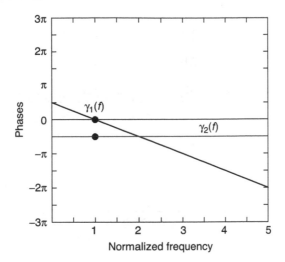

Figure 6.10 Phases for the reference sampling functions in Wyant's three-step algorithm.

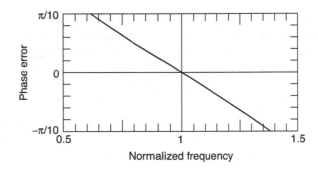

Figure 6.11 Phase error as a function of the normalized frequency for Wyant's three-step algorithm.

the phase is simply the ratio of the two measurements. Mendoza-Santoyo et al. (1988) determined the phase using this principle. This principle has also been applied to an interesting three-step method (Figure 6.12) suitable for systems with vibrations, such as in the testing of large astronomical mirrors (Angel and Wizinowich, 1988). The phase of one of the beams is rapidly switched between two values, separated by 90°. This

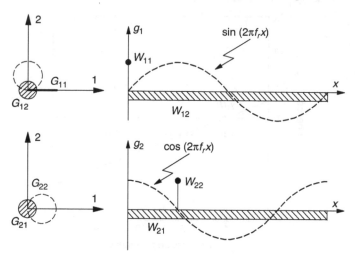

Figure 6.12   Sampling functions in the three-step (2 + 1) algorithm.

is done quickly enough to reduce the effects of vibration. Further readings are taken any time later to obtain the sum of the irradiance of the beams, independent of their relative phase. These later readings to find the irradiance sum can be performed in any of several possible ways, one of which is to take two readings separated by 180°. An alternative way is to use an integrating interval of $\Delta = 360°$. The Fourier analysis of this algorithm thus depends on the approach used to find this irradiance. Here, we consider the second method of integrating the signal in a period. Thus, we can write:

$$s_1 = a + b\cos\phi$$

$$s_2 = a + b\cos(\phi + 90°)$$

$$s_3 = \frac{1}{X_r}\int_0^{X_r} s(x)\mathrm{d}x = a$$

(6.24)

where $x = (X_r/2\pi)\phi$, which gives us the following for the phase:

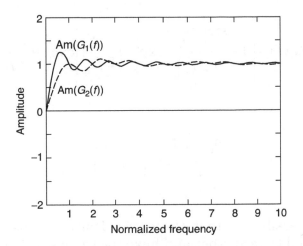

Figure 6.13 Amplitudes of Fourier transforms for reference sampling functions for the three-step (2 + 1) algorithm.

$$\tan\phi = -\frac{s_2 - s_3}{s_1 - s_3} \qquad (6.25)$$

The reference sampling functions are:

$$g_1(x) = \delta\left(x - \frac{X_r}{4}\right) - f(x) \qquad (6.26)$$

and

$$g_2(x) = \delta(x) - f(x) \qquad (6.27)$$

with

$$f(x) = 0, \quad \text{for } x \le 0$$

$$= \frac{1}{X_r}, \quad \text{for } 0 \le x \le X_r \qquad (6.28)$$

$$= 0, \quad \text{for } X_r \le x$$

Thus, the Fourier transforms of these sampling functions, as shown in Figure 6.13, are:

$$G_1(f) = \left(1 - \frac{\sin\left(\pi\dfrac{f}{f_r}\right)}{\left(\pi\dfrac{f}{f_r}\right)}\exp\left(-i\frac{\pi}{2}\frac{f}{f_r}\right)\right)\exp\left(-i\frac{\pi}{2}\frac{f}{f_r}\right) \quad (6.29)$$

and

$$G_2(f) = 1 - \frac{\sin\left(\pi\dfrac{f}{f_r}\right)}{\left(\pi\dfrac{f}{f_r}\right)}\exp\left(-i\pi\frac{f}{f_r}\right) \quad (6.30)$$

We can easily see that these two Fourier transforms are orthogonal to each other and have the same amplitude at the signal frequency and all of its harmonics. In other words, this algorithm is not insensitive to any of the signal harmonics. It is also sensitive to detuning. The value of $r(f)$, from Equation 5.77, is given by:

$$r(f) = \frac{\cos\left(\dfrac{\pi}{2}\dfrac{f}{f_r}+\phi\right) - \dfrac{\sin\left(\pi\dfrac{f}{f_r}\right)}{\left(\pi\dfrac{f}{f_r}\right)}\cos\left(\pi\dfrac{f}{f_r}+\phi\right)}{\cos\phi - \dfrac{\sin\left(\pi\dfrac{f}{f_r}\right)}{\left(\pi\dfrac{f}{f_r}\right)}\cos\left(\pi\dfrac{f}{f_r}+\phi\right)} \quad (6.31)$$

6.3 FOUR-STEP ALGORITHMS TO MEASURE THE PHASE

In principle, three steps are enough to determine the three unknown constants; however, small measurement errors can have a large effect in the results. Four-step methods can offer better results in this respect. With four steps, as noted earlier in this chapter, the sampling point distribution has an infinite number of solutions for the phase, and some of them are diagonal least-squares algorithm solutions.

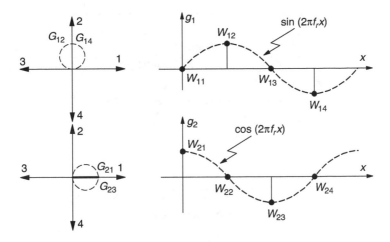

Figure 6.14 Four-step cross algorithm.

6.3.1 Four Steps in the Cross Algorithm

The values of the irradiance are measured using four different values of the phase: $\alpha_1 = 0°$, $\alpha_2 = 90°$, $\alpha_3 = 180°$, and $\alpha_4 = 270°$. Thus, as shown in Figure 6.14, we have:

$$s_1 = a + b\cos\phi$$
$$s_2 = a + b\cos(\phi + 90°)$$
$$s_3 = a + b\cos(\phi + 180°)$$
$$s_4 = a + b\cos(\phi + 270°)$$

(6.32)

From these expressions, one possible solution for the phase is:

$$\tan\phi = -\frac{s_2 - s_4}{s_1 - s_3}$$

(6.33)

The sampling weights have the values $W_{11} = 0$, $W_{12} = 1$, $W_{13} = 0$, $W_{14} = -1$, $W_{21} = 1$, $W_{22} = 0$, $W_{23} = -1$, and $W_{24} = 0$. We can see in Figure 6.14 that these sampling weights are described by Equation 5.19. Hence, this is a diagonal least-squares solution, with a diagonal system matrix. The reference sampling functions are:

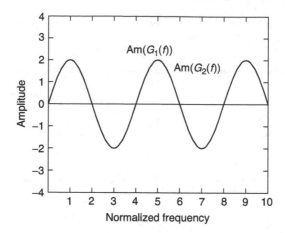

Figure 6.15 Amplitudes of Fourier transforms for reference sampling functions for the four-step cross algorithm.

$$g_1(x) = \delta\left(x - \frac{X_r}{4}\right) - \delta\left(x - \frac{3X_r}{4}\right) \qquad (6.34)$$

and

$$g_2(x) = \delta(x) - \delta\left(x - \frac{2X_r}{4}\right) \qquad (6.35)$$

Thus, the Fourier transforms of the sampling functions (Figure 6.15) are:

$$G_1(f) = 2\sin\left(\frac{\pi}{2}\frac{f}{f_r}\right)\exp\left[-i\pi\left(\frac{f}{f_r} - \frac{1}{2}\right)\right] \qquad (6.36)$$

and

$$G_2(f) = 2\sin\left(\frac{\pi}{2}\frac{f}{f_r}\right)\exp\left[-i\frac{\pi}{2}\left(\frac{f}{f_r} - 1\right)\right] \qquad (6.37)$$

The amplitudes of these functions are the same at all frequencies and are orthogonal at the reference frequency (f_r) and all

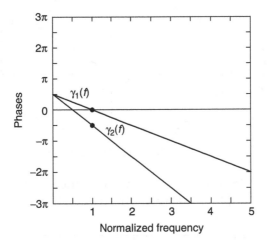

Figure 6.16 Phases for the reference sampling functions for the four-step cross algorithm.

its odd harmonics, as shown in Figure 6.16. Using Equation 5.77, the value of $r(f)$ is given by:

$$r(f) = -\frac{\sin\left(\phi + \pi\dfrac{f}{f_r}\right)}{\sin\left(\phi + \dfrac{\pi}{2}\dfrac{f}{f_r}\right)} \tag{6.38}$$

From Figure 6.15 we can see that this algorithm has the following properties:

1. It is quite sensitive to detuning error, because, as in Wyant's algorithm, the orthogonality of the Fourier transforms of the sampling functions is lost due to small detuning. The phase error as a function of the normalized frequency is shown in Figure 6.17 and as a function of the signal phase in Figure 6.18.
2. Phase errors can be introduced by the presence in the signal of all odd harmonics; however, it is insensitive to all even harmonics.

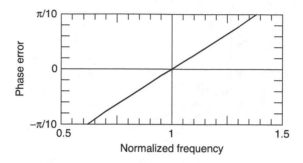

Figure 6.17 Phase error as a function of the normalized frequency for reference sampling functions in the four-step cross algorithm.

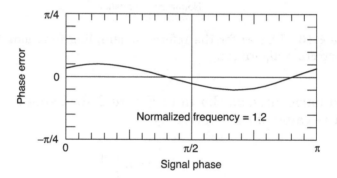

Figure 6.18 Phase error as a function of the signal frequency of the four-steps cross algorithm. The normalized frequency is equal to 1.2.

6.3.2 Algorithm for Four Steps in X

The values of the irradiance are measured at four different values of the phase: $\alpha_1 = 45°$, $\alpha_2 = 135°$, $\alpha_3 = 225°$, and $\alpha_4 = 315°$. Thus, as shown in Figure 6.19, we have:

$$s_1 = a + b\cos(\phi + 45°)$$
$$s_2 = a + b\cos(\phi + 135°)$$
$$s_3 = a + b\cos(\phi + 225°)$$
$$s_4 = a + b\cos(\phi + 315°)$$

$$(6.39)$$

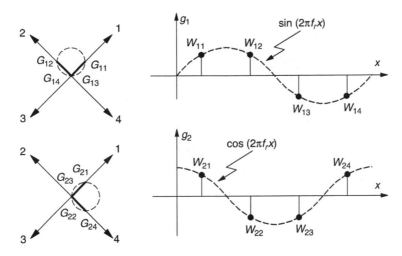

Figure 6.19 Four-step X algorithm.

From these equations, we can show that one solution for the phase is:

$$\tan\phi = -\frac{s_1 + s_2 - s_3 - s_4}{s_1 - s_2 - s_3 + s_4} \tag{6.40}$$

The sampling weights have the following values: $W_{11} = 1$, $W_{12} = 1$, $W_{13} = -1$, $W_{14} = -1$, $W_{21} = 1$, $W_{22} = -1$, $W_{23} = -1$, and $W_{24} = 1$. As in the preceding algorithm, we can see that these sampling weights are as described by Equation 5.19, thus this is another diagonal least-squares solution. The reference sampling functions, then, are:

$$g_1(x) = \delta\left(x - \frac{X_r}{8}\right) + \delta\left(x - \frac{3X_r}{8}\right) - \delta\left(x - \frac{5X_r}{8}\right) - \delta\left(x - \frac{7X_r}{8}\right) \tag{6.41}$$

and

$$g_2(x) = \delta\left(x - \frac{X_r}{8}\right) - \delta\left(x - \frac{3X_r}{8}\right) - \delta\left(x - \frac{5X_r}{8}\right) + \delta\left(x - \frac{7X_r}{8}\right) \tag{6.42}$$

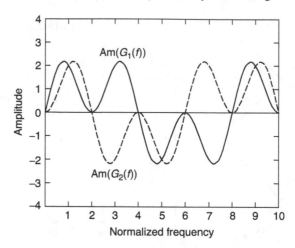

Figure 6.20 Amplitudes of Fourier transforms for reference sampling functions for the four-step X algorithm.

The Fourier transforms of the sampling functions (Figure 6.20) are:

$$G_1(f) = 2\sqrt{2}\sin\left(\frac{\pi}{2}\frac{f}{f_r}\right)\cos\left(\frac{\pi}{4}\frac{f}{f_r}\right)\exp\left[-i\pi\left(\frac{f}{f_r}-\frac{1}{2}\right)\right] \quad (6.43)$$

and

$$G_2(f) = 2\sqrt{2}\sin\left(\frac{\pi}{2}\frac{f}{f_r}\right)\sin\left(\frac{\pi}{4}\frac{f}{f_r}\right)\exp\left[-i\pi\left(\frac{f}{f_r}-1\right)\right] \quad (6.44)$$

These functions are orthogonal at all frequencies and have the same amplitude only at the reference frequency (f_r) and all of its odd harmonics. From Equation 5.75, the value of $r(f)$ can be shown to be given by:

$$r(f) = -\frac{\tan\phi}{\tan\left(\dfrac{\pi}{4}\dfrac{f}{f_r}\right)} \quad (6.45)$$

Thus, any detuning can be compensated, if the signal frequency is known, by dividing the calculated value of $r(f)$ by $\tan(\pi f/4f_r)$.

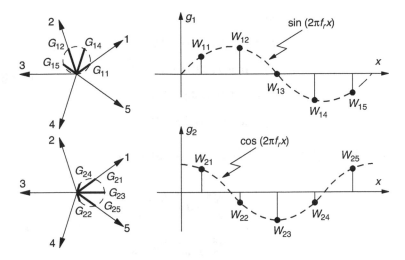

Figure 6.21 Five-step algorithm.

From Figure 6.20 we can see that this algorithm has the following properties:

1. It is quite sensitive to detuning error, as the amplitude of the Fourier transforms of the sampling functions are altered by small detunings.
2. Signals with frequencies f_r, $3f_r$, $5f_r$, $7f_r$, $9f_r$, etc. can be detected, as the amplitudes of the Fourier transforms are the same (even if of different sign) at these frequencies.
3. As in the preceding algorithm, phase errors can be introduced by the presence in the signal of all odd harmonics; also, it is insensitive to all even harmonics.

6.4 FIVE-STEP ALGORITHM

In this algorithm, the values of the irradiance are measured at five different values of the phase: $\alpha_1 = 36°$, $\alpha_2 = 108°$, $\alpha_3 = 180°$, $\alpha_4 = 252°$, and $\alpha_5 = 324°$. Thus, as shown in Figure 6.21, we have:

$$s_1 = a + b\cos(\phi + 36°)$$

$$s_2 = a + b\cos(\phi + 108°)$$

$$s_3 = a + b\cos(\phi + 180°)$$

$$s_4 = a + b\cos(\phi + 252°)$$

$$s_5 = a + b\cos(\phi + 324°)$$

(6.46)

Then, the diagonal least-squares solution is:

$$\tan\phi = -\frac{\displaystyle\sum_{n=1}^{6}\sin\left(\frac{2\pi n}{5}\right)s_n}{\displaystyle\sum_{n=1}^{6}\cos\left(\frac{2\pi n}{5}\right)s_n}$$

(6.47)

Thus, the reference sampling functions are:

$$g_1(x) = \delta\left(x - \frac{X_r}{10}\right) + \delta\left(x - \frac{3X_r}{10}\right) - \delta\left(x - \frac{7X_r}{10}\right) - \delta\left(x - \frac{9X_r}{10}\right)$$ (6.48)

and

$$g_2(x) = \delta\left(x - \frac{X_r}{10}\right) - \delta\left(x - \frac{3X_r}{10}\right) - \delta\left(x - \frac{5X_r}{10}\right)$$

$$- \delta\left(x - \frac{7X_r}{10}\right) + \delta\left(x - \frac{9X_r}{10}\right)$$

(6.49)

The Fourier transforms of the sampling functions (Figure 6.22) are:

$$G_1(f) = 2\left[\begin{array}{c}\sin\left(\dfrac{\pi}{5}\right)\sin\left(\dfrac{4\pi}{5}\dfrac{f}{f_r}\right) \\ + \sin\left(\dfrac{3\pi}{5}\right)\sin\left(\dfrac{2\pi}{5}\dfrac{f}{f_r}\right)\end{array}\right]\exp\left[-i\pi\left(\dfrac{f}{f_r} + \dfrac{1}{2}\right)\right]$$ (6.50)

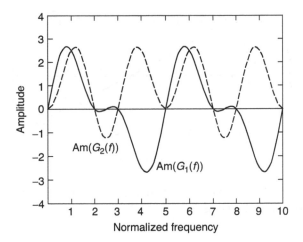

Figure 6.22 Amplitudes of the Fourier transforms for reference sampling functions of the five-step algorithm.

and

$$G_2(f) = 2 \left[\begin{array}{c} \dfrac{1}{2} - \cos\left(\dfrac{\pi}{5}\right)\cos\left(\dfrac{4\pi}{5}\dfrac{f}{f_r}\right) \\[2mm] + \cos\left(\dfrac{2\pi}{5}\right)\cos\left(\dfrac{2\pi}{5}\dfrac{f}{f_r}\right) \end{array} \right] \exp\left[-i\pi\left(\dfrac{f}{f_r}+1\right) \right] \quad (6.51)$$

These functions are orthogonal at all frequencies and have the same amplitude only at the reference frequency (f_r) and at the sixth harmonic. From Equation 5.77, we can see that the value of $r(f)$ is given by:

$$r(f) = \frac{\sin\left(\dfrac{\pi}{5}\right)\cos\left(\dfrac{4\pi}{5}\dfrac{f}{f_r}\right) + \sin\left(\dfrac{3\pi}{5}\right)\cos\left(\dfrac{2\pi}{5}\dfrac{f}{f_r}\right)\tan\left(\phi+\pi\dfrac{f}{f_r}\right)}{\dfrac{1}{2} - \cos\left(\dfrac{\pi}{5}\right)\cos\left(\dfrac{4\pi}{5}\dfrac{f}{f_r}\right) + \cos\left(\dfrac{2\pi}{5}\right)\cos\left(\dfrac{2\pi}{5}\dfrac{f}{f_r}\right)} \quad (6.52)$$

From Figure 6.22 we can see that this algorithm has the following properties:

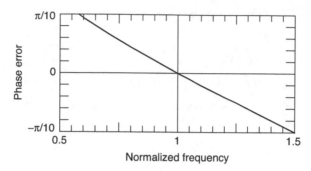

Figure 6.23   Phase error as a function of the normalized frequency for the five-step algorithm.

1. It is quite sensitive to detuning error, as the magnitudes of the Fourier transforms of the sampling functions are altered by small detunings. The phase error as a function of the normalized frequency is shown in Figure 6.23.
2. Signals with frequencies f_r, $4f_r$, $6f_r$, $9f_r$, etc. can be detected, as the amplitudes of the Fourier transforms are the same (even if of different sign) at these frequencies.
3. Phase errors can be introduced by the presence in the signal of fourth, sixth, and ninth harmonics. The signal is insensitive to the second, third, fifth, seventh, eighth, and tenth harmonics.

6.5 ALGORITHMS WITH SYMMETRICAL N + 1 PHASE STEPS

We have seen in Chapter 5 that any phase-detection algorithm must satisfy the condition that the reference sampling vectors G_1 and G_2 must be orthogonal to each other and must have the same magnitude. Also, the sums of their x and y components must be zero, as expressed by Equations 5.96 and 5.97. We have also seen in Chapter 5 that when we have N sampling points, equally and uniformly spaced, as described by:

$$x_n = \frac{(n-1)}{Nf_r} \tag{6.53}$$

then these conditions are satisfied if the sampling weights are given by:

$$W_{1n} = \sin\alpha_n \tag{6.54}$$

and

$$W_{2n} = \cos\alpha_n \tag{6.55}$$

where $\alpha_n = 2\pi f_r x_n$. Then, the signal phase becomes:

$$\tan\phi = -\left(\frac{\displaystyle\sum_{n=1}^{N} s(x_n)\sin\alpha_n}{\displaystyle\sum_{n=1}^{N} s(x_n)\cos\alpha_n} \right) \tag{6.56}$$

This expression is valid for all algorithms with N sampling points equally and uniformly spaced according to Equation 6.47. The first sampling point ($n = 1$) is located at a coordinate $x_n = 0$, and the last point is located at $x_N = (N-1)/Nf_r$. A point with $n = N + 1$ (which is not considered) would be located at $x_n = X_r = 1/f_r$ (that is, at a phase equal to 2π).

Let us now consider algorithms with $N + 1$ sampling points with the same separation as described earlier, such that the last point has a phase equal to 2π. This modification removes the orthogonality and equal magnitudes that are required from the reference sampling weights, but these conditions can be restored simply by splitting in half the magnitude of the first ($n = 1$) sampling weight (W_{21}) and setting the last ($n = N + 1$) sampling weight ($W_{2(N+1)}$) equal to this value. Thus, the modified sampling weights W_{21} and $W_{2(N+1)}$ have the same value:

$$W_{21} = W_{2(N+1)} = \frac{1}{2}\cos\alpha_1 = \frac{1}{2} \tag{6.57}$$

and all other sampling weights remain the same. These algorithms, first described by Larkin and Oreb (1992), are called symmetrical $N + 1$ sampling algorithms and have some interesting error-compensating properties.

The Fourier transforms of these reference sampling functions with $N + 1$ sampling points, from Equations 6.1 and 6.2, are given by:

$$G_m(f) = \sum_{n=1}^{N+1} W_{mn} \exp(-i2\pi f x_n) \tag{6.58}$$

With the sampling point distribution just described for these algorithms, its Fourier transforms become, after adding together terms symmetrically placed in the sampling interval,

$$G_m(f) = \sum_{n=1}^{\frac{(N+1)}{2}} \left[W_{mn} \exp(-i2\pi f x_n) + W_{m(N+2-n)} \exp(-i2\pi f x_{(N+2-n)}) \right] \tag{6.59}$$

for N odd, with no sampling point at the middle central position of the sampling interval as the total number of points ($N + 1$) is even; or

$$G_m(f) = \sum_{n=1}^{\frac{N}{2}} \left[W_{mn} \exp(-i2\pi f x_n) + W_{m(N+2-n)} \exp(-i2\pi f x_{(N+2-n)}) \right] + \\ + W_{m(N/2+1)} \exp(-i2\pi f x_{(N/2+1)}) \tag{6.60}$$

for N even. Because the total number of sampling points is odd, there is a point at the middle. The weights defined by Equations 6.54 and 6.55 are antisymmetrical, while the terms defined by Equation 6.57 are symmetrical. Then, we can show that $G_1(f)$ is given by:

$$G_1(f) = 2i \sum_{n=1}^{\frac{(N+1)}{2}} W_{1n} \sin\left(\pi \left(1 - \frac{2(n-1)}{N} \right) \frac{f}{f_r} \right) \exp\left(-i\pi \frac{f}{f_r} \right) \tag{6.61}$$

for N odd, and that

$$G_1(f) = 2i \sum_{n=1}^{\frac{N}{2}} W_{1n} \sin\left(\pi\left(1 - \frac{2(n-1)}{N}\right)\frac{f}{f_r}\right) \exp\left(-i\pi\frac{f}{f_r}\right) \quad (6.62)$$

for N even. The last term has disappeared, as the weight $(W_{1(N/2+1)})$ is equal to zero. In the same manner, $G_2(f)$ is given by:

$$G_2(f) = 2 \sum_{n=1}^{\frac{(N+1)}{2}} W_{2n} \cos\left(\pi\left(1 - \frac{2(n-1)}{N}\right)\frac{f}{f_r}\right) \exp\left(-i\pi\frac{f}{f_r}\right) \quad (6.63)$$

for N odd, and

$$G_2(f) = 2 \sum_{n=1}^{\frac{N}{2}} W_{2n} \cos\left(\pi\left(1 - \frac{2(n-1)}{N}\right)\frac{f}{f_r}\right) \exp\left(-i\pi\frac{f}{f_r}\right) +$$

$$+ W_{2(N/2+1)} \exp\left(-i\pi\frac{f}{f_r}\right) \quad (6.64)$$

for N even.

From Equations 6.54, 6.55, and 6.57 and because $\psi(f_r)$ is zero, the sampling weights, using the sampling point distribution in Equation 6.52, are:

$$W_{1n} = \sin\left(\frac{2\pi(n-1)}{N}\right) \quad (6.65)$$

for all values of n,

$$W_{2n} = \cos\left(\frac{2\pi(n-1)}{N}\right) \quad (6.66)$$

for $1 < n < N + 1$, and

$$W_{21} = \frac{1}{2} \quad (6.67)$$

for $n = 1$ and $n = N + 1$.

We can see that, due to their symmetry, these two functions are orthogonal at all frequencies. This is an important result, because we can conclude that, with detuning, the only condition that can fail is that requiring equal amplitudes of the Fourier transforms of the sampling functions.

The only requirement, then, for insensitivity to detuning, as studied in Chapter 5, is that the amplitude of the Fourier transforms must remain the same in a small frequency interval centered at f_r. As described in Chapter 4, this occurs when the two plots for $G_1(f)$ and $G_2(f)$ touch tangentially at the frequency f_r.

An important property of these symmetrical $N + 1$ algorithms is that they can be made insensitive to low-frequency detuning. The requirement that the slopes for $G_1(f)$ and $G_2(f)$ are equal, so that they touch tangentially, is satisfied in some of these algorithms (for some values of N) but not for all of them. When this happens, the algorithm can still be modified to obtain insensitivity to detuning.

Let us assume, as described by Larkin and Oreb (1992), that an additional term, $\Delta G_1(f)$, is added to the function $G_1(f)$, with the following conditions:

1. Its phase is equal to that of $G_1(f)$, so the orthogonality condition is not disturbed at any frequency.
2. Its amplitude at the frequency f_r is zero, so the condition of equal amplitudes is not disturbed at this frequency.
3. The sum of its sampling weights should be zero, so the condition for no DC bias is met.
4. Its amplitude is zero at the harmonics of the frequency f_r, so the absence of harmonics cross-talk is not altered by the presence of this extra term.
5. Its slope at the frequency f_r is not zero, so the final slope of the Fourier transform $G_1(f)$ can be changed as needed to make the algorithm insensitive to small detuning.

The sampling weights W_{11} and $W_{1(N+1)}$ have a zero value. Let us assume that the sampling weights for the additional term $\Delta G_1(f)$ are given nonzero values with the same amplitudes

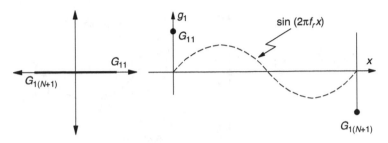

Figure 6.24 Sampling weights for the extra term $\Delta G_1(f)$.

but with opposite signs at these locations, as shown in Figure 6.24. The necessary conditions are satisfied, and the slope of the amplitude of the Fourier transform $G_1(f)$ at the signal frequency can be modified. Thus, we see that $\Delta G_1(f)$, as plotted in Figure 6.25, is:

$$\Delta G_1(f) = 2i W_{11} \sin\left(\pi \frac{f}{f_r}\right) \exp\left(-i\pi \frac{f}{f_r}\right) \qquad (6.68)$$

where $W_{11} = -W_{1(N+1)}$ is set to a value so that the two desired slopes become equal.

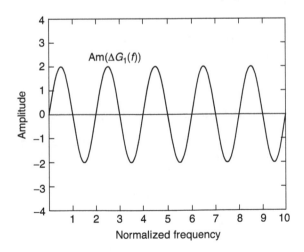

Figure 6.25 Amplitude of the Fourier transforms for the extra term $\Delta G_1(f)$.

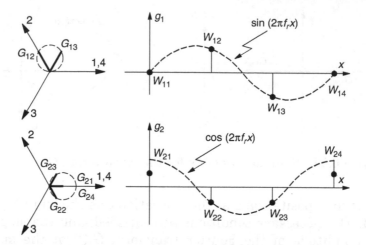

Figure 6.26 Symmetrical four-step (3 + 1) algorithm.

We will apply this extra term to some symmetrical algorithms, later in this chapter, to make them insensitive to detuning. Surrel (1993) developed symmetrical detuning-insensitive algorithms and showed that the sampling weights W_{11} and $W_{1(N+1)}$ must have the value:

$$W_{11} = W_{1(N+1)} = \frac{1}{2\tan\left(\dfrac{2\pi}{N}\right)} \tag{6.69}$$

6.5.1 Symmetrical Four-Step (3 + 1) Algorithm

For this algorithm, with $N = 3$, as illustrated in Figure 6.26, the four signal measurements are written as follows:

$$s_1 = a + b\cos\phi$$

$$s_2 = a + b\cos(\phi + 120°)$$

$$s_3 = a + b\cos(\phi + 240°) \tag{6.70}$$

$$s_4 = a + b\cos(\phi + 360°)$$

The first and last points have the same phase; thus, we can take the average of these points in order to reduce the number of equations to three. Then, from these equations we find:

$$\tan\phi = -\sqrt{3}\,\frac{s_2 - s_3}{s_1 - s_2 - s_3 + s_4} \tag{6.71}$$

It is interesting to note that this expression can be obtained from a three-point algorithm, such as the 120° three-step algorithm, with the first sampling point at zero degrees if s_1 is replaced by $(s_1 + s_4)/2$.

The sampling weights are $W_{11} = 0$, $W_{12} = \sqrt{3}/2$, $W_{13} = -\sqrt{3}/2$, $W_{21} = 0.5$, $W_{22} = -0.5$, $W_{23} = -0.5$, and $W_{24} = 0.5$. Then, the reference sampling functions are:

$$g_1(x) = \frac{\sqrt{3}}{2}\left(\delta\left(x - \frac{X_r}{3}\right) - \delta\left(x - \frac{2X_r}{3}\right)\right) \tag{6.72}$$

and

$$g_2(x) = \frac{1}{2}\left(\delta(x) - \delta\left(x - \frac{X_r}{3}\right) - \delta\left(x - \frac{2X_r}{3}\right) + \delta(x - X_r)\right) \tag{6.73}$$

The Fourier transforms of these sampling functions, plotted in Figure 6.27, are:

$$G_1(f) = \sqrt{3}\sin\left(\frac{\pi}{3}\frac{f}{f_r}\right)\exp\left[-i\pi\left(\frac{f}{f_r} + \frac{1}{2}\right)\right] \tag{6.74}$$

and

$$G_2(f) = 2\sin\left(\frac{2\pi}{3}\frac{f}{f_r}\right)\sin\left(\frac{\pi}{3}\frac{f}{f_r}\right)\exp\left[-i\pi\left(\frac{f}{f_r} - 1\right)\right] \tag{6.75}$$

The value of $r(f)$, from Equation 5.77, is given by:

$$r(f) = \frac{\sqrt{3}\tan\left(\phi + \pi\dfrac{f}{f_r}\right)}{2\sin\left(\dfrac{2\pi}{3}\dfrac{f}{f_r}\right)} \tag{6.76}$$

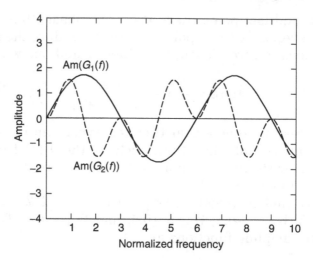

Figure 6.27 Amplitudes of the Fourier transforms for reference sampling functions for the symmetrical four-step (3 + 1) algorithm.

These Fourier transforms are orthogonal at all frequencies. We can see that the two curves do not touch each other tangentially at the reference frequency (f_r). In order to have detuning insensitivity, to the function $G_1(f)$ we must add the additional term $\Delta G_1(f)$, with the proper amplitude σ. Then, the value of W_{11} that makes the slope of $\Delta G_1(f)$ equal to minus this value is equal to $W_{11} = 1/(2\sqrt{3})$. The sampling weights for the final algorithm are shown in Figure 6.28.

The plots of the amplitudes of the Fourier transforms are shown in Figure 6.29, where we can see that this algorithm has the following properties:

1. It is insensitive to small detuning errors, as the two plots for the Fourier transform magnitudes touch each other tangentially at the reference frequency.
2. Signals with frequencies f_r, $2f_r$, $4f_r$, $5f_r$, $7f_r$, etc. can be detected, as the amplitudes of the Fourier transforms are the same (even if of different sign) at these frequencies.

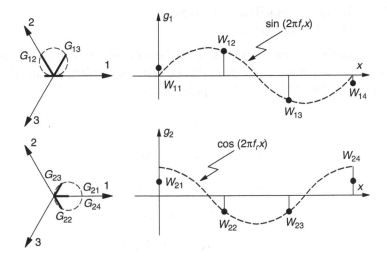

Figure 6.28 Symmetrical four-step $(3 + 1)$ algorithm with an extra term to obtain detuning insensitivity.

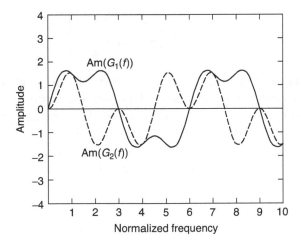

Figure 6.29 Amplitudes of the Fourier transforms for reference sampling functions for the symmetrical four-step $(3 + 1)$ algorithm with an extra term.

3. Phase errors can be introduced by the presence in the signal of second, fourth, fifth, seventh, and eighth harmonics. It is insensitive to third, sixth, and ninth harmonics.

6.5.2 Schwider–Hariharan Five-Step (4 + 1) Algorithm

This algorithm was described by Schwider et al. (1983) and later by Hariharan et al. (1987). The irradiance measurements for the five sampling points are:

$$s_1 = a + b\cos\phi$$

$$s_2 = a + b\cos(\phi + 90°)$$

$$s_3 = a + b\cos(\phi + 180°) \tag{6.77}$$

$$s_4 = a + b\cos(\phi + 270°)$$

$$s_5 = a + b\cos(\phi + 360°)$$

From these equations, the phase can be obtained as follows:

$$\tan\phi = -\left(\frac{s_2 - s_4}{\dfrac{1}{2}s_1 - s_3 + \dfrac{1}{2}s_5}\right) \tag{6.78}$$

This expression can be obtained from the four steps of the $n\pi/2$ algorithm by substituting the measurement s_1 with the average of the measurements s_1 and s_5. The sampling weights, as shown in Figure 6.30, have the values $W_{11} = 0$, $W_{12} = 1$, $W_{13} = 0$, $W_{14} = -1$, $W_{15} = 0$, $W_{21} = 1/2$, $W_{22} = 0$, $W_{23} = -1$, $W_{24} = 0$, and $W_{25} = 1/2$. Then, the reference sampling functions are:

$$g_1(x) = \delta\left(x - \frac{X_r}{4}\right) - \delta\left(x - \frac{3X_r}{4}\right) \tag{6.79}$$

and

$$g_2(x) = \frac{1}{2}\delta(x) - \delta\left(x - \frac{X_r}{2}\right) + \frac{1}{2}\delta(x - X_r) \tag{6.80}$$

The amplitudes of the Fourier transforms of the sampling functions, shown in Figure 6.31, are:

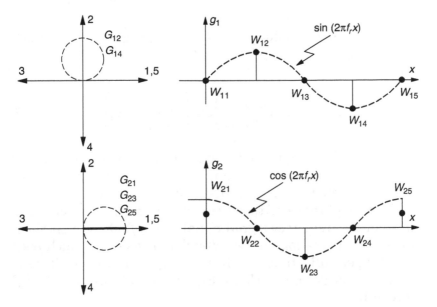

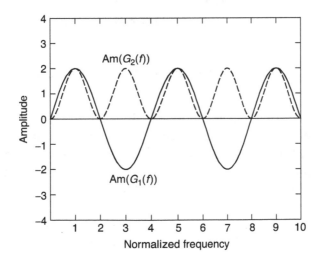

Figure 6.30 Schwider–Hariharan symmetrical five-step (4 + 1) algorithm.

Figure 6.31 Amplitudes of the Fourier transforms for reference sampling functions for the symmetrical five-step (4 + 1) algorithm.

$$G_1(f) = 2\sin\left(\frac{\pi}{2}\frac{f}{f_r}\right)\exp\left[-i\pi\left(\frac{f}{f_r} - \frac{1}{2}\right)\right] \qquad (6.81)$$

and

$$G_2(f) = 2\sin^2\left(\frac{\pi}{2}\frac{f}{f_r}\right)\exp\left[-i\pi\left(\frac{f}{f_r} - 1\right)\right] \qquad (6.82)$$

As illustrated in Figure 6.32, these functions are orthogonal at all frequencies and their amplitudes are equal only at the reference frequency (f_r) and at its odd harmonics.

The amplitudes of these two functions become equal at values of the frequency signal equal to f_r, $5f_r$, $9f_r$, etc. At these points, the curves for the two Fourier transforms touch each other tangentially, thus making the algorithm insensitive to low-frequency detuning. Using Equation 5.77, the value of $r(f)$ is given by:

$$r(f) = -\frac{\tan\left(\phi + \pi\frac{f}{f_r}\right)}{\sin\left(\pi\frac{f}{f_r}\right)} \qquad (6.83)$$

From Figure 6.31 we can see that this algorithm has the following properties:

1. It is insensitive to small detuning errors, as the two plots for the Fourier transform magnitude touch each other tangentially at the reference frequency. The phase error as a function of the normalized frequency is illustrated in Figures 6.33 and 6.34.
2. Signals with frequencies f_r, $3f_r$, $5f_r$, $7f_r$, $9f_r$, etc. can be detected, as the amplitudes of the Fourier transforms are the same (even if of different sign) at these frequencies.
3. Phase errors can be introduced by the presence of odd harmonics in the signal, but it is insensitive to even harmonics.

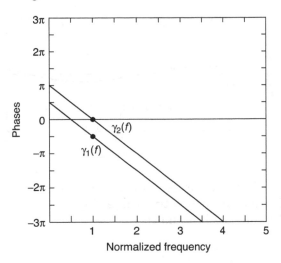

Figure 6.32 Phases for the sampling functions in the Schwider–Hariharan symmetrical five-step (4 + 1) algorithm.

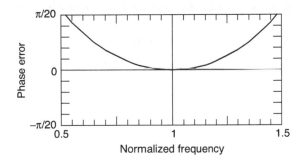

Figure 6.33 Phase error as a function of the normalized frequency for the Schwider–Hariharan symmetrical five-step (4 + 1) algorithm.

Hariharan et al. (1987) derived this algorithm by assuming that the phase separation between the five sampling points was not known and algebraically represented it by α in Equation 6.70. In this case, the value of α is found by equating to zero the derivative of $\tan(\phi_0)$ with respect to angle α; thus, angle α equal to 90° is found. In this algorithm, a symmetrical sampling point distribution from $-\pi$ to π is used.

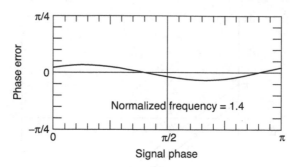

Figure 6.34 Phase error as a function of the signal phase for the Schwider–Hariharan symmetrical five-step (4 + 1) algorithm. The normalized frequency is equal to 1.4.

6.5.3 Symmetrical Six-Step (5 + 1) Algorithm

In this algorithm, the irradiance measurements for the six sampling points, as illustrated in Figure 6.35, are:

$$
\begin{aligned}
s_1 &= a + b\cos\phi \\
s_2 &= a + b\cos(\phi + 72°) \\
s_3 &= a + b\cos(\phi + 144°) \\
s_4 &= a + b\cos(\phi + 216°) \\
s_5 &= a + b\cos(\phi + 288°) \\
s_6 &= a + b\cos(\phi + 360°)
\end{aligned}
\tag{6.84}
$$

From these equations, the phase can be shown to be:

$$
\tan\phi = -\frac{\displaystyle\sum_{n=1}^{6}\sin\left(\frac{2\pi(n-1)}{5}\right)s_n}{\dfrac{1}{2}s_1 + \displaystyle\sum_{n=2}^{5}\cos\left(\frac{2\pi(n-1)}{5}\right)s_n + \dfrac{1}{2}s_6}
\tag{6.85}
$$

The reference sampling functions are:

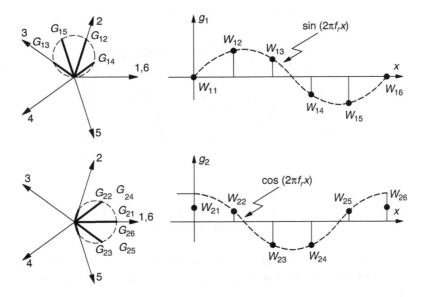

Figure 6.35 Symmetrical six-step (5 + 1) algorithm.

$$g_1(x) = \sum_{n=1}^{6} \sin\left(\frac{2\pi(n-1)}{5}\right)\delta\left(x - \frac{(n-1)}{5}X_r\right) \tag{6.86}$$

and

$$g_2(x) = \frac{1}{2}\delta(x) + \sum_{n=2}^{5} \cos\left(\frac{2\pi(n-1)}{5}\right)\delta\left(x - \frac{n}{5}X_r\right) + \frac{1}{2}\delta(x - X_r) \tag{6.87}$$

The Fourier transforms of the sampling functions (Figure 6.36) are:

$$G_1(f) = 2\left(\begin{array}{c} \sin\left(\dfrac{2\pi}{5}\right)\sin\left(\dfrac{3\pi}{5}\dfrac{f}{f_r}\right) \\ + \sin\left(\dfrac{\pi}{5}\right)\sin\left(\dfrac{\pi}{5}\dfrac{f}{f_r}\right) \end{array}\right)\exp\left[-i\pi\left(\dfrac{f}{f_r} - \dfrac{1}{2}\right)\right] \tag{6.88}$$

and

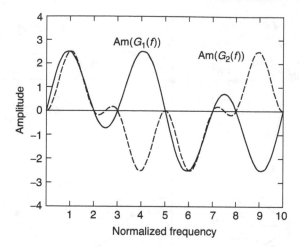

Figure 6.36 Amplitudes of the Fourier transforms for reference sampling functions for the symmetrical six-step (5 + 1) algorithm.

$$G_2(f) = 2 \begin{pmatrix} \cos\left(\dfrac{\pi}{5}\right)\cos\left(\dfrac{\pi}{5}\dfrac{f}{f_r}\right) \\[2mm] -\cos\left(\dfrac{2\pi}{5}\right)\cos\left(\dfrac{3\pi}{5}\dfrac{f}{f_r}\right) \\[2mm] -\dfrac{1}{2}\cos\left(\pi\dfrac{f}{f_r}\right) \end{pmatrix} \exp\left[-i\pi\left(\dfrac{f}{f_r}-1\right)\right] \quad (6.89)$$

These functions are orthogonal at all frequencies, as expected. The amplitudes of these two functions become equal at values of the frequency signal equal to f_r, $6f_r$, etc. Using Equation 5.77, the value of $r(f)$ is given by:

$$r(f) = \frac{\left(\sin\left(\dfrac{2\pi}{5}\right)\sin\left(\dfrac{3\pi}{5}\dfrac{f}{f_r}\right)+\sin\left(\dfrac{\pi}{5}\right)\sin\left(\dfrac{\pi}{5}\dfrac{f}{f_r}\right)\right)\tan\left(\phi-\pi\dfrac{f}{f_r}\right)}{\left(\dfrac{1}{2}\cos\left(\pi\dfrac{f}{f_r}\right)+\cos\left(\dfrac{2\pi}{5}\right)\cos\left(\dfrac{3\pi}{5}\dfrac{f}{f_r}\right)-\cos\left(\dfrac{\pi}{5}\right)\cos\left(\dfrac{\pi}{5}\dfrac{f}{f_r}\right)\right)} \quad (6.90)$$

From Figure 6.36 we can see that this algorithm has the following properties:

1. It is not insensitive to small detuning errors, as the two plots for the Fourier transform magnitude do not touch each other tangentially at the reference frequency, as desired.
2. Signals with frequencies f_r, $4f_r$, $6f_r$, $9f_r$, etc. can be detected, as the amplitudes of the Fourier transforms are the same (even if of different sign) at these frequencies.
3. Phase errors can be introduced by the presence in the signal of fourth, sixth, and ninth harmonics. It is insensitive to second, third, fifth, seventh, eighth, and tenth harmonics.

6.5.4 Symmetrical Seven-Step (6 + 1) Algorithm

This algorithm was first described by Larkin and Oreb (1992). The irradiance measurements for the seven sampling points, as illustrated in Figure 6.37, are:

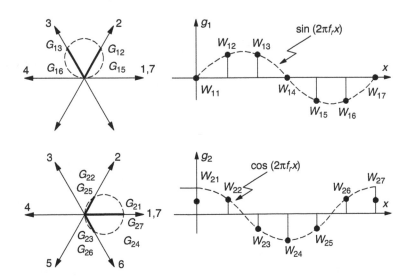

Figure 6.37 Symmetrical seven-step (6 + 1) algorithm.

$$s_1 = a + b\cos\phi$$

$$s_2 = a + b\cos(\phi + 60°)$$

$$s_3 = a + b\cos(\phi + 120°)$$

$$s_4 = a + b\cos(\phi + 180°) \tag{6.91}$$

$$s_5 = a + b\cos(\phi + 240°)$$

$$s_6 = a + b\cos(\phi + 300°)$$

$$s_7 = a + b\cos(\phi + 360°)$$

From these equations, the desired solution for the phase is:

$$\tan\phi = -\sqrt{3}\,\frac{s_2 + s_3 - s_5 - s_6}{s_1 + s_2 - s_3 - 2s_4 - s_5 + s_6 + s_7} \tag{6.92}$$

The sampling weights have the values: $W_{11} = 0$, $W_{12} = \sqrt{3}/2$, $W_{13} = \sqrt{3}/2$, $W_{14} = 0$, $W_{15} = -\sqrt{3}/2$, $W_{16} = -\sqrt{3}/2$, $W_{17} = 0$, $W_{21} = 1/2$, $W_{22} = 1/2$, $W_{23} = -1/2$, $W_{24} = -1$, $W_{25} = -1/2$, $W_{26} = 1/2$, and $W_{27} = 1/2$. Thus, the reference sampling functions are:

$$g_1(x) = \sqrt{3}\left(\begin{array}{c}\delta\left(x - \dfrac{X_r}{6}\right) + \delta\left(x - \dfrac{2X_r}{6}\right) \\[2mm] -\delta\left(x - \dfrac{4X_r}{6}\right) - \delta\left(x - \dfrac{5X_r}{6}\right)\end{array}\right) \tag{6.93}$$

and

$$g_2(x) = \delta(x) + \delta\left(x - \frac{X_r}{6}\right) - \delta\left(x - \frac{2X_r}{6}\right) - 2\delta\left(x - \frac{3X_r}{6}\right) - $$
$$\qquad\qquad - \delta\left(x - \frac{4X_r}{6}\right) + \delta\left(x - \frac{5X_r}{6}\right) + \delta\left(x - \frac{6X_r}{6}\right) \tag{6.94}$$

The Fourier transforms of the sampling functions, shown in Figure 6.38, are:

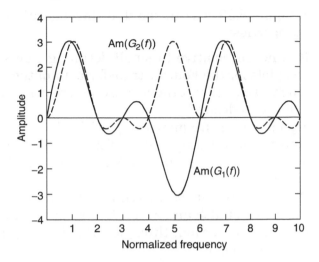

Figure 6.38 Amplitudes of the Fourier transforms for reference sampling functions for the symmetrical seven-step (6 + 1) algorithm.

$$G_1(f) = \sqrt{3}\left[\sin\left(\frac{2\pi}{3}\frac{f}{f_r}\right) + \sin\left(\frac{\pi}{3}\frac{f}{f_r}\right)\right]\exp\left[-i\pi\left(\frac{f}{f_r} - \frac{1}{2}\right)\right] \qquad (6.95)$$

and

$$G_2(f) = \left[1 - \cos\left(\pi\frac{f}{f_r}\right) - \cos\left(\frac{2\pi}{3}\frac{f}{f_r}\right) + \cos\left(\frac{\pi}{3}\frac{f}{f_r}\right)\right]\exp\left(-i\pi\frac{f}{f_r}\right) \qquad (6.96)$$

These functions are orthogonal at all frequencies, as expected. The amplitudes of these two functions become equal at values of the frequency signal equal to f_r, $7f_r$, etc. Using Equation 5.77, the value of $r(f)$ is given by:

$$r(f) = \frac{\left(\sin\left(\frac{2\pi}{3}\frac{f}{f_r}\right) + \sin\left(\frac{\pi}{3}\frac{f}{f_r}\right)\right)\tan\left(\phi + \pi\frac{f}{f_r}\right)}{\cos\left(\pi\frac{f}{f_r}\right) + \cos\left(\frac{2\pi}{3}\frac{f}{f_r}\right) - \cos\left(\frac{\pi}{3}\frac{f}{f_r}\right) - 1} \qquad (6.97)$$

From Figure 6.38, we can see that this algorithm has the following properties:

1. It is not insensitive to small detuning errors, as the two plots for the Fourier transform amplitudes do not touch each other tangentially at the reference frequency, as desired.
2. Signals with frequencies f_r, $5f_r$, $7f_r$, etc. can be detected, as the amplitudes of the Fourier transforms are the same (even if of different sign) at these frequencies.
3. Phase errors can be introduced by the presence in the signal of fifth and seventh harmonics. It is insensitive to the second, third, fourth, sixth, eighth, and ninth harmonics.

6.6 COMBINED ALGORITHMS IN QUADRATURE

We saw at the beginning of this chapter that, if the reference function $g_1(f)$ is symmetric and $g_2(f)$ is antisymmetric, or vice versa, the two functions are orthogonal at all frequencies. Then, as shown in Chapter 5, in this case the phase error due to detuning oscillates sinusoidally with the value of the phase $(\phi + \psi(f_r))$, as expressed by Equation 5.154. Thus, if we use two different sampling algorithms of this kind, but with two different values of this phase $(\phi + \psi(f_r))$, the phase errors upon detuning will have the same magnitudes but opposite sign. If the two phase results are averaged, as follows, the phase error due to detuning will cancel out:

$$\phi' = \frac{\tan^{-1}\phi_a + \tan^{-1}\phi_b}{2} \tag{6.98}$$

Another possibility is to superimpose the two algorithms, as proposed by Schwider et al., 1983, 1993). Let us assume that the basic reference sampling functions are $g_1(x)$ and $g_2(x)$. The only requirement is that the phase separation between the sampling points must be a submultiple of $\pi/2$. Thus, the shifted algorithm will have the same sampling points, with only a few points being added to the final algorithm. For the initial algorithm the phase equation is:

$$\tan\phi_a = -\frac{\displaystyle\sum_{n=1}^{N} g_1(x_n)s(x_n)}{\displaystyle\sum_{n=1}^{N} g_2(x_n)s(x_n)} \tag{6.99}$$

and for the shifted algorithm, from Equations 5.217 and 5.218, the phase equation is:

$$\tan\phi_b = -\frac{\displaystyle\sum_{n=M}^{N+M} g_2\left(x_n - \frac{X_r}{4}\right)s(x_n)}{\displaystyle\sum_{n=M}^{N+M} -g_1\left(x_n - \frac{X_r}{4}\right)s(x_n)} \tag{6.100}$$

Then, the phase equation for the combined algorithm is:

$$\tan\phi' = -\frac{\displaystyle\sum_{n=1}^{M} g_1'(x_n)s(x_n)}{\displaystyle\sum_{n=1}^{M} g_2'(x_n)s(x_n)} \tag{6.101}$$

where $x_n = f_r/4$. The reference sampling functions for this combined algorithm are:

$$g_1'(x) = g_1(x) + g_2\left(x - \frac{X_r}{4}\right) \tag{6.102}$$

and

$$g_2'(x) = g_2(x) - g_1\left(x - \frac{X_r}{4}\right) \tag{6.103}$$

The Fourier transforms of these functions are:

$$G_1'(f) = G_1(f) + G_2(f)\exp\left(-i\frac{\pi}{2}\frac{f}{f_r}\right) \tag{6.104}$$

and

$$G_2'(f) = G_2(f) - G_1(f)\exp\left(-i\frac{\pi}{2}\frac{f}{f_r}\right) \tag{6.105}$$

but this last expression can be transformed into:

$$G_2'(f) = \left(G_1(f) + G_2(f)\exp i\left(\frac{\pi}{2}\frac{f}{f_r} - \pi\right)\right)\exp\left[-i\pi\left(\frac{1}{2}\frac{f}{f_r} - 1\right)\right] \tag{6.106}$$

Then, writing the Fourier transforms in terms of their magnitudes and phases, we find:

$$G_1'(f) = \left(|G_1(f)| + |G_2(f)|\exp\left[-i\left(\frac{\pi}{2}\frac{f}{f_r} - \gamma_2 + \gamma_1\right)\right]\right)\exp(i\gamma_1) \tag{6.107}$$

and

$$G_2'(f) = \left(|G_1(f)| + |G_2(f)|\exp\left[-i\left(\frac{\pi}{2}\frac{f}{f_r} + \gamma_2 - \gamma_1 - \pi\right)\right]\right) \times$$

$$\times \exp\left[i\left(\gamma_1 - \frac{\pi}{2}\frac{f}{f_r} + \pi\right)\right] \tag{6.108}$$

where γ_1 and γ_2 are the phases of the complex functions $G_1(f)$ and $G_2(f)$, respectively.

This is a general expression for the combined algorithm, formed by the base algorithm and its 90° shifted version. Here, we have two possible cases. The first case is when, in the base algorithm, the magnitudes of the Fourier transforms $G_1(f)$ and $G_2(f)$ are equal at all frequencies but are orthogonal only at the reference frequency (f_r). In this case, we can show that:

$$G_1'(f) = 2|G_1(f)|\cos\left(\frac{\pi}{4}\frac{f}{f_r} - \frac{(\gamma_2 - \gamma_1)}{2}\right) \times$$

$$\times \exp\left[i\left(-\frac{\pi}{4}\frac{f}{f_r} + \frac{(\gamma_2 - \gamma_1)}{2}\right)\right]$$

(6.109)

and

$$G_1'(f) = 2|G_1(f)|\sin\left(\frac{\pi}{4}\frac{f}{f_r} + \frac{(\gamma_2 - \gamma_1)}{2}\right) \times$$

$$\times \exp\left[i\left(-\frac{\pi}{4}\frac{f}{f_r} + \frac{(\gamma_2 + \gamma_1)}{2} + \frac{\pi}{2}\right)\right]$$

(6.110)

We can see that these Fourier transforms are orthogonal at all frequencies, but their magnitudes are equal only at the reference frequency (f_r).

A second particular case is when the orthogonality condition in the original algorithm is satisfied at all frequencies ($\gamma_2 = \gamma_1 + \pi/2$), but the magnitudes of $G_1(f)$ and $G_2(f)$ are equal only at the reference frequency. In this case, we have:

$$G_1'(f) = \left(|G_1(f)| + |G_2(f)|\exp\left[-i\frac{\pi}{2}\left(\frac{f}{f_r} - 1\right)\right]\right)\exp(i\gamma_1) \quad (6.111)$$

and

$$G_2'(f) = \left(|G_1(f)| + |G_2(f)|\exp\left[i\frac{\pi}{2}\left(\frac{f}{f_r} - \frac{1}{2}\right)\right]\right) \times$$

$$\times \exp\left[i\left(\gamma_1 - \frac{\pi}{2}\frac{f}{f_r} + \pi\right)\right]$$

(6.112)

We can see that the two reference sampling functions of the combined algorithm have equal magnitudes at all frequencies, but they are orthogonal only at the signal frequency. The square magnitude is equal to:

$$\left|G_2'(f)\right|^2 = \left|G_1(f)\right|^2 + \left|G_2(f)\right|^2 + 2\left|G_1(f)\right|\left|G_2(f)\right|\cos\left(\frac{\pi}{2}\frac{f}{f_r}\right) \quad (6.113)$$

In both cases, as expected, the combined algorithm is insensitive to a small detuning. The formal mathematical proof is left to the reader as an exercise.

Schmit and Creath (1995) extended this averaging concept to multiple steps. Combining two detuning, uncompensated algorithms provides an algorithm that is insensitive to small detuning (that is, in a relatively small frequency range). By repeating the same process in sequence and combining an already compensated algorithm and its 90° shifted version, a better compensated algorithm is obtained. These algorithms (class B), are detuning insensitive in a wider frequency range.

Instead of multiple sequential applications of an algorithm and its shifted version in a process referred to as the *multiple sequential technique*, Schmit and Creath (1996) proposed a method in which several shifted algorithms are combined at the same time, in a process they call the *multiple averaging technique*. Equations 6.102 and 6.103 then become:

$$g_1'(x) = g_1(x) + g_2\left(x - \frac{X_r}{4}\right) - g_1\left(x - \frac{X_r}{2}\right)\ldots \quad (6.114)$$

and

$$g_2'(x) = g_2(x) + g_1\left(x - \frac{X_r}{4}\right) - g_2\left(x - \frac{X_r}{2}\right)\ldots \quad (6.115)$$

6.6.1 Schwider Algorithm

Schwider et al. (1983, 1993) described an algorithm with four sampling points separated by 90° that can be considered as the sum of two three-point algorithms separated by 90°. The first algorithm, shown in Figure 6.39a, is the three-step inverted T algorithm described previously, for which the phase equation is:

$$\tan\phi_a = -\frac{-s_1 + 2s_2 - s_3}{s_1 - s_3} \quad (6.116)$$

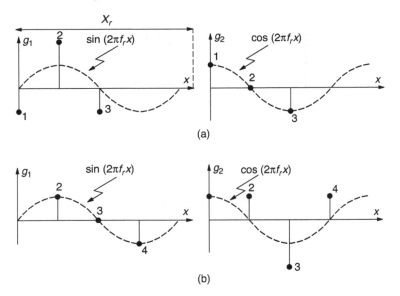

Figure 6.39 Sampling with two combined algorithms in quadrature: (a) three-steps inverted T algorithm, and (b) inverted T algorithm for $\pi/2$ shifted three steps.

The second algorithm is identical, but shifted by $\varepsilon = \pi/2$, as described in Section 5.7.2 and illustrated in Figure 6.39b. Then, the reference functions for the second algorithm, as described by Equations 5.217 and 5.218, are as follows:

$$\tan\phi_b = -\frac{s_2 - s_4}{s_2 - 2s_3 + s_4} \qquad (6.117)$$

Let us now superimpose the two algorithms to obtain the combined reference functions shown in Figure 6.40:

$$g_1(x) = -\delta(x) + 3\delta\left(x - \frac{X_r}{4}\right) - \delta\left(x - \frac{2X_r}{4}\right) - \delta\left(x - \frac{3X_r}{4}\right) \qquad (6.118)$$

and

$$g_2(x) = \delta(x) + \delta\left(x - \frac{X_r}{4}\right) - 3\delta\left(x - \frac{2X_r}{4}\right) + \delta\left(x - \frac{3X_r}{4}\right) \qquad (6.119)$$

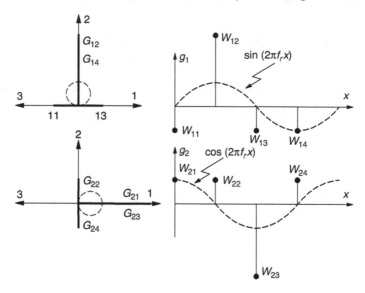

Figure 6.40 Sampling functions for the Schwider algorithm obtained by combining two algorithms in quadrature.

The phase is now given by:

$$\tan\phi = -\frac{-s_1 + 3s_2 - s_3 - s_4}{s_1 + s_2 - 3s_3 + s_4} \qquad (6.120)$$

and the sampling points are located at $\alpha_1 = 0°$, $\alpha_2 = 90°$, $\alpha_3 = 180°$, and $\alpha_4 = 270°$. The Fourier transforms of the sampling functions become:

$$G_1(f) = 4\sin\left(\frac{\pi}{4}\frac{f}{f_r}\right)\left(\sin\left(\frac{\pi}{2}\frac{f}{f_r}\right) + i\right)\exp\left(-i\frac{3\pi}{4}\frac{f}{f_r}\right) \qquad (6.121)$$

and

$$G_2(f) = 4\sin\left(\frac{\pi}{4}\frac{f}{f_r}\right)\left(\sin\left(\frac{\pi}{2}\frac{f}{f_r}\right) - i\right)\exp\left[-i\pi\left(\frac{3}{4}\frac{f}{f_r} - 1\right)\right] \qquad (6.122)$$

We can see that the amplitudes of these functions are equal at all frequencies, as the orthogonality condition in the original three-point algorithm was preserved at all frequencies

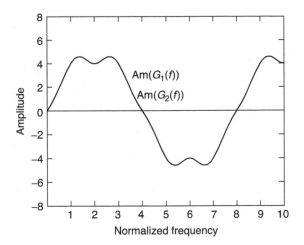

Figure 6.41 Fourier transform amplitudes of sampling functions for the Schwider algorithm obtained by combining two algorithms in quadrature.

(see Figure 6.41). These Fourier transforms are orthogonal only at the reference frequency (f_r) and all odd harmonics, as shown in Figure 6.42.

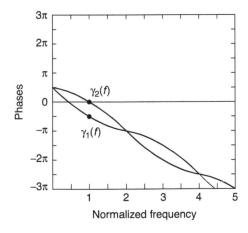

Figure 6.42 Phases for the two reference functions in the Schwider algorithm.

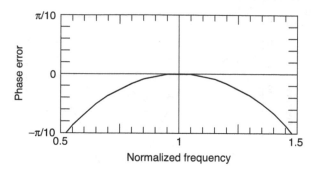

Figure 6.43 Phase error as a function of the normalized frequency for the Schwider algorithm.

We can also note in this figure that, at the signal frequency and all its odd harmonics, the slope of this phase difference is zero. Thus, we see that this algorithm has a low detuning sensitivity, as shown in the phase error illustrated in Figure 6.43. It has no sensitivity to the fourth and eight harmonics.

Another equivalent algorithm with low sensitivity to detuning can be obtained from this one by shifting the sampling points $\pi/2 + \pi/4$ to the left, which is equal to $-3\pi/4$, as shown in Section 5.10. Then, by applying the corresponding relations, we obtain:

$$\tan\phi = -2\frac{-s_2 + s_3}{-s_1 + s_2 + s_3 - s_4} \qquad (6.123)$$

A singularity and indetermination are observed when $\phi = 0°$ ($s_1 = -s_4$ and $s_2 = -s_3$). The sampling weights have the values $W_{11} = 0$, $W_{12} = -2$, $W_{13} = 2$, $W_{14} = 0$, $W_{21} = -1$, $W_{22} = 1$, $W_{23} = 1$, and $W_{24} = -1$. The reference sampling functions for this algorithm (Figure 6.44) are:

$$g_1(x) = 2\left[-\delta\left(x + \frac{X_r}{8}\right) - \delta\left(x - \frac{X_r}{8}\right)\right] \qquad (6.124)$$

and

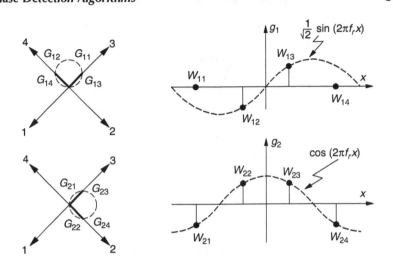

Figure 6.44 Reference sampling functions for the shifted Schwider algorithm.

$$g_2(x) = \begin{bmatrix} -\delta\left(x + \dfrac{3X_r}{8}\right) + \delta\left(x + \dfrac{X_r}{8}\right) \\ + \delta\left(x - \dfrac{X_r}{8}\right) - \delta\left(x - \dfrac{3X_r}{8}\right) \end{bmatrix} \tag{6.125}$$

and the sampling points are located at $\alpha_1 = -135°$, $\alpha_2 = -45°$, $\alpha_3 = 45°$, and $\alpha_4 = 135°$.

These Fourier transforms, shown in Figure 6.45, are thus given by:

$$G_1(f) = 4\sin\left(\frac{\pi}{4}\frac{f}{f_r}\right)\exp\left(i\frac{\pi}{2}\right) \tag{6.126}$$

and

$$G_2(f) = 8\cos\left(\frac{\pi}{4}\frac{f}{f_r}\right)\sin^2\left(\frac{\pi}{4}\frac{f}{f_r}\right)\exp(i\pi) \tag{6.127}$$

As we expected, these two functions are orthogonal at all frequencies, as the original algorithm had the same amplitudes

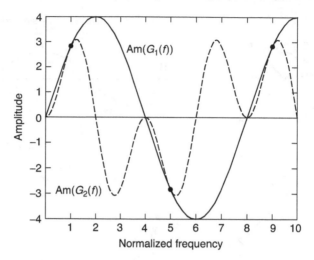

Figure 6.45 Amplitudes of the Fourier transforms of reference sampling functions for the shifted Schwider algorithm.

of the Fourier transforms at all frequencies. Because the two Fourier transform plots touch each other tangentially at the reference frequency, the algorithm has detuning insensitivity. As for the original algorithm, this one has no sensitivity to the fourth and eighth harmonics. The value of $r(f)$, using Equation 5.77, is given by:

$$r(f) = \frac{2\tan\phi}{\sin\left(\dfrac{\pi}{2}\dfrac{f}{f_r}\right)} \qquad (6.128)$$

With this procedure more complex algorithms can be generated by linearly combining several inverted T algorithms instead of only two, each one shifted with respect to the preceding algorithm by 90°. It must be noted, however, that the insensitivity to detuning is obtained only when they are added in such a manner that the sum of all odd coefficients of the linear combination is equal to the sum of all even coefficients.

6.6.2 Schmit and Creath Algorithm

This class B algorithm with five sampling points was described by Schmit and Creath (1995). The base algorithm is the Schwider algorithm (Equation 6.123):

$$\tan \phi_a = -\frac{2s_2 - 2s_3}{s_1 - s_2 - s_3 + s_4} \quad (6.129)$$

and the 90° shifted algorithm is:

$$\tan \phi_b = -\frac{s_2 - s_3 - s_4 + s_5}{-2s_3 + 2s_4} \quad (6.130)$$

Hence, the combined algorithm is:

$$\tan \phi = -\frac{3s_2 - 3s_3 - s_4 + s_5}{s_1 - s_2 - 3s_3 + 3s_4} \quad (6.131)$$

with the reference sampling functions shown in Figure 6.46; the sampling functions are located at $\alpha_1 = -135°$, $\alpha_2 = -45°$, $\alpha_3 = 45°$, $\alpha_4 = 135°$, and $\alpha_5 = 225°$. The Fourier transforms of these reference sampling functions, illustrated in Figure 6.47, are:

$$G_1(f) = 4\sin\left(\frac{\pi}{4}\frac{f}{f_r}\right)\left[\cos\left(\frac{\pi}{2}\frac{f}{f_r}\right) + 2i\sin\left(\frac{\pi}{2}\frac{f}{f_r}\right)\right] \times$$
$$\times \exp{-i\left(\frac{5\pi}{4}\frac{f}{f_r} + \frac{\pi}{2}\right)} \quad (6.132)$$

and

$$G_2(f) = -4i\sin\left(\frac{\pi}{4}\frac{f}{f_r}\right)\left[\cos\left(\frac{\pi}{2}\frac{f}{f_r}\right) - 2i\sin\left(\frac{\pi}{2}\frac{f}{f_r}\right)\right] \times$$
$$\times \exp{-i\left(\frac{3\pi}{4}\frac{f}{f_r}\right)} \quad (6.133)$$

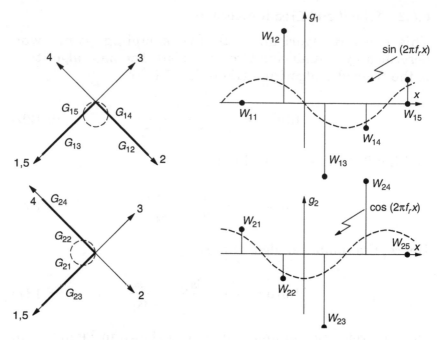

Figure 6.46 Reference sampling functions for the Schmit and Creath algorithm.

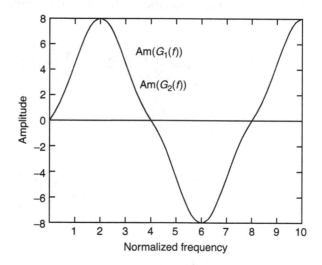

Figure 6.47 Fourier transforms of reference sampling functions for the Schmit and Creath algorithm.

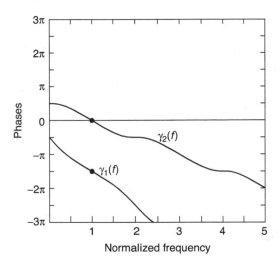

Figure 6.48 Phase for the reference sampling functions for the Schmit and Creath algorithm.

The amplitudes of these Fourier transforms are equal at all frequencies. The orthogonality condition is valid in a small region about the reference frequency (Figure 6.48), making the algorithm insensitive to small detunings. As the figure shows, it has insensitivity to only the fourth and eighth harmonics.

The phase error with detuning for this algorithm is shown in Figure 6.49. If we shift the sampling points of this

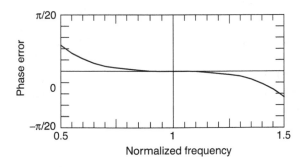

Figure 6.49 Phase error vs. the normalized frequency for the Schmit and Creath algorithm.

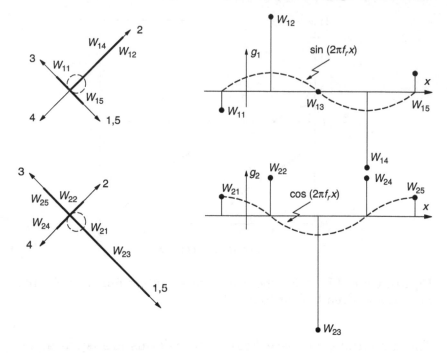

Figure 6.50 Reference sampling functions for the shifted Schmit and Creath algorithm.

algorithm by $\pi/4$ to the left and apply Equations 5.223 and 5.224, we obtain:

$$\tan\phi = -\frac{-s_1 + 4s_2 - 4s_4 + s_5}{s_1 + 2s_2 - 6s_3 + 2s_4 + s_5} \qquad (6.134)$$

with the reference sampling functions as illustrated in Figure 6.50 and the sampling points at $\alpha_1 = -45°$, $\alpha_2 = 45°$, $\alpha_3 = 135°$, $\alpha_4 = 225°$, and $\alpha_5 = 315°$.

The Fourier transforms of these reference sampling functions, illustrated in Figure 6.51, are:

$$G_1(f) = 2\left(4\sin\left(\frac{\pi}{2}\frac{f}{f_r}\right) - \sin\left(\pi\frac{f}{f_r}\right)\right)\exp\left[-i\pi\left(\frac{f}{f_r} + \frac{1}{2}\right)\right] \qquad (6.135)$$

and

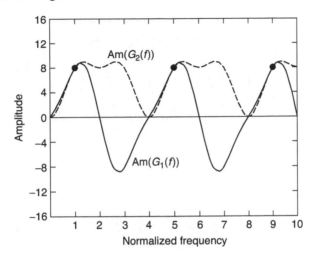

Figure 6.51 Fourier transforms of reference sampling functions for the shifted Schmit and Creath algorithm.

$$G_2(f) = \left[6 - 4\cos\left(\frac{\pi}{2}\frac{f}{f_r}\right) - 2\cos\left(\pi\frac{f}{f_r}\right) \right] \exp\left[-i\pi\left(\frac{f}{f_r} + 1\right) \right] \quad (6.136)$$

These Fourier transforms are orthogonal at all signal frequencies. The slope of these functions is the same at the reference frequency, where we also have the same amplitudes, thus making the algorithm insensitive to small detuning. As for the original algorithm, this one is insensitive to the fourth and eighth signal harmonics.

6.6.3 Other Detuning-Insensitive Algorithms

Many other detuning-insensitive algorithms have been designed, some of which have the additional important characteristic that they are also insensitive to harmonics (that is, to distorted signals). An interesting algorithm with great detuning insensitivity was designed by Servín et al. (1997) using an optimization procedure as described in Chapter 5. This algorithm was designed with seven equally spaced sampling points with a phase interval of $\pi/2$ and optimized for detuning, using the following weights:

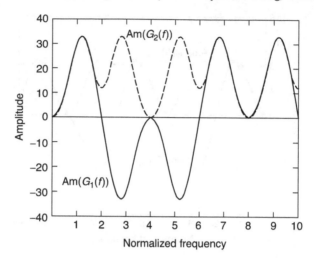

Figure 6.52 Fourier transforms of reference sampling functions for the optimized seven-sample algorithm designed by Servín et al.

$$\rho_0 = \rho_1 = 1$$

$$\rho_3 = 0.01$$

$$\rho_2 = \rho_4 = \rho_5 = \rho_6 \ldots = 0 \qquad (6.137)$$

$$\Delta_1 = 0.8$$

$$\Delta_2 = 0.1$$

With these parameters, we can define an algorithm with attenuation in the third harmonic. The solution of the linear system with seven phase steps (α_i) at $-3\pi/2$, $-\pi$, $-\pi/2$, 0, $\pi/2$, π, and $3\pi/2$ produce the phase equation:

$$\tan\phi = -\frac{1s_1 + 4.3s_2 - 14s_3 + 14s_5 - 4.3s_6 - 1s_7}{1.5s_1 - 6s_2 - 4.5s_3 + 18s_4 - 4.5s_5 - 6s_6 + 1.5s_7} \quad (6.138)$$

Figure 6.52 shows the Fourier transforms of the reference sampling functions, illustrating the frequency response and detuning insensitivity of this algorithm. Figure 6.53 shows the detuning insensitivity of this algorithm. For comparison

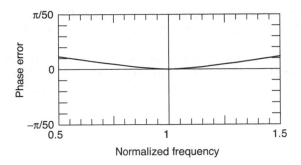

Figure 6.53 Detuning sensitivity of the optimized seven-sample algorithm.

purposes, this figure shows the detuning insensitivity of the Schwider–Hariharan algorithm compared with this algorithm. It should be pointed out that the detuning insensitivity obtained in the algorithms presented here has been obtained at the expense of any possible harmonic leaks.

6.7 DETUNING-INSENSITIVE ALGORITHMS FOR DISTORTED SIGNALS

When a signal is distorted and, as a consequence, harmonics are present, a detuning-insensitive algorithm must also be insensitive to the signal harmonics. The reason is that, when detuning is present, not only is the fundamental frequency detuned but also its harmonics. This problem, first studied by Hibino et al. (1995) and a little later by Surrel (1996) and Zhao and Surrel (1995), has been described in Section 5.9.

In order to have an algorithm with detuning sensitivity up the mth harmonic we need enough sampling points to determine the signal bias, the amplitudes of all harmonic components (i.e., S_0, S_1, S_2, ..., S_m), their phases (ϕ_1, ϕ_2, ..., ϕ_m) in Equation 5.57, and the magnitude of the linear phase error. This results in a total of $2m + 2$ unknowns; thus, a minimum of $2m + 2$ sampling points is needed. It should be pointed out here that Hibino et al. (1995) found that a minimum of $2m + 3$ points was necessary, but this value was later corrected by Surrel (1996).

TABLE **6.1**   Minimum Number of Sampling Points for Detuning-Insensitive Algorithms with Harmonically Distorted Signals

Minimum Number of Samples ($N = 2m + 2$)	Maximum Harmonic (m) with Detuning Insensitivity	Maximum Phase Interval ($2\pi/(m + 2)$)
4	1	120°
6	2	90°
8	3	72°
10	4	60°
12	5	51.14°
14	6	45°

Source: Data from Hibino et al. (1995) and Surrel (1996).

An algorithm with detuning insensitivity up to the mth harmonic, as pointed out before, requires that:

1. The phase interval between sampling points is smaller than $2\pi/(m+2)$.
2. When the maximum phase interval is used, the minimum number of sampling points is $2m + 2$. With a smaller phase interval the number of required sampling points would be larger.

For example, as described in Table 6.1, an algorithm that is detuning insensitive only up to the second harmonic using the maximum phase interval of 90° must have at least six sampling points. If this phase interval is reduced, more than six points are needed.

6.7.1   Zhao and Surrel Algorithm

Let us now consider the six-sample algorithm (Zhao and Surrel, 1995; Surrel, 1996), which takes six signal measurements at constant phase intervals equal to 90°, as follows:

$$s_1 = a + b\cos\phi$$
$$s_2 = a + b\cos(\phi + 90°)$$
$$s_3 = a + b\cos(\phi + 180°)$$
$$s_4 = a + b\cos(\phi + 270°)$$
$$s_5 = a + b\cos(\phi + 360°)$$
$$s_6 = a + b\cos(\phi + 450°)$$

(6.139)

From these equations, the desired solution for the phase that satisfies the conditions described earlier, is:

$$\tan\phi = -\frac{s_1 + 3s_2 - 4s_4 - s_5 + s_6}{s_1 - s_2 - 4s_3 + 3s_5 + s_6}$$

(6.140)

Thus, the reference sampling functions (Figure 6.54) are:

$$g_1(x) = \delta\left(x - \frac{X_r}{4}\right) + 3\delta\left(x - \frac{X_r}{2}\right) - 4\delta(x - X_r) -$$
$$- \delta\left(x - \frac{5X_r}{4}\right) + \delta\left(x - \frac{3X_r}{2}\right)$$

(6.141)

and

$$g_2(x) = \delta\left(x - \frac{X_r}{4}\right) - \delta\left(x - \frac{X_r}{2}\right) - 4\delta\left(x - \frac{3X_r}{4}\right) +$$
$$+ 3\delta\left(x - \frac{5X_r}{4}\right) + \delta\left(x - \frac{3X_r}{2}\right)$$

(6.142)

The Fourier transforms for these reference sampling functions (Figure 6.55) are:

$$G_1(f) = 2\left[\begin{array}{l} \cos\left(\dfrac{5\pi}{4}\dfrac{f}{f_r}\right) - \cos\left(\dfrac{3\pi}{2}\dfrac{f}{f_r}\right) \\ +4\sin\left(\dfrac{\pi}{2}\dfrac{f}{f_r}\right) \times \exp\left[i\left(\dfrac{\pi}{4}\dfrac{f}{f_r} - \dfrac{\pi}{2}\right)\right] \end{array}\right]\exp\left[i\left(\dfrac{5\pi}{4}\dfrac{f}{f_r}\right)\right]$$ (6.143)

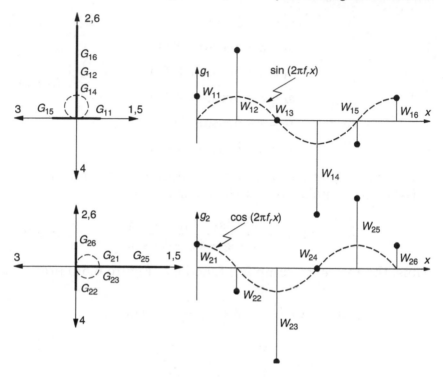

Figure 6.54 Reference sampling functions for the six-sample detuning-insensitive algorithm.

and

$$
G_2(f) = 2 \left[\begin{array}{l} \cos\!\left(\dfrac{5\pi}{4}\dfrac{f}{f_r} \right) - \cos\!\left(\dfrac{3\pi}{2}\dfrac{f}{f_r} \right) \\[2ex] + 4\sin\!\left(\dfrac{\pi}{2}\dfrac{f}{f_r} \right) \times \exp\!\left[-i\!\left(\dfrac{\pi}{4}\dfrac{f}{f_r} - \dfrac{\pi}{2} \right) \right] \end{array} \right] \exp\!\left[-i\!\left(\dfrac{5\pi}{4}\dfrac{f}{f_r} \right) \right] \quad (6.144)
$$

These Fourier transforms have the same amplitudes at all frequencies, but they are orthogonal in the vicinity of the reference frequency and the second harmonic, as illustrated in Figure 6.56. This algorithm is shifted π/4 with respect to the one described in the articles by Zhao and Surrel (1995) and Surrel (1996) which is orthogonal to all frequencies, but

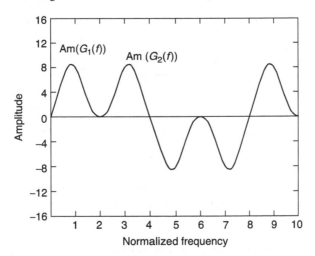

Figure 6.55 Fourier transforms for the six-sample detuning-insensitive algorithm.

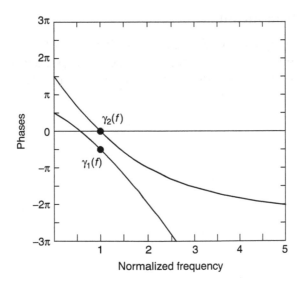

Figure 6.56 Phases for the reference functions in the Zhao–Surrel six-sample detuning-insensitive algorithm.

their magnitudes are equal in the vicinity of the reference frequency and its second harmonic. When shifting, the algorithm properties are preserved. This algorithm is detuning

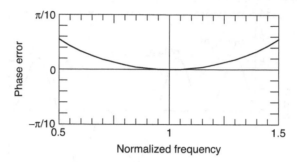

Figure 6.57 Phase error vs. the normalized frequency in the Zhao–Surrel six-sample detuning-insensitive algorithm.

insensitive up to the second harmonic, but it is not insensitive to the third harmonic. The phase error in the presence of detuning is shown in Figure 6.57.

6.7.2 Hibino Algorithm

Another algorithm with small sensitivity to the second harmonic, even when detuning is present, uses seven sampling points and has been described by Hibino et al. (1995). The phase is calculated by:

$$\tan\phi = -\frac{s_2 - 2s_4 + s_6}{0.5s_1 - 1.5s_3 + 1.5s_5 - 0.5s_7} \qquad (6.145)$$

and the reference sampling functions (Figure 6.58) are:

$$g_1(x) = \delta\left(x - \frac{X_r}{4}\right) - 2\delta\left(x - \frac{3X_r}{4}\right) + \delta\left(x - \frac{5X_r}{4}\right) \qquad (6.146)$$

and

$$g_2(x) = \frac{1}{2}\delta(x) - 1.5\delta\left(x - \frac{X_r}{2}\right) + 1.5\delta(x - X_r) - \qquad (6.147)$$

$$- \frac{1}{2}\delta\left(x - \frac{3X_r}{2}\right)$$

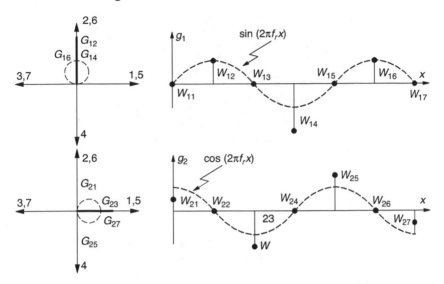

Figure 6.58 Reference sampling functions for the seven-sample detuning-insensitive algorithm.

The Fourier transforms for the reference sampling functions (Figure 6.59) are:

$$G_1(f) = 2\left[\cos\left(\pi\frac{f}{f_r}\right) - 1\right]\exp\left(-i\frac{3\pi}{2}\frac{f}{f_r}\right) \tag{6.148}$$

and

$$G_2(f) = \left[\sin\left(\frac{3\pi}{2}\frac{f}{f_r}\right) + 3\sin\left(\frac{\pi}{2}\frac{f}{f_r}\right)\right]\exp\left[-i\left(\frac{3\pi}{2}\frac{f}{f_r} - \frac{\pi}{2}\right)\right] \tag{6.149}$$

An interesting property of this algorithm is that it is insensitive to all even harmonics as well as to small detuning of these harmonics; however, it is sensitive to odd harmonics. The phase error for this algorithm in the presence of detuning is illustrated in Figure 6.60.

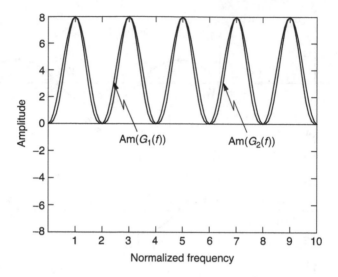

Figure 6.59 Fourier transforms for the seven-sample, detuning-insensitive algorithm.

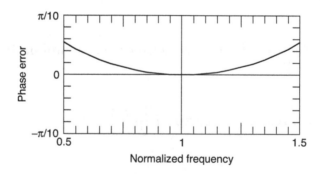

Figure 6.60 Phase error vs. the normalized frequency in the seven-sample, detuning-insensitive algorithm.

6.7.3 Six-Sample, Detuning-Insensitive Algorithm

By using the graphical method described in Section 5.5.4, some other detuning-insensitive algorithms have been designed. As an example, let us consider the one designed by Malacara-Doblado and Vazquez-Dorrío (2000) that has six sampling points. The phase is given by:

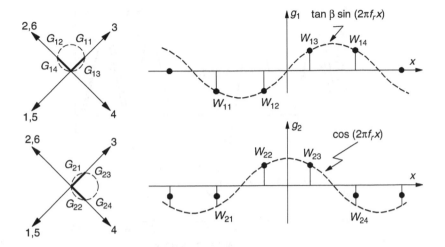

Figure 6.61 Reference sampling functions for the six-sample, detuning-insensitive algorithm designed by Malacara-Doblado and Vazquez-Dorrío (2000).

$$\tan\phi = -\frac{s_2 + s_3 - s_4 - s_5}{-0.5s_1 - 0.5s_2 + s_3 + s_4 - 0.5s_5 - 0.5s_6} \quad (6.150)$$

and the reference sampling functions (Figure 6.61) are given by:

$$g_1(x) = -\delta\left(x + \frac{3X_r}{4}\right) - \delta\left(x + \frac{X_r}{4}\right) + \delta\left(x - \frac{X_r}{4}\right) + \delta\left(x - \frac{3X_r}{4}\right) \quad (6.151)$$

and

$$g_2(x) = -\frac{1}{2}\delta\left(x + \frac{5X_r}{4}\right) - \frac{1}{2}\delta\left(x + \frac{3X_r}{2}\right) + \delta\left(x + \frac{X_r}{4}\right) +$$

$$+ \delta\left(x - \frac{X_r}{4}\right) - \frac{1}{2}\delta\left(x - \frac{3X_r}{2}\right) - \frac{1}{2}\delta\left(x - \frac{5X_r}{4}\right) \quad (6.152)$$

The Fourier transforms for the reference sampling functions (Figure 6.62) are:

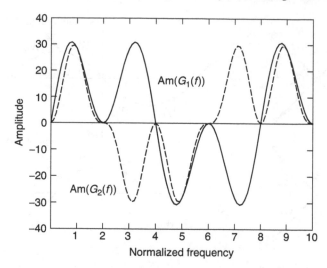

Figure 6.62 Fourier transforms of the reference sampling functions for the six-sample, detuning-insensitive algorithm.

$$G_1(f) = 2\left[\sin\left(\frac{3\pi}{4}\frac{f}{f_r}\right) + \sin\left(\frac{\pi}{4}\frac{f}{f_r}\right)\right]\exp\left(-i\frac{\pi}{2}\right) \quad (6.153)$$

and

$$G_2(f) = 2\cos\left(\frac{\pi}{4}\frac{f}{f_r}\right) - \cos\left(\frac{3\pi}{4}\frac{f}{f_r}\right) - \cos\left(\frac{5\pi}{4}\frac{f}{f_r}\right) \quad (6.154)$$

This algorithm is detuning insensitive at the fundamental frequency as well as at the second, sixth, and eighth harmonics. It is insensitive to all even harmonics. The detuning phase error is illustrated in Figure 6.63.

6.8 ALGORITHMS CORRECTED FOR NONLINEAR PHASE-SHIFTING ERROR

In Chapter 5, we described how algorithms can be designed for insensitivity to high-order nonlinear phase shifting in the presence of signal harmonic distortion (Hibino, 1997; Surrel, 1998; Hibino, 1999; Hibino and Yamauchi, 2000). It was

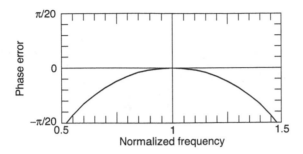

Figure 6.63 Phase error as a function of the normalized frequency for the six-sample, detuning-insensitive algorithm.

shown that the minimum number of samples necessary to compensate for these errors is six and that a very good correction can be achieved with eleven points. In this section, we describe three of these algorithms.

The first algorithm uses six sampling points. The reference sampling functions for the six-sample algorithm with correction for nonlinear phase errors are shown in Figure 6.64. The Fourier transforms of the reference sampling functions for this six-sample algorithm with correction for nonlinear phase errors are shown in Figure 6.65. The phase errors as a function of the normalized frequency for the six-sample algorithm with correction for nonlinear phase errors are illustrated in Figure 6.66.

The second algorithm uses nine sampling points. The reference sampling functions for the nine-sample algorithm with correction for nonlinear phase errors are shown in Figure 6.67. The Fourier transforms of the reference sampling functions for the nine-sample algorithm with correction for nonlinear phase errors are shown in Figure 6.68. The phase errors as a function of the normalized frequency for the nine-sample algorithm with correction for nonlinear phase errors are illustrated in Figure 6.69.

The last example is an algorithm that uses eleven sampling points. The reference sampling functions for the eleven-sample algorithm with correction for nonlinear phase errors are shown in Figure 6.70. The Fourier transforms of the

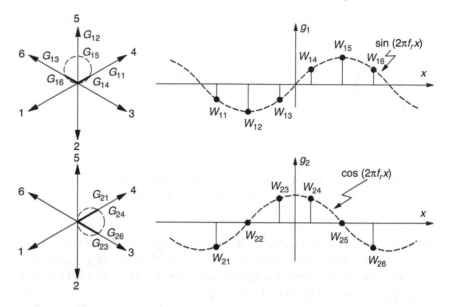

Figure 6.64 Reference sampling functions for the six-sample algorithm with correction for nonlinear phase error designed by Hibino et al. (1997).

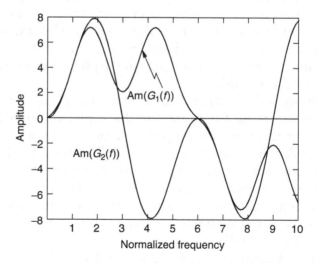

Figure 6.65 Fourier transforms of the reference sampling functions for the six-sample algorithm with correction for nonlinear phase error designed by Hibino et al. (1997).

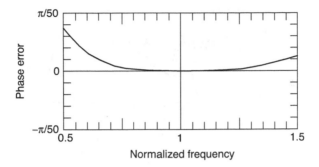

Figure 6.66 Phase error as a function of the normalized frequency for the six-sample algorithm with correction for nonlinear phase error designed by Hibino et al. (1997).

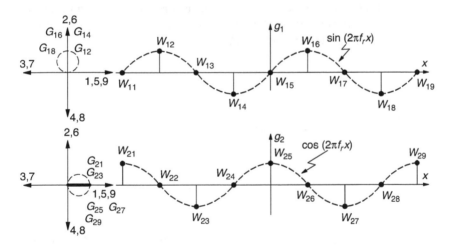

Figure 6.67 Reference sampling functions for the nine-sample algorithm with correction for nonlinear phase error designed by Hibino et al. (1997).

reference sampling functions for the eleven-sample algorithm with correction for nonlinear phase errors are shown in Figure 6.71. The phase errors as a function of the normalized frequency for the eleven-sample algorithm with correction for nonlinear phase errors are illustrated in Figure 6.72.

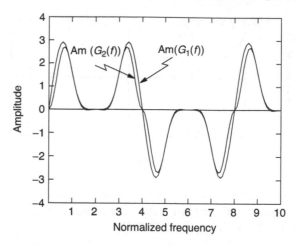

Figure 6.68 Fourier transforms of the reference sampling functions for the nine-sample algorithm with correction for nonlinear phase error designed by Hibino et al. (1997).

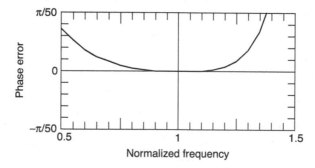

Figure 6.69 Phase error as a function of the normalized frequency for the nine-sample algorithm with correction for nonlinear phase error designed by Hibino et al. (1997).

6.9 CONTINUOUS SAMPLING IN A FINITE INTERVAL

When sampling a sinusoidal signal with a finite aperture or a finite sampling interval, this aperture or finite interval acts as a filtering window. This problem has been studied by Nakadate (1988a,b) but with a different approach than that presented here. Here, we will use a similar but slightly simpler approach, using the Fourier theory just developed.

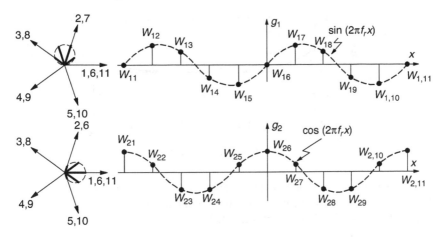

Figure 6.70 Reference sampling functions for the eleven-sample algorithm with correction for nonlinear phase error designed by Hibino et al. (1997).

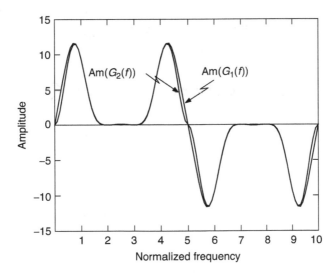

Figure 6.71 Fourier transforms of the reference sampling functions for the eleven-sample algorithm with correction for nonlinear phase error designed by Hibino et al. (1997).

The tentative sampling functions using a finite interval of size X can be written as:

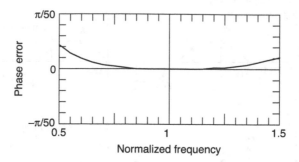

Figure 6.72 Phase error as a function of the normalized frequency for the eleven-sample algorithm with correction for nonlinear phase error designed by Hibino et al. (1997).

$$g_1(x) = \sin(2\pi f_r x), \quad \text{for } -\frac{X}{2} \le x \le \frac{X}{2}$$
$$= 0, \quad\quad\quad\quad \text{for } |x| > \frac{X}{2} \tag{6.155}$$

and

$$g_2(x) = \cos(2\pi f_r x), \quad \text{for } -\frac{X}{2} \le x \le \frac{X}{2}$$
$$= 0, \quad\quad\quad\quad \text{for } |x| > \frac{X}{2} \tag{6.156}$$

Then, the Fourier transforms of these functions (Figure 6.73) can be written as:

$$G_1(f) = i\Big[\text{sinc}\big(\pi(f + f_r)X\big) - \text{sinc}\big(\pi(f - f_r)X\big)\Big] \tag{6.157}$$

and

$$G_2(f) = \Big[\text{sinc}\big(\pi(f + f_r)X\big) + \text{sinc}\big(\pi(f - f_r)X\big)\Big] \tag{6.158}$$

We can see, as shown in Figure 6.73, that the separation between these two sinc functions is equal to twice the reference frequency (f_r). When the reference frequency is large compared to $1/X$, the two sinc functions are quite separated from each other, and the side lobes of one will not overlap the

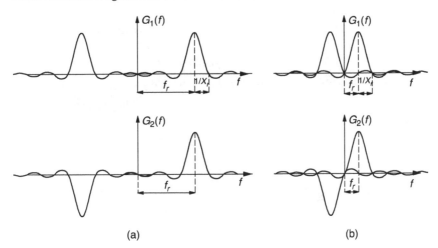

Figure 6.73 Fourier transforms of functions $g_1(x)$ and $g_2(x)$ with continuous sampling in a finite interval: (a) with $X \gg X_r$ and (b) $X = X_r$.

other (Figure 6.73a). On the other hand, if the reference frequency is low as compared to $1/X$, the side lobes of one sinc function will overlap the other sinc function (Figure 6.73b), where $X = X_r = 1/f_r$.

Because the functions $G_i(f)$ are the sum of the two sinc functions, the $G_i(f_r)$ will not change and will remain equal to each other when:

$$f_r X = \frac{n}{2} \qquad (6.159)$$

where n is any positive integer. In this case, no error is present in the phase detection. This result means that the sampling interval (or aperture) should be an integral number of half the spatial period of the fringes (refer to Section 5.2). This property was used by Morimoto and Fujisawa (1994). A peak in the error will occur, however, at intermediate positions given by:

$$f_r X = \frac{n}{2} + \frac{1}{4} \qquad (6.160)$$

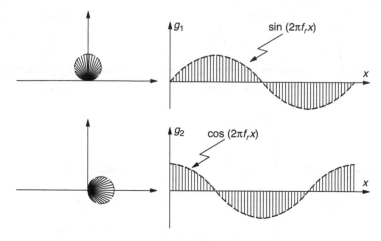

Figure 6.74   Reference sampling functions $g_1(x)$ and $g_2(x)$ for a continuous sampling interval $X_r = 1/f_r$.

If a phase-detecting algorithm uses the sampling interval X_r, then the phase ϕ is given by:

$$\tan\phi = -\frac{\displaystyle\int_{x=0}^{X_r} s(x)\sin(2\pi fx)\,dx}{\displaystyle\int_{x=0}^{X_r} s(x)\cos(2\pi fx)\,dx} \qquad (6.161)$$

with the reference sampling functions as shown in Figure 6.74. The Fourier transforms of the reference sampling functions are:

$$G_1(f) = i\left[\operatorname{sinc}\left(\pi\left(\frac{f}{f_r}+1\right)\right) - \operatorname{sinc}\left(\pi\left(\frac{f}{f_r}-1\right)\right)\right] \qquad (6.162)$$

and

$$G_2(f) = \left[\operatorname{sinc}\left(\pi\left(\frac{f}{f_r}+1\right)\right) + \operatorname{sinc}\left(\pi\left(\frac{f}{f_r}-1\right)\right)\right] \qquad (6.163)$$

which are illustrated in Figure 6.75. The Fourier transforms shown in this figure are orthogonal at all signal frequencies, but they have the same amplitude only at the reference

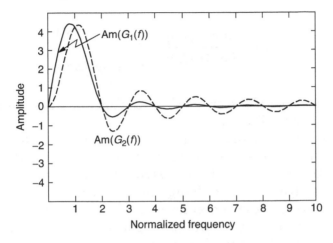

Figure 6.75 Fourier transforms of functions $g_1(x)$ and $g_2(x)$ for a continuous sampling interval $X_r = 1/f_r$.

frequency. Thus, this algorithm is sensitive to detuning. It is quite interesting to note the lack of sensitivity to any harmonics in the absence of detuning. Insensitivity to small detuning can be obtained if the additional sampling points at the ends of the sampling interval, as described in Section 6.5, are used. This is a limit case for discrete sampling algorithms, when the number of sampling steps tends to infinity.

6.10 ASYNCHRONOUS PHASE-DETECTION ALGORITHMS

In synchronous detection we have assumed that the frequency of the detected signal and the phase steps taken during the measurements are known; however, at times the phase steps or frequency of the measured signal are unknown. In that case, before calculating the phase the signal frequency must be determined. To do so, we need a minimum of four sampling points.

If we examine the expression for $r(f)$ in Equation 5.62, we see that, if we require that the two Fourier transforms $G_1(f)$ and $G_2(f)$ have the same phase ϕ instead of being orthogonal to each other and if we also remove the condition that their magnitudes are equal, using Equation 5.77 we obtain:

$$r(f) = \frac{Am\big(G_1(f_r)\big)}{Am\big(G_2(f_r)\big)} = \frac{\displaystyle\int_{-\infty}^{\infty} s(x)g_1(x)\mathrm{d}x}{\displaystyle\int_{-\infty}^{\infty} s(x)g_2(x)\mathrm{d}x} \qquad (6.164)$$

This is possible if the two reference functions are both anti-symmetric and different.

Then, we can see that the value of $r(f)$ is not a function of the signal phase ϕ as before. Instead, it is a function of the signal frequency. The value of $r(f)$ *can* be calculated for a given sampling algorithm satisfying this condition, thus allowing determination of the signal frequency. A simple way to obtain Fourier transforms with the same phase is to require that the reference sampling functions $g_1(x)$ and $g_2(x)$ are both antisymmetrical or both symmetrical. Thus, they must have different frequencies, normally equal to f_r and $2f_r$, respectively.

We can see that if the reference functions $g_1(x)$ and $g_2(x)$ are antisymmetrical and the signal is symmetrical, or vice versa, both integrals in this expression become equal to zero. Then, with symmetric reference functions the value of $r(f)$ becomes undetermined when the signal is symmetrical (that is, when the phase has a value equal to $n\pi$, n being an integer). On the other hand, with antisymmetric reference functions, the value of $r(f)$ becomes undetermined when the signal is antisymmetrical (that is, when the phase has a value equal to $n\pi/2$, n being an odd integer).

6.10.1 Carré Algorithm

This is the classic asynchronous algorithm, developed by Carré (1966), where four measurements of the signal are taken at equally spaced phase increments. The sampling points are symmetrically placed with respect to the origin, as expressed by:

$$\begin{aligned}
s_1 &= a + b\cos(\phi - 3\beta) \\
s_2 &= a + b\cos(\phi - \beta) \\
s_3 &= a + b\cos(\phi + \beta) \\
s_4 &= a + b\cos(\phi + 3\beta)
\end{aligned} \qquad (6.165)$$

where the phase increment is 2β. If the reference frequency (f_r) and signal frequency (f) are different, the phase increments would have a different value when referred to the reference function or to the signal phase scales. When measured with respect to the signal phase scale, its value is β, but if measured with respect to the reference function phase scale its value is α. In synchronous phase detection, we have $\alpha = \beta$, but in general we have:

$$\beta = \alpha \frac{f}{f_r} \qquad (6.166)$$

The value of β is unknown, either because the value of α or the frequency (f) of the signal is unknown. The most common phase step used in this algorithm is $\alpha = \pi/4$. The value of β can be calculated by using the following expression obtained from Equation 6.165:

$$\tan^2 \beta = \frac{3(s_2 - s_3) - (s_1 - s_4)}{(s_1 - s_4) + (s_2 - s_3)} \qquad (6.167)$$

or, alternatively, by defining a value of $r\beta(f)$ given by:

$$r_\beta(f) = -\frac{\sin 2\beta \cos \beta \sin \phi}{\cos 2\beta \sin \beta \sin \phi} = -\frac{\tan 2\beta}{\tan \beta} = \frac{-s_1 - s_2 + s_3 + s_4}{s_1 - s_2 + s_3 - s_4} \qquad (6.168)$$

with the reference functions for which the sampling weights have the values $W_{11} = -1$, $W_{12} = -1$, $W_{13} = 1$, $W_{14} = 1$, $W_{21} = 1$, $W_{22} = -1$, $W_{23} = 1$, and $W_{24} = -1$. Singularity and indetermination are observed when $\sin\phi = 0$, because then $s_2 = s_3$ and $s_1 = s_4$. Singularity and indetermination also occur when $\beta = \pi/2$. The reference sampling functions for $\alpha = \pi/4$ (Figure 6.76) are:

$$g_1(x) = -\delta\left(x + \frac{3X_r}{8}\right) - \delta\left(x + \frac{X_r}{8}\right) + \delta\left(x - \frac{X_r}{8}\right) + \delta\left(x - \frac{3X_r}{8}\right) \qquad (6.169)$$

and

$$g_2(x) = \delta\left(x + \frac{3X_r}{8}\right) - \delta\left(x + \frac{X_r}{8}\right) + \delta\left(x - \frac{X_r}{8}\right) - \delta\left(x - \frac{3X_r}{8}\right) \qquad (6.170)$$

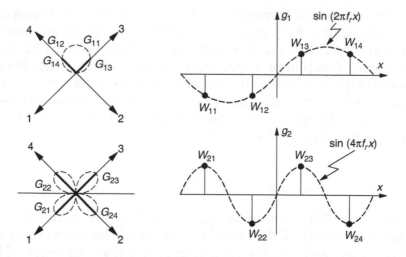

Figure 6.76 Sampling in the Carré algorithm, with $\alpha = \pi/4$, to obtain the signal frequency.

The Fourier transforms of the sampling functions for $\alpha = \pi/4$ (Figure 6.77) are:

$$G_1(f) = 4\cos\left(\frac{\pi}{4}\frac{f}{f_r}\right)\sin\left(\frac{\pi}{2}\frac{f}{f_r}\right)\exp-i\left(\frac{\pi}{2}\right) \qquad (6.171)$$

and

$$G_2(f) = 4\sin\left(\frac{\pi}{4}\frac{f}{f_r}\right)\cos\left(\frac{\pi}{2}\frac{f}{f_r}\right)\exp-i\left(\frac{\pi}{2}\right) \qquad (6.172)$$

We can observe in this figure that these functions are symmetrical about the value of the normalized frequency equal to 2, which corresponds to $\beta = \pi/2$. Hence, the measurement of β can be performed without uncertainty only if it is in the range $0 < \beta < \pi/2$. Hence, the value of the reference frequency (f_r) should in principle be chosen so that the values of α and β are as close as possible to each other. In other words, the reference frequency should be higher than half the signal frequency but as close as possible to this value. This condition can also be expressed by saying that the four sampling points

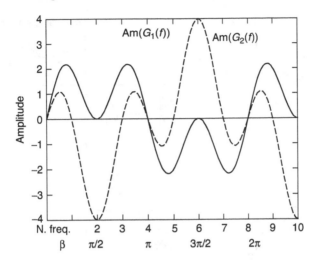

Figure 6.77 Amplitudes of the Fourier transforms of the reference functions for the Carré algorithm for $\alpha = \pi/4$, to obtain the signal frequency.

must be separated by at least a fourth of the period of the signal. Nevertheless, if we take into account the presence of additive noise in the measurements, it can be shown that the noise influence is minimized when $\beta = 110°$, as pointed out by Carré (1966) and Freischlad and Koliopoulos (1990).

Figure 6.77 illustrates the singularity and indetermination that occur when $\beta = \pi$, as both Fourier transform amplitudes are zero. This algorithm is quite sensitive to signal harmonics.

Once the value of β has been calculated, the signal phase ϕ can be found using another algorithm with the same sampling points and, hence, the same measured values:

$$\tan\phi = \frac{(s_1 - s_4) + (s_2 - s_3)}{(s_2 + s_3) - (s_1 + s_4)}\tan\beta \qquad (6.173)$$

As in the previous algorithm, indetermination occurs when $\phi = 0$, as $s_1 = s_3$ and $s_1 = s_4$. Hence, when ϕ is small, large errors can occur.

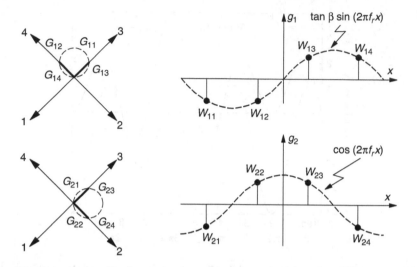

Figure 6.78 Sampling in the reference function for the Carré algorithm with $\alpha = \pi/4$ and a constant value of β, to find the phase.

Having calculated the value of β with a set of four sampling points, the same value of β can be used to calculate the phase for several signal points with different locations, if the frequency for the signal is the same everywhere. This is the case of temporal phase shifting, where the signal frequency is frequently the same for all points in the interferogram. Alternatively, if the frequency is not constant, such as in space phase shifting, when the wavefront is not aberration free the value of β has to be calculated for every point where the phase is to be determined.

Let us consider the first case in which the value of β is a constant. We can write Equation 6.161 as:

$$\tan \phi = -\tan \beta \frac{s_1 + s_2 - s_3 - s_4}{s_1 - s_2 - s_3 + s_4} \qquad (6.174)$$

with the sampling weight values $W_{11} = \tan\beta$, $W_{12} = \tan\beta$, $W_{13} = -\tan\beta$, $W_{14} = -\tan\beta$, $W_{21} = 1$, $W_{22} = -1$, $W_{23} = -1$, and $W_{24} = 1$. The reference sampling functions (Figure 6.78) are:

$$g_1(x) = \delta\left(x + \frac{3X_r}{8}\right) + \delta\left(x + \frac{X_r}{8}\right) - \delta\left(x - \frac{X_r}{8}\right) - \delta\left(x - \frac{3X_r}{8}\right) \quad (6.175)$$

and

$$g_2(x) = \delta\left(x + \frac{3X_r}{8}\right) - \delta\left(x + \frac{X_r}{8}\right) - \delta\left(x - \frac{X_r}{8}\right) + \delta\left(x - \frac{3X_r}{8}\right) \quad (6.176)$$

The Fourier transforms of the sampling functions with $\alpha = \pi/4$ are thus given by:

$$G_1(f) = 4\sin\left(\frac{\pi}{2}\frac{f}{f_r}\right)\cos\left(\frac{\pi}{4}\frac{f}{f_r}\right)\tan\beta\exp i\left(\frac{\pi}{2}\right) \quad (6.177)$$

and

$$G_2(f) = 4\sin\left(\frac{\pi}{2}\frac{f}{f_r}\right)\sin\left(\frac{\pi}{4}\frac{f}{f_r}\right) \quad (6.178)$$

which are illustrated in Figure 6.79.

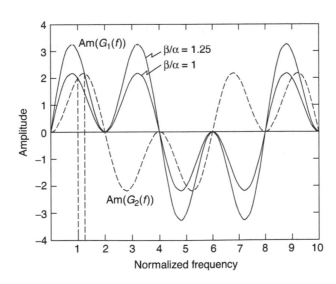

Figure 6.79 Amplitudes of the Fourier transforms of the reference functions in the Carré algorithm using $\alpha = \pi/4$ and two different constant values of β.

We can see that this algorithm is insensitive to all even harmonics, only if $\beta/\alpha = 1$, which is not frequent, and it is always quite sensitive to all odd harmonics. It must be pointed out here that this is for the second part, after β has been calculated, but errors due to the presence of harmonics can also appear in the calculation of β, as we pointed out before. We can also see that it is quite sensitive to detuning, but that is not a serious problem, as the frequency has been previously calculated in the first step. Notice that this algorithm is identical to the four points in the X algorithm, described previously, when $\beta/\alpha = 1$.

A problem arises, however, if the value of β is not a constant for all locations where it is measured. Then, the frequency is not a constant, and it is better to recalculate β every time the phase is to be obtained. Then, we can combine Equations 6.156 and 6.161, with the result:

$$\tan\phi = \frac{\left[3(s_2 - s_3)^2 - (s_1 - s_4)^2 + 2(s_1 - s_4)(s_2 - s_3)\right]^{1/2}}{(s_2 + s_3) - (s_1 + s_4)} \quad (6.179)$$

thus removing the indetermination.

We can see that, in this case, by substituting the value of β in Equation 6.166 into Equation 6.177 for $G_1(f)$, the two Fourier transforms, $G_1(f)$, and $G_2(f)$, become equal at all frequencies. This is to be expected, because we now have no detuning error, as the algorithm is self calibrating.

One problem with this algorithm is that the numerator in this expression is the square of a number; thus, the sign of $\sin\phi$ is lost. As a consequence, the phase is wrapped modulo π instead of modulo 2π as for most phase-detecting algorithms. Figure 6.80 shows the phase wrapping in the Carré algorithm compared with phase wrapping in other algorithms. The Carré algorithm has been adapted by Rastogi (1993) to the study of four-wave holographic interferometry.

6.10.2 Schwider Asynchronous Algorithm

This asynchronous algorithm (Schwider et al., 1983; Cheng and Wyant, 1985) has four sampling points at phases -2β, $-\beta$,

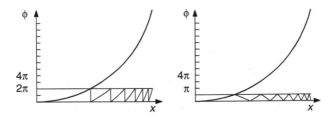

Figure 6.80 Phase wrapping in the Carré algorithm compared with that for other phase-detecting algorithms.

β, and 2β (with β as defined in Equation 6.166) and a value of $\alpha = \pi/4$. The cosine of the phase increment becomes:

$$r_\beta(f) = \cos\beta = \frac{-s_1 + s_4}{-2s_2 + 2s_3} \qquad (6.180)$$

and the reference sampling functions (Figure 6.81) are:

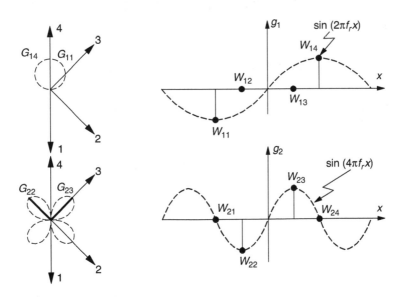

Figure 6.81 Reference sampling functions for the Schwider asynchronous algorithm.

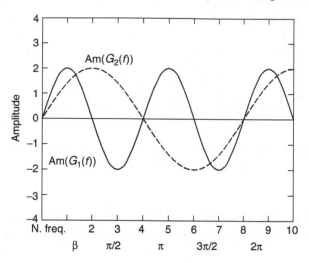

Figure 6.82   Amplitudes of the Fourier transforms of the reference sampling functions for the Schwider asynchronous algorithm.

$$g_1(x) = -\delta\left(x - \frac{X_r}{2}\right) + \delta\left(x + \frac{X_r}{2}\right) \tag{6.181}$$

and

$$g_2(x) = -2\delta\left(x - \frac{X_r}{4}\right) + 2\delta\left(x + \frac{X_r}{4}\right) \tag{6.182}$$

The Fourier transforms of these reference sampling functions (Figure 6.82) are:

$$G_1(f) = 2\sin\left(\frac{\pi}{2}\frac{f}{f_r}\right)\exp i\left(\frac{\pi}{2}\right) \tag{6.183}$$

and

$$G_2(f) = 2\sin\left(\frac{\pi}{4}\frac{f}{f_r}\right)\exp i\left(\frac{\pi}{2}\right) \tag{6.184}$$

In this algorithm the reference frequency can be as low as one eighth of the signal frequency; however, singularities and indeterminations are observed at f/f_r equal to 4 and 8. Ideally, the reference frequency should be as close as possible to the signal frequency. This algorithm has a large sensitivity to the presence of signal harmonics.

6.10.3 Two Algorithms in Quadrature

We have seen in Section 6.6 that two algorithms in quadrature produce phases with opposite errors in the phase; hence, by averaging their phases, as in Equation 6.85, the error-free phase can be calculated. The error in the phase can be obtained if, instead of averaging the two phases, their difference is taken:

$$\delta\phi = \frac{\tan^{-1}\phi_a - \tan^{-1}\phi_b}{2} \tag{6.185}$$

Now, from Equation 5.154, if the base (nonshifted) algorithm is orthogonal at all frequencies, we have:

$$\frac{\text{Am}(G_1(f))}{\text{Am}(G_2(f))} = \rho(f) = 1 + \frac{2\delta\phi}{\sin 2\phi} \tag{6.186}$$

where the phase ϕ is calculated with Equation 6.98.

Once the value of $\rho(f)$ (which is different from 1) has been obtained, the normalized frequency f/f_r can be calculated, because, for these algorithms, from Equation 5.77 we have $r(f) = \pm\rho(f)\tan\phi$. For example, if the inverted T algorithm has been used, we have:

$$\frac{f}{f_r} = \frac{4}{\pi}\tan^{-1}\rho \tag{6.187}$$

6.10.4 An Algorithm for Zero Bias and Three Sampling Points

We have seen that four measurements are necessary to determine the four parameters of a sinusoidal signal (i.e., a, b, ϕ_0, and ω). Ransom and Kokal (1986) and later Servín and Cuevas (1995) described a method in which the DC (bias) term is first eliminated from the signal by means of a convolution with a high-pass filter, as described in Section 2.4.1. Then, the only problem remaining is that the entire signal interval must be sampled and processed before sampling the phase-measuring points. Thus, after eliminating the bias (coefficient a), the signal can be expressed by:

$$s(x) = b\cos(\omega x + \phi)$$ (6.188)

If three sampling points at x positions x_0, 0, and $-x_0$ are used, we have:

$$s_1 = b\cos(-\omega x_0 + \phi)$$ (6.189)

$$s_2 = b\cos\phi$$ (6.190)

and

$$s_3 = b\cos(\omega x_0 + \phi)$$ (6.191)

But, these three expressions can also be written as:

$$s_1 = b\cos(\omega x_0)\cos\phi + b\sin(\omega x_0)\sin\phi$$ (6.192)

$$s_2 = b\cos\phi$$ (6.193)

and

$$s_3 = b\cos(\omega x_0)\cos\phi - b\sin(\omega x_0)\sin\phi$$ (6.194)

Then, it is easy to see that

$$\frac{s_1 + s_3}{2s_2} = \cos(\omega x_0)$$ (6.195)

and

$$\frac{s_1 - s_3}{2s_2} = \sin(\omega x_0)\tan\phi$$ (6.196)

Now, from Equation 6.195:

$$\sin(\omega x_0) = \left[1 - \left(\frac{s_1 + s_3}{2s_2}\right)^2\right]^{1/2}$$ (6.197)

Thus, it is easy to show from Equations 6.196 and 6.197 that

$$\tan\phi = \frac{s_1 - s_3}{(\text{sign } s_2)\left[4s_2^2 - (s_1 + s_3)^2\right]^{1/2}}$$ (6.198)

We can see that this phase expression is insensitive to the signal frequency; hence, the result is not affected by detunings. The unknown signal frequency can then be found with:

$$\omega = \frac{1}{x_0} \cos^{-1}\left(\frac{s_1 + s_3}{2s_2}\right) \qquad (6.199)$$

6.10.5 Correlation with Two Sinusoidal Signals in Quadrature

In Chapter 5, we studied the synchronous detection method utilizing multiplication of the signal by two orthogonal sinusoidal reference functions with the same frequency as the signal. Let us now assume that the two reference orthogonal functions have a different frequency (ω_r) than the signal. The parameters S and C are not constants; instead, we now have:

$$S(x) = s(x)\sin(\omega_r x) = a\sin(\omega_r x) + b\cos(\phi + \omega x)\sin(\omega_r x)$$

$$= a\sin(\omega_r x) + \frac{b}{2}\sin\big(\phi + (\omega + \omega_r)x\big) - \frac{b}{2}\sin\big(\phi + (\omega - \omega_r)x\big) \qquad (6.200)$$

and

$$C(x) = s(x)\cos(\omega_r x) = a\cos(\omega_r x) + b\cos(\phi + \omega x)\cos(\omega_r x)$$

$$= a\cos(\omega_r x) + \frac{b}{2}\cos\big(\phi + (\omega + \omega_r)x\big) + \frac{b}{2}\cos\big(\phi + (\omega - \omega_r)x\big) \qquad (6.201)$$

These two functions contain three spatial frequencies, the reference frequency, the sum of the reference and the signal frequencies, and their difference. If we apply a low-pass filter, so that only the term with the frequency difference remains, we obtain the filtered versions of $S(x)$ and $C(x)$ as:

$$\overline{S}(x) = S(x)\sin\omega_{rx} \qquad (6.202)$$

and

$$\overline{C}(x) = C(x)\cos\omega_{rx} \qquad (6.203)$$

Thus, we can obtain:

$$\tan\left(\phi + (\omega_S - \omega_r)x\right) = -\frac{\overline{S}(x)}{\overline{C}(x)} \qquad (6.204)$$

which is possible only if the reference frequency (f_r) is higher than half the signal frequency:

$$\omega_r > \frac{\omega}{2} \qquad (6.205)$$

but ideally both frequencies should be equal.

The low-pass filtering process is performed by means of a convolution with a filtering function, $h(x)$. Then, the values of $S(x)$ and $C(x)$ can be expressed by:

$$\overline{S}(x) = \int_{-\infty}^{\infty} s(\alpha)\sin(\omega_r\alpha)h(x-\alpha)d\alpha \qquad (6.206)$$

and

$$\overline{C}(x) = \int_{-\infty}^{\infty} s(\alpha)\cos(\omega_r\alpha)h(x-\alpha)d\alpha \qquad (6.207)$$

The filtering function must be selected so the term with the lowest frequency (the difference term) remains; hence, we can also write:

$$\phi + (\omega - \omega_r)x = -\tan^{-1}\left(\frac{S(x)}{C(x)}\right) \qquad (6.208)$$

6.11 ALGORITHM SUMMARY

In this section, we describe some of the main properties of phase-detecting algorithms.

6.11.1 Detuning Sensitivity

We have seen in Chapter 4 that by shifting the sampling point locations we can obtain an algorithm in which the Fourier transforms of the reference sampling functions are either orthogonal or have the same magnitudes at all frequencies. We have also seen that the sensitivity to detuning is not affected by this shifting of the sampling points.

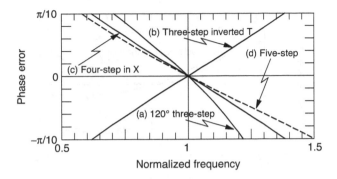

Figure 6.83 Detuning sensitivity for four algorithms: (a) 120° three-step, (b) three-step inverted T, (c) four-step in X, and (d) five-step.

The detuning sensitivity for some of the main algorithms described in this chapter are now described. In the following figures, the peak phase error is represented by the quantity in front of the sine function in Equation 5.154. Figure 6.83 illustrates the detuning errors for four algorithms. The first plot (Figure 6.83a) is for the 120° three-step algorithm. This is the algorithm with the largest error. The second plot (Figure 6.83b) is for the three-step inverted T algorithm. In this case, the sign of the error is opposite the sign of that in Figure 6.83a. The third plot (Figure 6.83c) is for the four-step X algorithm. The fourth plot (Figure 6.83d) is for the five-step algorithm. This phase error is the smallest of the four algorithms, but not by much.

Figure 6.84 shows the detuning phase error for some symmetrical (N +1) algorithms. The first plot (Figure 6.84a) is for the four-step (3 +1) algorithm, and we can detect sensitivity to detuning in the plot. If this algorithm is compensated with the extra sampling weights described before (Figure 6.84b), the sensitivity to detuning is reduced, as the slope of the curve is zero at the origin. The next plot (Figure 6.84c) is for the popular Schwider–Hariharan five-step (4+1) algorithm, where the insensitivity to detuning is clearly seen to be better than in the four-step (3 + 1) algorithm. The six-step (5 + 1) algorithm is not compensated by the extra sampling weights; thus, some

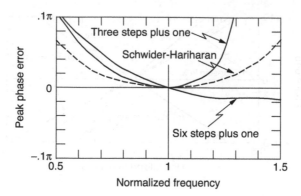

Figure 6.84 Detuning sensitivity for five symmetrical $N + 1$ algorithms: (a) uncompensated four-step $(3 + 1)$, (b) compensated four-step, (c) Schwider–Hariharan five-step $(4 + 1)$, (d) uncompensated six-step $(5 + 1)$, and (e) uncompensated seven-step $(6 + 1)$.

detuning sensitivity is present. Finally, the seven-step $(6 + 1)$ algorithm also has some detuning sensitivity because it is also uncompensated. If compensated, this algorithm features the lowest detuning sensitivity. Figure 6.85 shows the detuning sensitivities for the Schwider–Hariharan, Schmit–Creath, Servín, and Malacara-Dorrío algorithms.

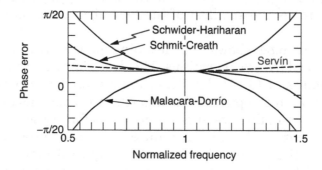

Figure 6.85 Detuning sensitivities for the Schwider–Hariharan, Schmit–Creath, Servín, and Malacara–Dorrío algorithms.

TABLE 6.2 Sensitivity to Signal Harmonics of Some Algorithms

Algorithm	Harmonics Being Suppressed								
	2	3	4	5	6	7	8	9	10
Three-points (120° or T)	—	Y	—	—	Y	—	—	Y	—
Three-point (Wyant's)	—	—	Y	—	—	—	Y	—	—
Four-point (X or cross)	Y	—	Y	—	Y		Y	—	—
Five-point	Y	Y	—	Y	—	Y	Y	—	Y
Symmetrical four-point (3 + 1)	—	Y	—	—	Y	—	—	Y	—
Symmetrical five-point (4 + 1)	Y	Y	Y	—	Y	—	Y	Y	—
Symmetrical six-point (5 + 1)	Y	Y	Y	—	Y	—	Y	Y	—
Symmetrical seven-point (6 + 1)	Y	Y	Y	—	Y	—	Y	Y	Y
Schwider	—	—	Y	—	—	—	Y	—	—
Schmit–Creath	—	—	Y	—	—	—	Y	—	—

6.11.2 Harmonic Sensitivity

The harmonic sensitivities for some of the algorithms described in this chapter are summarized in Table 6.2.

REFERENCES

Angel, J.R.P. and Wizinowich, P.L., A method of phase shifting in the presence of vibration, *Eur. Southern Obs. Conf. Proc.*, 30, 561, 1988.

Bhushan, B., Wyant, J.C., and Koliopoulos, C.L., Measurement of surface topography of magnetic tapes by Mirau interferometry, *Appl. Opt.*, 24, 1489–1497, 1985.

Carré, P., Installation et Utilisation du Comparateur Photoelectrique et Interferentiel du bureau International des Poids et Measures, *Metrologia*, 2, 13–23, 1966.

Cheng, Y.-Y. and Wyant, J.C., Phase shifter calibration in phase-shifting interferometry, *Appl. Opt.*, 24, 30–49, 1985.

Creath, K., Comparison of phase measuring algorithms, *Proc. SPIE*, 680, 19–28, 1986.

Creath, K., Phase measuring interferometry: beware these errors, *Proc. SPIE*, 1553, 213–220, 1991.

de Groot, P., Derivation of algorithms for phase shifting interferometry using the concept of a data-sampling window, *Appl. Opt.*, 34, 4723–4730, 1995.

Freischlad, K. and Koliopoulos, C. L., Fourier description of digital phase measuring interferometry, *J. Opt. Soc. Am. A*, 7, 542–551, 1990.

Greivenkamp, J.E. and Bruning, J.H., Phase shifting interferometers, in *Optical Shop Testing*, Malacara, D., Ed., John Wiley & Sons, New York, 1992.

Hariharan, P., Areb, B.F., and Eyui, T., Digital phase-shifting interferometry: a simple error-compensating phase calculation algorithm, *Appl. Opt.*, 26, 2504– 2505, 1987.

Hibino, K., Phase-shifting algorithms for nonlinear spatially nonuniform phase shifts, *J. Opt. Soc. Am.*, 14, 919–930, 1997.

Hibino, K., Error-compensating phase measuring algorithms in a Fizeau interferometer, *Opt. Review*, 6, 529–538, 1999.

Hibino, K. and Yamauchi, M., Phase-measuring algorithms to suppress spatially nonuniform phase modulation in a two beam interferometer, *Opt. Rev.*, 7, 543–549, 2000.

Hibino, B., Oreb, F., and Farrant, D.I., Phase shifting for non-sinusoidal waveforms with phase shift errors, *J. Opt. Soc. Am. A*, 12, 761–768, 1995.

Joenathan, C., Phase measuring interferometry: new methods and error analysis, *Appl. Opt.*, 33, 4147–4155, 1994.

Larkin, K.G., New seven sample symmetrical phase-shifting algorithm, *Proc. SPIE*, 1755, 2–11, 1992.

Larkin, K.G. and Oreb, B.F., Design and assessment of symmetrical phase-shifting algorithm, *J. Opt. Soc. Am.*, 9, 1740–1748, 1992.

Malacara-Doblado, D. and Vazquez-Dorrío, B., Family of detuning insensitive phase shifting algorithms, *J. Opt. Soc. Am. A*, 17, 1857–1863, 2000.

Mendoza-Santoyo, F., Kerr, D., and Tyrer, J.R., Interferometric fringe analysis using a single phase step technique, *Appl. Opt.*, 27, 4362–4364, 1988.

Morimoto, Y. and Fujisawa, M., Fringe pattern analysis by a phase-shifting method using Fourier transform, *Opt. Eng.*, 33, 3709–3714, 1994.

Nakadate, S., Phase detection of equidistant fringes for highly sensitive optical sensing, I. Principle and error analysis, *J. Opt. Soc. Am. A*, 5, 1258–1264, 1988a.

Nakadate, S., Phase detection of equidistant fringes for highly sensitive optical sensing, II. Experiments, *J. Opt. Soc. Am. A*, 5, 1265–1269, 1988b.

Parker, D.H., Moiré patterns in three-dimensional Fourier space, *Opt. Eng.*, 30, 1534–1541, 1991.

Ransom, P.L. and Kokal, J.B., Interferogram analysis by a modified sinusoid fitting technique, *Appl. Opt.*, 25, 4199–4204, 1986.

Rastogi, P.K., Modification of the Carré phase-stepping method to suit four-wave hologram interferometry, *Opt. Eng.*, 32, 190–191, 1993.

Schmit, J. and Creath, K., Extended averaging technique for derivation of error-compensating algorithms in phase-shifting interferometry, *Appl. Opt.*, 34, 3610–3619, 1995.

Schmit, J. and Creath, K., Window function influence on phase error in phase-shifting algorithms, *Appl. Opt.*, 35, 5642–5649, 1996.

Schwider, J., Burow, R., Elssner, K.-E., Grzanna, J., Spolaczyk, R., and Merkel, K., Digital wave-front measuring interferometry: some systematic error sources, *Appl. Opt.*, 22, 3421–3432, 1983.

Schwider, J., Falkenstörfer, O., Schreiber, H., Zöller, A., and Streibl, N., New compensating four-phase algorithm for phase-shift interferometry, *Opt. Eng.*, 32, 1883–1885, 1993.

Servín, M. and Cuevas, F.J., A novel technique for spatial phase-shifting interferometry, *J. Mod. Opt.*, 42, 1853–1862, 1995.

Servín, M., Malacara, D., Marroquín, J.L., and Cuevas, F.J., Complex linear filters for phase shifting with very low detuning sensitivity, *J. Mod. Opt.*, 44, 1269–1278, 1997.

Surrel, Y., Phase stepping: a new self-calibrating algorithm, *Appl. Opt.*, 32, 3598–3600, 1993.

Surrel, Y., Design of algorithms for phase measurements by the use of phase stepping, *Appl. Opt.*, 35, 51–60, 1996.

Surrel, Y., Phase-shifting algorithms for nonlinear and spatially nonuniform phase shifts [comment], *J. Opt. Soc. Am. A*, 15, 1227–1233, 1998.

Wizinowich, P.L., Phase shifting interferometry in the presence of vibration: a new algorithm and system, *Appl. Opt.*, 29, 3271–3279, 1990.

Wyant, J.C., Koliopoulos, C.L., Bushan, B., and George, D.E., An optical profilometer for surface characterization of magnetic media, *ASLE Trans.*, 27, 101, 1984.

Zhao, B. and Surrel, Y., Phase shifting: six-sample self-calibrating algorithm insensitive to the second harmonic in the fringe signal, *Opt. Eng.*, 34, 2821–2822, 1995.

7

Phase-Shifting Interferometry

7.1 PHASE-SHIFTING BASIC PRINCIPLES

Early phase-shifting interferometric techniques can be traced back to Carré (1966), but their further development and application were later reported by Crane (1969), Moore (1973), and Bruning et al. (1974), among others. These techniques have also been applied to speckle-pattern interferometry (Creath, 1985; Nakadate and Saito, 1985; Robinson and Williams, 1986) and to holographic interferometry (Nakadate et al., 1986; Stetson and Brohinski, 1988), and many reviews of this field have been published (e.g., Greivenkamp and Bruning, 1992).

In phase-shifting interferometers, the reference wavefront is moved along the direction of propagation with respect to the wavefront being analyzed, thus changing the phase differences. By measuring the irradiance changes for various phase shifts, it is possible to determine the phase for a wavefront, relative to the reference wavefront, for the measured point on that wavefront. The irradiance signal, $s(x,y)$, at point (x,y) in the detector changes with the phase:

$$s(x, y, \alpha) = a(x, y) + b(x, y)\cos(\alpha + \phi(x, y)) \qquad (7.1)$$

where $\phi(x,y)$ is the phase at the origin, and α is a known phase shift with respect to the origin. By measuring the phase for many points over the wavefront, the complete wavefront shape is thus determined.

If we consider any fixed point in the interferogram, the phase difference between the two wavefronts must be changed. We might wonder, though, how this is possible, because relativity does not permit either of the two wavefronts to move faster than the other, as the phase velocity is c for both waves. It has been shown (Malacara et al., 1969), however, that the Doppler effect occurs, producing a shift in both frequency and wavelength. The two beams, with different wavelengths, interfere with each other, producing beats. These beats can also be interpreted as changes in irradiance due to the continuously changing phase difference. These two conceptually different models are physically equivalent.

The change in the phase, then, can be accomplished if the frequency of one of the beams is modified during the process. This is possible in a continuous fashion using some devices, but for only a relatively short period of time with other devices. This fact has led to the following problem in semantics: When the frequency can be modified in a permanent way, some people refer to such instruments as AC, heterodyne, or frequency-shift interferometers; otherwise, the instrument is considered a phase-shifting interferometer. Here, we will refer to *all* of these instruments as phase-shifting interferometers.

7.2   AN INTRODUCTION TO PHASE SHIFTING

The procedure just described can be implemented using almost any kind of two-beam interferometer, such as, for example, Twyman–Green or Fizeau interferometers. The phase can be shifted in several different ways, as reviewed by Creath (1988).

7.2.1   Moving Mirror with a Linear Transducer

One method is to move the mirror for the reference beam along the light trajectory by means of an electromagnetic or piezoelectric transducer, as shown in Figure 7.1 for a Twyman–

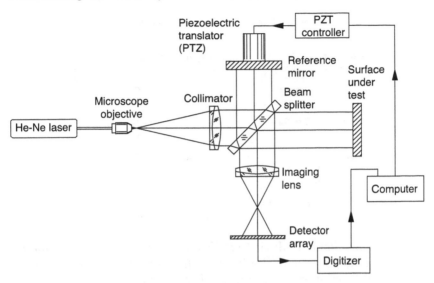

Figure 7.1 Twyman–Green interferometer with a phase-shifting transducer.

Green interferometer. The transducer moves the mirror so the phase is changed to a new value, as shown in Figure 7.2a. Alternatively, one can think of the reflected light as Doppler-shifted light. A piezoelectric transducer (PZT) typically has a linear displacement of over 1 μm (2λ). Voltages ranging from zero to a few hundred volts are used to produce the displacement.

7.2.2 Rotating Glass Plate

Another method for shifting the phase is to insert a plane-parallel glass plate in the light beam (Wyant and Shagam, 1978), as shown in Figure 7.2b. The phase shift (α) introduced by this glass plate, when tilted by angle θ with respect to the optical axis, is given by:

$$\alpha = \frac{t}{k}(n\cos\theta' - \cos\theta) \tag{7.2}$$

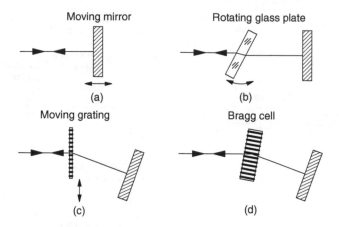

Figure 7.2 Some methods to shift the phase in an interferometer: (a) mirror moving along the light path, (b) rotating glass plate, (c) moving diffraction grating, and (d) Bragg cell.

where t is the plate thickness, n is its refractive index, and $k = 2\pi/\lambda$. The angles θ and θ' are the angles between the normal to the glass plate and the light rays outside and inside the plate, respectively. A rotation of the plate that increases angle θ also increases the optical path difference; thus, if the plate is rotated a small angle $(\Delta\theta)$, the phase shift (α) is given by:

$$\alpha = \frac{t}{k}\left(1 - \frac{1}{n}\frac{\cos\theta}{\cos\theta'}\right)\sin\theta\,\Delta\theta \qquad (7.3)$$

An important requirement in this method is that the plate must be inserted in a collimated light beam to avoid introducing aberrations.

7.2.3 Moving Diffraction Grating

Another way to shift the phase is to use a diffraction grating or ruling moving perpendicularly to the light beam (Suzuki and Hioki, 1967; Stevenson, 1970; Bryngdahl, 1976; Srinivasan et al., 1985) as shown in Figure 7.2c. It is easy to see that the phase of the diffracted light beam is shifted $n \times 2\pi$ the number of slits that pass through a fixed point, where n

represents the order of diffraction. Thus, the shift in the frequency is equal to n times the number of slits in the grating that pass through a fixed point within a unit of time. Put differently, the shift in the frequency is equal to the speed of the grating divided by period d of the grating. It is interesting to note that the frequency is increased for the light beams diffracted in the same direction as the movement of the grating. Light beams diffracted in the direction opposite that of the movement of the grating decrease in frequency. As expected, the direction of the beam is changed because the first-order beam must be used and the zero-order beam must be blocked by means of a properly placed diaphragm.

If the diffraction grating is moved a small distance (Δy), then the phase changes by an amount (α) given by:

$$\alpha = \frac{2\pi n}{d} \Delta y \tag{7.4}$$

where d is the period of the grating and n is the order of diffraction.

A Ronchi ruling moving perpendicularly to its lines in the Ronchi test is a particular case of a moving diffraction grating. This method has been used by several researchers (e.g., Indebetow, 1978) under the name of *running projection fringes.*

A similar method utilizes diffraction of light by means of an acoustic optic Bragg cell (Massie and Nelson, 1978; Wyant and Shagam, 1978; Shagam, 1983), as shown in Figure 7.2d. An acoustic transducer produces ultrasonic vibrations in the liquid of the cell. These vibrations produce periodic changes in the refractive index, inducing the cell to act as a thick diffraction grating. This thickness effect makes this diffraction device an efficient one for the desired order of diffraction.

7.2.4 Rotating Phase Plate

The phase can also be shifted by means of a rotating plane-parallel glass plate (Crane, 1969; Okoomian, 1969; Bryngdahl, 1972; Sommargren, 1975; Shagam and Wyant, 1978; Hu, 1983; Zhi, 1983; Kothiyal and Delisle, 1984, 1985; Salbut

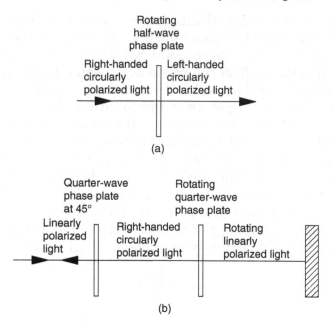

Figure 7.3   Polarized light device to shift the phase.

and Patorski, 1990) as shown in Figure 7.3. If a beam of circularly polarized light goes through a half-wave phase plate, the direction of the circular polarization is reversed, as shown in Figure 7.3a. If the half-wave phase plate rotates, the frequency of the light changes. If the plate rotates in a continuous manner, the frequency change ($\Delta\nu$) is equal to twice the frequency of rotation of the plate. If the phase plate is rotated a small angle ($\Delta\theta$), the phase changes by α as follows:

$$\alpha = 2\Delta\theta \tag{7.5}$$

This arrangement works if the light passes through the phase plate only once; however, in a Twyman–Green interferometer, the light passes through the system twice, so the configuration shown in Figure 7.3b is used. The first quarter-wave retarding plate is stationary, with its slow axis located at 45° with respect to the plane of polarization of the incident linearly polarized light. This plate also transforms the returning circularly polarized light back to being linearly polarized. The second phase retarder is also a quarter-wave plate, but it rotates and the light passes through it twice, so it really acts as a half-wave plate.

7.2.5 Moiré in an Interferogram with a Linear Carrier

Let us consider an interferogram with a large linear carrier — that is, with many fringes produced by means of a reference wavefront tilt. If a Ronchi ruling or a similar linear ruling with about the same number of fringes is placed on top of this interferogram, a moiré fringe appears (see Chapter 9). This moiré represents the interferogram with the linear carrier removed. The phase of this interferogram can be changed by moving the superimposed linear ruling. The phase changes by an amount equal to 2π if the linear ruling is moved perpendicular to the fringes a distance equal to its period. This phase-shifting scheme has been described by Kujawinska et al. (1991) and Dorrío et al. (1995a,b). The Ronchi ruling is placed on top of the interferogram to produce multiplication of the interferogram irradiance by the ruling transmission. In principle, this ruling can be implemented by computer software, but information about very high spatial frequencies must be stored in the computer memory, thus making the system quite inefficient. It is advisable, then, to use a real Ronchi ruling and perform spatial filtering of the high frequencies before the light detector. The low-pass filtering can be performed by defocusing the lens to form an interferogram image on the light detector.

7.2.6 Frequency Changes in the Laser Light Source

Another method for producing the phase shift is to shift the frequency of the laser light source. This shift can be done in two possible ways, one of which is to illuminate the interferometer with a Zeeman frequency split laser line. The frequency of the laser is split into two orthogonally polarized output frequencies by means of a DC magnetic field (Burgwald and Kruger, 1970). The frequency separation of the two spectral lines is of the order of 2 to 5 MHz in a helium–neon laser. In the interferometer system, the two lines travel different paths and the plane of polarization of one of them is rotated to produce the interference. Another method is to use an unbalanced interferometer (i.e., one with a large optical path difference) and a laser diode for which the frequency is controlled by an injected electrical current, as proposed by Ishii et al. (1991) and later studied by Onodera and Ishii (1996). This method is based on the fact that the phase difference in an interferometer is proportional to the product of the optical path difference (OPD) and its temporal frequency and that varying one of them will produce a piston phase change.

7.2.7 Simultaneous Phase-Shift Interferometry

Phase-shifting methods in an environment with vibrations cannot give good results due to the long time required to take all the measurements. This problem has been avoided by the use of interferometer systems in which all the necessary interferometer frames are taken at the same time (Kujawinska, 1987, 1993; Kujawinska and Robinson, 1988, 1989; Kujawinska et al., 1990). One approach is to use multichannel interferometers (Kwon, 1984); an interferometer in a Mach–Zehnder configuration produces three frames at the same time by means of a diffraction grating. Kwon and Shough (1985) and Kwon et al. (1987) used radial shear interferometers, also in Mach–Zehnder or triangular configurations, with a diffraction grating. Bareket (1985) and Koliopoulos (1991) have also designed other simultaneous or multiple-channel phase-shift interferometers. The great disadvantage of these arrangements is the complicated and expensive hardware that is required. Also, exact pixel-to-pixel correlation between the images is required.

7.3 PHASE-SHIFTING SCHEMES AND PHASE MEASUREMENT

We have seen in Chapter 1 that the signal is a sinusoidal function of the phase, as shown in Figure 1.2. In phase-shifting interferometers, the wavelength of the signal to be detected is

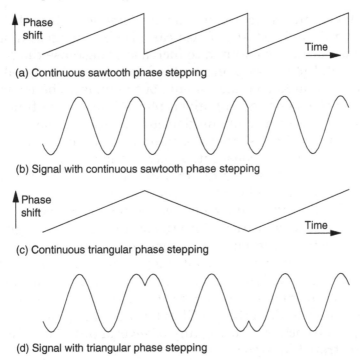

(a) Continuous sawtooth phase stepping

(b) Signal with continuous sawtooth phase stepping

(c) Continuous triangular phase stepping

(d) Signal with triangular phase stepping

Figure 7.4 Signals obtained in phase-shifting interferometry.

equal to the wavelength of the illuminating light. The basic problem is to determine the nonshifted phase difference between the two waves with the highest possible precision. This can be done by any of several procedures described here.

The best method for determining the phase depends on many factors, but primarily on how the phase shift was performed. The phase can be changed in a continuous manner by introducing a permanent frequency shift in the reference beam. Some authors refer to this as a *heterodyne interferometer*. As described by Moore (1973), heterodyne interferometry has three possible basic approaches: (1) the frequency is permanently shifted, and the signal output is continuous; (2) the phase is changed in a sinusoidal manner (Figure 7.4a) to obtain the signal shown in Figure 7.4b; or (3) the phase is changed in a triangular manner (Figure 7.4c) to obtain the symmetrical signal shown in Figure 7.4d.

When the synchronous phase-detection algorithms in Chapter 5 are used, the phase can also be changed in steps, in a discontinuous manner, to increase or decrease the phase. The digital phase-stepping method measures the signal values at several known increments of the phase. The measurement of the signal at any given phase takes some time, due to the time response of the detector; hence, the phase must be stationary for a short time in order to take the measurement. Between two consecutive measurements, the phase can change as quickly as desired in order to get to the next phase with the smallest delay. One problem with the phase-stepping method is that the sudden changes in the mirror position can introduce some vibrations into the system. In the integrating bucket method, the phase changes continuously, not by discrete steps. The detector continuously measures the irradiance during a fixed time interval, without stopping the mirror; hence, an average value during the measuring time interval is measured, as described in Chapter 3. A change of the phase, thus, can be achieved using any of several different schemes, as illustrated in Figure 7.5.

Some analog methods can also be used to measure the relative irradiance phase at different interferogram points — for example, detection of the zero crossing point of the phase (Crane, 1969) or the phase-lock method (Moore et al., 1978). In the zero crossing method, the phase is detected by locating the phase point where the signal passes through the axis of symmetry of the function, not really zero, which has a signal value equal to a. The points crossing the axis of symmetry can be found by amplifying the signal function to saturation levels so the sinusoidal signal becomes a square function. Digital phase-stepping methods are used more extensively than analog methods, however.

7.4 HETERODYNE INTERFEROMETRY

When the phase shift is continuous, we speak of heterodyne or DC interferometry. As pointed out before, two equivalent models can describe the phase shift: (1) a change in the optical path difference, or (2) a change in the frequency of one of the

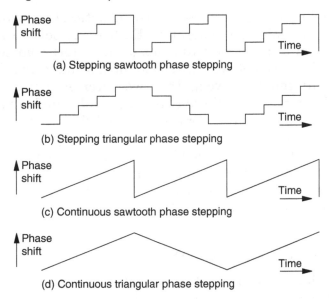

(a) Stepping sawtooth phase stepping

(b) Stepping triangular phase stepping

(c) Continuous sawtooth phase stepping

(d) Continuous triangular phase stepping

Figure 7.5 Four different ways to shift the phase periodically.

two interfering light beams. In this case, the most common interpretation is that of two different interfering frequencies, and we consider heterodyning beats. If we measure the relative phase of these beats at different points over the wavefront, we obtain the wavefront deformations. The phase of the detected beats is measured in real time using electronics hardware instead of by sampling the irradiance (Wyant, 1975; Massie, 1978, 1980, 1987; Massie and Nelson, 1978; Massie et al., 1979; Sommargren, 1981; Hariharan et al., 1983; Hariharan, 1985; Thalmann and Dändliker, 1985). The great advantage of this approach is that a fast measurement is achieved which is important in many applications, such as dynamical systems. Beat frequencies of the order of 1 MHz can be obtained, so a high-speed detector is necessary. A standard television camera cannot be used; instead, a high-frame-rate image tube (also called an *image dissector tube*) can be used.

Smythe and Moore (1983, 1984) proposed an alternative heterodyne interferometric system in which the beats are not measured; instead, by means of an optical procedure (not described here) that utilizes polarizing optics, two orthogonal bias-free signals are generated. Each of these two signals comes from each of the two arms of the interferometer. The phase difference between these two orthogonal signals is the phase difference between the two interferometer optical paths. If we represent these two orthogonal signals in a polar diagram, one along the vertical axis and the other along the horizontal axis, the path described in this diagram when the phase is continually changed is a circle. The angle with respect to the optical axis is the phase. This heterodyning procedure can be easily implemented to measure wavefront deformations in two dimensions.

7.5 PHASE-LOCK DETECTION

In the phase-lock method for detecting a signal, the phase reference wave is phase modulated with a sinusoidally oscillating mirror (Moore, 1973; Moore et al., 1978; Johnson et al., 1979; Moore and Truax, 1979). Two phase components — δ_0 and $\delta_1 \sin(\omega t)$ — are added to the signal phase, $\phi(x,y)$. One of the additional phase components being added has a fixed value and the other a sinusoidal time oscillation. Both components are independent and can have any desired value. Omitting the x,y dependence for notational simplicity, the total time-dependent phase is:

$$\phi + \delta_0 + \delta_1 \cos(2\pi ft) \tag{7.6}$$

thus, the signal is:

$$s(t) = a + b\cos(\phi + \delta_0 + \delta_1 \cos(2\pi ft)) \tag{7.7}$$

The phase modulation is carried out only in an interval smaller than π, as illustrated in Figure 7.6. The output signal can be interpreted as the phase-modulating signal, after being harmonically distorted by the signal to be detected. This harmonic distortion is a function of the phase (ϕ), as shown in Figure 7.7. This function is periodic and symmetrical; thus, to find the harmonic distortion using Equations 2.6 and 2.7, this function can now be expanded in series as:

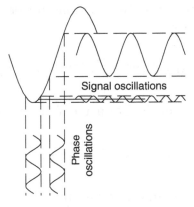

Figure 7.6 Phase lock detection of the signal phase.

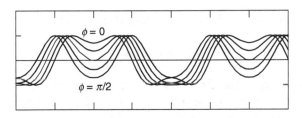

Figure 7.7 Output of an harmonically distorted signal, where $\delta = 0.75\pi$.

$$s(t) = \frac{c_0}{2} + \sum_{n=1}^{\infty} c_n \cos(2\pi nft) \qquad (7.8)$$

where:

$$c_n = \frac{1}{t_0} \int_{-t_0}^{t_0} s(t)\cos(2\pi nft)dt \qquad (7.9)$$

Then, making the variable substitution $\theta = 2\pi ft$, we can show that:

$$c_n = \frac{b}{\pi} e^{i(\phi+\delta_0)} \int_0^{\pi} e^{i(\delta_1 \cos\theta)} \cos(n\theta)d\theta +$$

$$+ \frac{b}{\pi} e^{-i(\phi+\delta_0)} \int_0^{\pi} e^{-i(\delta_1 \cos\theta)} \cos(n\theta)d\theta \qquad (7.10)$$

On the other hand, the Bessel function of the first kind, of order n, is given by:

$$J_n(\delta) = \frac{1}{\pi} e^{-in\pi/2} \int_0^{\pi} e^{i(\delta_1 \cos\theta)} \cos(n\theta) d\theta \tag{7.11}$$

Using this expression in Equation 7.10, we obtain:

$$c_n = 2b J_n(\delta) \cos\left(\phi + n\frac{\pi}{2}\right) \tag{7.12}$$

Hence, the output signal is given by:

$$s(x,y) = a +$$

$$+ b\cos(\phi(x,y) + \delta_0)\left[J_0(\delta_1) - 2J_2(\delta_1)\cos(2\omega t) + ...\right] \tag{7.13}$$

$$+ b\sin(\phi(x,y) + \delta_0)\left[2J_1(\delta_1)\sin(\omega t) - 2J_3(\delta_1)\sin(3\omega t) + ...\right]$$

where $\omega = 2\pi f$. The first part of this expression represents harmonic components of even order, and the second part represents harmonic components of odd order.

Let us now assume that the amplitudes of the phase oscillation component $\delta_1 \sin(\omega t)$ are much smaller than π. Then, if we adjust the δ_0 component to a value such that $\phi + \delta_0 = n\pi$, then $\sin(\phi + \delta_0)$ is zero and only even harmonics remain. This effect is illustrated in Figure 7.6, near one of the minima of the signal $s(x,y)$. This is done in practice by slowly changing the value of the phase component δ_0 while maintaining the oscillation $\delta_1 \sin(\omega t)$ until the minimum amplitude of the first harmonic (fundamental frequency) is obtained. We now have $\phi + \delta_0 = n\pi$, and because the value of δ_0 is known the value of ϕ has been determined.

This method can also be used at the inflection point for the sinusoidal signal function (Figure 7.7) by changing the fixed phase component until the first harmonic reaches its maximum amplitude. From Equation 7.12 we obtain:

$$\tan\phi = \frac{c_1}{c_2}\frac{J_2(\delta)}{J_1(\delta)} \tag{7.14}$$

Thus, because the Bessel function values are known, if the value of δ is also known, the signal phase can be determined if the ratio of the amplitudes of the fundamental component to the second harmonic component is measured. This measurement can be performed analogically by means of electronic hardware. Matthews et al. (1986) used this method with a null detection method instead of a maximum detection procedure. One disadvantage of this method is that a two-dimensional array of detectors cannot be used. A single detector must move to scan the entire picture.

7.6 SINUSOIDAL PHASE OSCILLATION DETECTION

Sasaki and Okasaki (1986a,b) and Sasaki et al. (1987) proposed a sinusoidal phase-modulating interferometer in which the reference wave is phase modulated with a sinusoidally oscillating mirror, as in the phase-lock method just described. The main difference is that the phase determination is performed with a digital sampling procedure. The modulated phase is:

$$\phi + \delta \cos(2\pi ft + \theta) \tag{7.15}$$

which differs from Equation 7.6 in that the constant phase value is not present and an extra term (θ) has been added. The value of θ is the phase of the phase-shifter oscillation at $t = 0$. It will be shown later that $\theta = 0$ is not the best value. Sasaki and Okasaki (1986a) added an extra random phase term $n(t)$ to this expression to consider the presence of multiplicative noise due to disturbing effects such as system vibrations. They derived the optimum values of the amplitude (δ) and phase (θ) of the oscillating driving signal by considering minimization of the effects of noise. For notational simplicity, we did not add this term here; thus, the modulated signal to be measured is:

$$s(t) = a + b \cos(\phi + \delta \cos(2\pi ft + \theta)) \tag{7.16}$$

This function is periodic but asymmetric ($\theta = 0$) and can be written as:

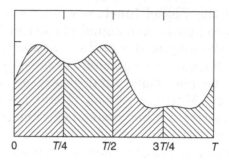

Figure 7.8 Interval integrating sampling of harmonic distorted signal at four points.

$$s(t) = a + b\cos\phi\cos(\delta\cos(2\pi ft + \theta)) - $$
$$-b\sin\phi\sin(\delta\cos(2\pi ft + \theta)) \tag{7.17}$$

This signal contains a large amount of signal harmonics. A phase-detecting sampling algorithm different from those studied in Chapter 6 can be used to take into account the presence of these harmonics. Four sampling measurements with 90° separation and interval averaging (as described in Chapter 2) are used. The integrating interval has a width of 90°, equal to the sampling point separation. This integration eliminates most harmonic content above the third harmonic. The associated filter function has its first zero at the frequency of the fourth harmonic. The second and third harmonic remain. As shown in Figure 7.8, the averaged signal measurements are:

$$\bar{s}_i = a + (b\cos\phi)C_i - (b\sin\phi)S_i \tag{7.18}$$

with

$$C_i = \frac{1}{4T}\int_{(i-1)T/4}^{iT/4} \cos(\delta\cos(2\pi ft + \theta)) \tag{7.19}$$

and

$$S_i = \frac{1}{4T}\int_{(i-1)T/4}^{iT/4} \sin(\delta\cos(2\pi ft + \theta)) \tag{7.20}$$

where T is the signal period.

Sasaki and Okasaki (1986a) found the expressions for C_i to be:

$$C_1 = C_3 = J_0(\delta) + \frac{4}{\pi} \sum_{n=1}^{\infty} \frac{J_{2n}(\delta)}{2n} \left[1 - (-1)^n\right] \sin(2n\pi) \quad (7.21)$$

and

$$C_2 = C_4 = J_0(\delta) - \frac{4}{\pi} \sum_{n=1}^{\infty} \frac{J_{2n}(\delta)}{2n} \left[1 - (-1)^n\right] \sin(2n\pi) \quad (7.22)$$

and the values of S_i to be:

$$S_1 = -S_3$$
$$= -\frac{4}{\pi} \sum_{n=1}^{\infty} \frac{J_{2n-1}(\delta)}{(2n-1)} \left[(-1)^n \sin(2(n-1)\pi) + \cos(2(n-1)\pi)\right] \quad (7.23)$$

and

$$S_2 = -S_4$$
$$= -\frac{4}{\pi} \sum_{n=1}^{\infty} \frac{J_{2n-1}(\delta)}{(2n-1)} \left[(-1)^n \sin(2(n-1)\pi) - \cos(2(n-1)\pi)\right] \quad (7.24)$$

The signal phase can then be proved to be:

$$\tan\phi = \frac{(C_1 - C_2)}{(S_1 + S_2)} \frac{\bar{s}_1 - \bar{s}_2 + \bar{s}_3 - \bar{s}_4}{\bar{s}_1 + \bar{s}_2 - \bar{s}_3 - \bar{s}_4} \quad (7.25)$$

and the optimum values of δ and θ are $\delta = 0.78\pi = 2.45$ and $\theta = 56°$.

According to Sasaki et al. (1987), this interferometric phase demodulation system yields a measurement accuracy of the order of 1.0 to 1.5 nm. Sasaki et al. (1990a) used a laser diode as a light source with a reference fringe pattern and electronic feedback to the laser current. In this manner, they eliminated noise due to variations in the laser intensity and to object vibrations. Zhao et al. (2004) used a charged-coupled device (CCD) as an image sensor to integrate the light. By changing the injection current in the laser diode light source, its frequency can be shifted to change the interference phase. Sinusoidal phase-modulating schemes can be implemented in Twyman–Green and Fizeau interferometers (Sasaki et al., 1990b).

7.7 PRACTICAL SOURCES OF PHASE ERROR

In Chapter 5, we studied some sources of systematic and random error produced by algorithm calculations when some important sources of instrument error must be taken into account. In this section, we describe some other practical sources of phase error that might be present in phase-shifting interferometers.

7.7.1 Vibration and Air Turbulence

Two important sources of error in phase-shifting interferometry are vibration and air turbulence. Their nature and consequences have been studied by many researchers (e.g., Kinnstaetter et al., 1988; Crescentini, 1989; Wingerden et al., 1991; de Groot, 1995; de Groot and Deck, 1996; Deck, 1996). It is desirable to apply as many preventive measures as possible in order to the reduce these two disturbing factors to a minimum. If the vibration frequency is high enough, with an average period higher than the integration time of the detector (which is of the order of 1/60th of a second), then the interference fringes are washed out, their contrast reduced.

Using an approach similar to the mathematical treatment for phase-lock and sinusoidal phase oscillation detection, de Groot and Deck (1996) studied the effects of noise by considering the signal to be phase modulated with the noise, as follows:

$$s(t) = a + b\cos(\alpha + \phi + n(t)) \qquad (7.26)$$

This expression is not restricted to any particular case of vibrational noise; however, some insight can be gained by

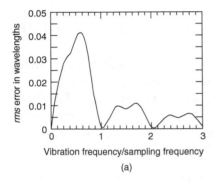
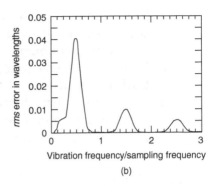

Figure 7.9 Vibrational root mean square (*rms*) error for two different algorithms: (a) three sampling points algorithm, and (b) seven sampling points algorithm. (From de Groot de, P. and Deck, L.L., *Appl. Opt.*, 35, 2173–2181, 1996. With permission.)

assuming that the noise is of a sinusoidal nature, with amplitude δ and phase offset θ, as follows:

$$s(t) = a + b\cos(\alpha + \phi + \delta\cos(2\pi ft + \theta)) \qquad (7.27)$$

In a linear approximation, if the noise is not sinusoidal but the amplitudes are small, we can sum the contributions from each of the Fourier components of the vibration (de Groot and Deck, 1996). When the noise amplitudes are not small, nonlinear couplings between these components can occur. In general, the phase of the noise vibration is not coherent but varies at random; thus, it is more logical to express the phase error as the root mean square (*rms*) value of the disturbed phase. This *rms* error varies sinusoidally with the phase of the signal and has twice the frequency of the signal.

Numerical simulations have been performed by de Groot and Deck (1996) to calculate the effect of vibrational noise for several phase-detecting algorithms. Figure 7.9 shows the *rms* error for two of these algorithms. In the figure, we can observe the following general, interesting facts that are valid for most algorithms:

1. The maximum vibrational sensitivity occurs when the vibration has a frequency equal to one half the sampling frequency.
2. Zeros of the sensitivity occur at vibration frequencies that are multiples of the sampling frequency.
3. The sensitivity decreases exponentially for high vibrational frequencies. If the frequency is extremely high, only the contrast is reduced, but its dependence on the signal phase is lost.

Brophy (1990) studied the effect of additive noise, particularly mechanical vibrations with frequencies that were extremely high or of the order of the sampling rate.

An immediate practical consequence of these findings is that, to reduce the effect of the vibrations, the sampling rate has to be as high as possible with respect to the vibration frequency. Unfortunately, high sampling rates require light detectors with a low integration time, which are quite expensive. As an alternative, Deck (1996) proposed an interferometer with two light detectors, one with a fast integration time and the other with a low integration time, to reduce the interferometer sensitivity to vibrations. Another approach to eliminating the effect of vibrations is to take the necessary irradiance samples at the same time, not in sequence (Kwon, 1984; Kwon and Shough, 1985; Kujawinska, 1987; Kujawinska and Robinson, 1988, 1989; Kujawinska et al., 1990).

7.7.2 Multiple-Beam Interference and Frequency Mixing

Signal harmonics can also occur in the interference process if more than two beams are interfering. In many cases, this effect is due to the nature of the interferometer; in other cases, it is accidental. Typical examples of multiple-beam interferometers include the Ronchi test and Newton or Fizeau interferometers with high-reflection beam splitters; however, even if the beam splitter in the Fizeau interferometer has a very low reflectance, it is impossible to reduce multiple reflections to absolute zero. Multiple reflections can occur by accident, due to spurious unwanted reflections. The influence of these spurious reflections has been considered by several authors (e.g., Bruning et al., 1974; Schwider et al., 1983; Hariharan et al., 1987; Ai and Wyant, 1988; Dorrío et al., 1996).

In Chapter 1, we studied a signal (irradiance) due to two beams with amplitudes A_1 and A_2. If, following Schwider et al. (1983) and Ai and Wyant (1988), we add a third coherent beam with amplitude B due to the coherent noise, we obtain:

$$E = A_1 \exp(i\phi) + A_2 \exp(i\alpha) + B\exp(i\beta) \tag{7.28}$$

where ϕ is the signal phase, α is the sampling reference function phase, and β is the extraneous coherent wave phase. The phases of these beams are referred to the same origin as the sampling reference functions. We also assume an absence of detuning, so the reference wavefront can be considered to have the same phase as the reference sampling function. Thus, the signal (irradiance) in the presence of coherent noise is given by:

$$s' = E \cdot E^* = A_1^2 + A_2^2 + B^2 + 2A_1A_2 \cos(\phi - \alpha) + \\ + 2A_1B\cos(\phi - \beta) + 2A_2B\cos(\beta - \alpha) \tag{7.29}$$

or

$$s' = s + B^2 + 2A_1B\cos(\phi - \beta) + 2A_2B\cos(\beta - \alpha)$$

$$= s + B^2 + 2A_1B\cos(\phi - \beta) + 2A_2B\cos\beta\cos(\alpha) + \tag{7.30}$$

$$+ 2A_2B\sin\beta\sin\alpha$$

Now we will study the particular case of algorithms with equally and uniformly spaced sampling points. In this case, the phase of the signal without coherent noise, from Equation 5.19, is:

$$\tan\phi = \frac{\sum\limits_{n=1}^{N} s\sin(\alpha_n)}{\sum\limits_{n=1}^{N} s\cos(\alpha_n)} \tag{7.31}$$

where α_n is the value of phase α for sampling point n. Taking into account the presence of the coherent noise, we have:

$$\tan\phi' = \frac{\sum_{n=1}^{N} s' \sin(\alpha_n)}{\sum_{n=1}^{N} s' \cos(\alpha_n)} \tag{7.32}$$

Thus, using Equations 5.11, 5.13, and 5.14, we find:

$$\tan\phi' = \frac{\sin\phi + \dfrac{B}{A_1}\sin\beta}{\cos\phi + \dfrac{B}{A_1}\cos\beta} \tag{7.33}$$

and the phase error is given by:

$$\tan(\phi' - \phi) = -\frac{\dfrac{B}{A_1}\sin(\phi - \beta)}{1 + \dfrac{B}{A_1}\cos(\phi - \beta)} \tag{7.34}$$

This phase error is a periodic, although not exactly sinusoidal, function of the signal phase. Its period is equal to that of the signal frequency. This phase error is illustrated in Figure 7.10.

This phase error can thus be substantially reduced by averaging two sets of measurements with a phase difference $(\phi - \beta)$ of π between them. This is possible only if another phase shifter is placed in the object beam. A phase shift in the reference beam does not change the phase difference $\phi - \beta$. Ai and Wyant (1988) pointed out that, if the spurious light comes from the reference arm in the interferometer or from the test surface, this method does not work, and they proposed an alternative way to eliminate the error.

In a Fizeau interferometer, as explained by Hariharan et al. (1987), the spurious light appears to be due to multiple reflections between the object being analyzed and the reference surface (beam splitter). In this case, the error can be

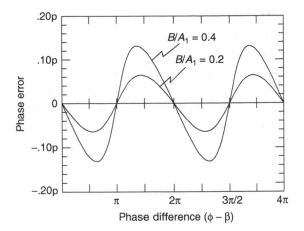

Figure 7.10   Phase error due to the presence of spurious coherent light beams.

minimized by proper selection of the sampling algorithm to eliminate the signal harmonics being generated.

Speckle noise is another kind of coherent noise that can become important in some applications, such as, for example, speckle interferometry. This kind of noise can also be reduced in some cases (Creath, 1985; Slettemoen and Wyant, 1986).

7.7.3   Spherical Reference Wavefronts

If the reference wavefront in phase-shifting interferometry is not planar it is spherical, as in the spherical Fizeau interferometer, where the spherical surface being analyzed is shifted to introduce the phase shift. If the phase shift at the center of the fringe pattern is 90°, the phase shift at the edge of the pupil would be slightly smaller. A phase error is introduced, as pointed out by Moore and Slaymaker (1980) and Schwider et al. (1983); nevertheless, this error is not large. For spherical test surfaces with numerical apertures smaller than 0.8, the phase error introduced can be smaller than one hundredth of a wavelength. If this error becomes important, it can be minimized using Carré's algorithm.

7.7.4 Quantization Noise

As we studied in Section 3.4, in the digitization of images the number of bits used to digitize the image defines the number of gray levels. A simple method to evaluate the quantization error has been provided by Brophy (1990), who demonstrated a correlation between signal samples taken 90° apart. He showed that, for algorithms for which samples are taken at 90° intervals, the *rms* error (σ) due to quantization into Q gray levels is given by:

$$\sigma = \frac{a}{\sqrt{3}bQ} \qquad (7.35)$$

where a and b are the bias and amplitude, respectively, of the signal. For example, if 8 bits are used, Q is equal to 256 gray levels. Then, if a/b is equal to one, the *rms* quantization error (σ) is equal to 0.00036 wavelengths, or about $\lambda/2777$. This value is so small that it is difficult to reach this limit. Zhao and Surrel (1997) made a detailed study of quantization noise for several algorithms.

Of course, the fringe contrast is not always perfect, and the ratio of a/b can be much larger than one. To avoid this error, the signal must cover as much of the detector dynamic range as possible.

7.7.5 Photon Noise Phase Errors

Other random phase errors include, for example, photon noise (Koliopoulos, 1981; Brophy, 1990; Freischlad and Koliopoulos, 1991). This error occurs due to fluctuations in the arrival frequency of the photons to the light detector when the number of photons is not large. In other words, this noise appears where the signal is relatively small.

7.7.6 Laser Diode Intensity Modulation

When a phase shift is produced by phase current modulation of a laser diode in an unbalanced interferometer an amplitude modulation also occurs simultaneously with the phase modulation, as described in Section 7.2.6. The phase error introduced by this undesired intensity modulation has been studied by Onodera and Ishii (1996) and by Surrel (1997), assuming that the irradiance variation is linear with the phase shift.

7.8 SELECTION OF THE REFERENCE SPHERE IN PHASE-SHIFTING INTERFEROMETRY

When digitizing an interferogram with a detector array, the sampling theorem requires the minimum local fringe spacing or period to be greater than twice the pixel separation; thus, each detector has a minimum fringe period that can be allowed. This minimum period, in turn, is set by the wavefront asphericity and the testing method. This section discusses the optimum defocusing and tilt necessary to test aspherical wavefronts for which the asphericity is as large as possible in a non-null-test configuration (Malacara-Hernández et al., 1996).

A general expression for an aspherical wavefront deformation, $W(S)$, for different focus shifts and only a primary spherical aberration is:

$$W(S) = aS^2 + bS^4 \tag{7.36}$$

where a is the defocusing term and b is the primary spherical aberration coefficient. Figure 7.11 shows the wavefront deformation (W) values for three different focus settings to be described later. The first derivative, $W'(S)$, with respect to S is the radial slope of this wavefront, as given by:

$$W'(S) = \frac{dW(S)}{dS} = 2aS + 4bS^3 \tag{7.37}$$

These radial derivatives for the three focus positions are illustrated in Figure 7.12. If we plot this wavefront slope, $W'(S)$, any change in the focus or in the amount of tilt can be easily represented in this graph. As shown in Figure 7.13, a tilt is a vertical displacement of the curve, and a change in the focus is represented by a small rotation of the graph about the origin. The wavefront can be measured with respect to many reference spheres by selection of the defocusing coefficient a. Here, we will study the three main possibilities.

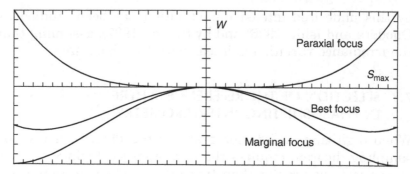

Figure 7.11 Aspherical wavefront deformations at paraxial focus, best focus, and marginal focus with primary spherical aberration.

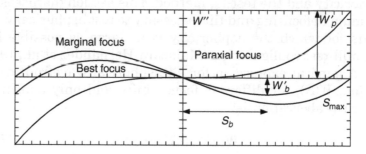

Figure 7.12 Wavefront radial slopes at the paraxial focus, best focus, and marginal focus for a wavefront with primary spherical aberration. The maximum radial slope for the best focus is at S_b and at the edge of the pupil.

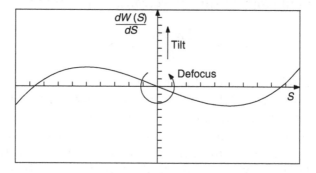

Figure 7.13 Tilt and defocus effect on the derivative of a wavefront. A defocus rotates the curve about the origin, and a tilt displaces the curve vertically.

7.8.1 Paraxial Focus

The paraxial focus is defined by a zero defocusing coefficient ($a = 0$), and the slope of the wavefront measured with respect to a sphere with its center at the paraxial focus is:

$$W_p'(S) = 4bS^3 \qquad (7.38)$$

Then, the maximum slope of the wavefront at the paraxial focus $W_{p\text{max}}'$ occurs at the edge of the pupil; that is, $S = S_{\text{max}}$. Thus,

$$W_{p\text{max}}' = W_p'(S_{\text{max}}) = 4bS_{\text{max}}^3 \qquad (7.39)$$

where S_{max} is the semidiameter of the wavefront.

7.8.2 Best Focus

The best focus is defined as the focus setting that minimizes the absolute value of the maximum radial slope over the pupil. This maximum slope occurs at the edge (S_{max}) of the pupil and at some intermediate pupil radius (S_b) but with opposite values. Opposite signs but the same magnitude for the radial slope means that the transverse aberrations $TA(S_b)$ and $TA(S_{\text{max}})$ are also equal in magnitude but with opposite signs. This is the condition for the waist of the caustic; hence, the optimum or best focus occurs when the center of the reference sphere is located at the waist of the caustic, as illustrated in Figure 7.14. Thus, we can write:

$$W_{b\text{max}}' = -W'(S_b) = W'(S_{\text{max}}) \qquad (7.40)$$

After some algebraic manipulation using this condition for the first derivative as well as the condition that the second derivative of W is zero, it is possible to show that at this focus setting the defocusing coefficient (a) is related to the primary aberration coefficient (b) by the expression:

$$2\left(\frac{b}{a}\right)S^3 + \frac{2}{3}\left(-\frac{a}{6b}\right)^{1/2} + S = 0 \qquad (7.41)$$

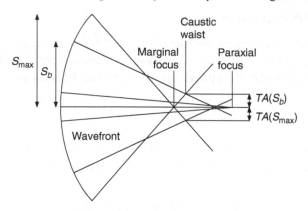

Figure 7.14 Aspherical wavefront and its caustic, showing the paraxial, marginal, and best focus.

Solving this equation, it is possible to find that at the best focus the defocusing coefficient is given by:

$$a = -\left(\frac{3}{2}\right)bS_{\max}^2 \tag{7.42}$$

Then, it is easy to see that the ratio between the maximum wavefront deformation at the paraxial focus and at the best focus positions is a constant given by:

$$\frac{W'_{p\max}}{W'_{b\max}} = 4 \tag{7.43}$$

7.8.3 Marginal Focus

The wavefront slope $W'_m(S_{\max})$ at the marginal focus and the edge of the pupil has to be zero; thus,

$$W'_m(S_{\max}) = 2aS_{\max} + 4bS_{\max}^3 = 0 \tag{7.44}$$

Hence, the defocusing coefficient (a) at the marginal focus is:

$$a = -2bS_{\max}^2 \tag{7.45}$$

and the first radial derivative of the wavefront at the marginal focus is:

$$W_b'(S) = 4b\left(S^3 - S_{\max}^2 S\right) \tag{7.46}$$

Then, the maximum slope value of this wavefront deformation is given by equating to zero the second radial derivative with respect to S. Thus, we obtain a value for the radial position (S_m) of this maximum wavefront deformation at the marginal focus:

$$S_{m\,\max} = \frac{S_{\max}}{\sqrt{3}} \tag{7.47}$$

so that

$$W_{b\,\max}' = W'(S_b) = -\frac{8}{3\sqrt{3}} b S_{\max}^3 \tag{7.48}$$

The ratio between the slope maxima at the paraxial and at the marginal foci can be shown to be:

$$\frac{W_{p\,\max}'}{W_{m\,\max}'} = -2.6 \tag{7.49}$$

7.8.4 Optimum Tilt and Defocusing in Phase-Shifting Interferometry

The optimum tilt magnitude and reference sphere (defocusing) for the different interferogram analysis methods can now be estimated using these results. The sampling theorem requires the minimum local fringe spacing or period to be greater than twice the pixel separation. Thus, each detector has a minimum fringe period that can be allowed (see Table 7.1). This minimum period, in turn, is set by the wavefront asphericity and the testing method, as pointed out by Creath and Wyant (1987). The fringe period, $s(S)$, or its fringe frequency, $f(S)$, in the interferogram is related to the wavefront slope by the relation:

TABLE 7.1 Relative Minimum Fringe Periods for Wavefronts and Three Methods for Interferometric Analysis

Interferometric Analysis Method	Wavefront Focus	Wavefront Tilt	Relative Minimum Fringe Period
Temporal phase-shifting techniques	Paraxial	None	1.0
	Best	None	4.0
	Marginal	None	2.6
Spatial linear carrier demodulation	Paraxial	Yes	0.5
	Best	Yes	2.0
	Marginal	Yes	1.3
Circular spatial circular carrier demodulation	Marginal	None	2.6

Note: The relative fringe period is defined as the ratio of the minimum fringe spacing for the focus setting to that of the paraxial focus setting.

$$f(S) = \frac{1}{s(S)} = \frac{W'(S)}{\lambda} \qquad (7.50)$$

On the other hand, from geometrical optics, the slope, $W'(S)$, of the wavefront is related to the ray transverse aberration by:

$$W'(S) = \frac{TA(S)}{r} \qquad (7.51)$$

where r is the radius of curvature of the reference wavefront. The maximum wavefront slope, $W'_{p\max}$, with a paraxial focus setting is related to the maximum wavefront deformation with the focus setting $W_{p\ \max}$ by means of the relation:

$$W'_{p\max} = 4\frac{W_{p\max}}{S_{\max}} \qquad (7.52)$$

The maximum fringe frequency and the minimum fringe period (spacing) at this paraxial focus (without any tilt) occurs at the edge of the fringe pattern and is given by:

$$f_{p\max} = \frac{1}{s_{p\min}} = \frac{W'_{p\max}}{\lambda} = 4\frac{W_{p\max}}{\lambda S_{\max}} = 4\frac{n_p}{S_{\max}} \qquad (7.53)$$

where $sp_{\min}$ is the period with $f_{p\max}$, and n_p is the number of fringes at the paraxial focus without any tilt.

The condition to maximize the minimum fringe period is equivalent to minimizing the peak ray transverse aberration, which occurs at the best focus position. On the other hand, the best focus position is obtained when the center of the reference sphere is at the center of the waist of the caustic. In this case, the maximum fringe frequency and the minimum fringe spacing are given by:

$$f_{b\max} = \frac{1}{s_{b\min}} = \frac{W'_{b\max}}{\lambda} \qquad (7.54)$$

The ratio $s_{b\max}/s_{p\max}$ is:

$$\frac{s_{b\max}}{s_{p\min}} = 4 \qquad (7.55)$$

This result tells us that at the best focus position the minimum fringe period or fringe spacing is increased by a factor of four with respect to the paraxial focus setting. The relative fringe period will be defined as the ratio of the minimum fringe spacing for the focus setting under consideration to that of the paraxial focus setting. This is a useful advantage when testing aspheric wavefronts.

7.8.4.1 Temporal Phase-Shifting Techniques

In this case, no tilt is necessary but the focus can be adjusted with any value. Let us consider the following three focus possibilities:

1. *Paraxial focus* — In this case, the minimum fringe period is defined as the unit ($\eta = 1$). A phase-shifting method can be used, but to obtain the maximum asphericity capacity this focus setting is not the optimum.

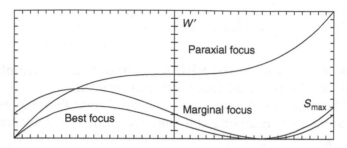

Figure 7.15 Effect on the radial wavefront slope of introducing tilt in a wavefront until the derivative of the wavefront is positive everywhere.

  2. *Best focus* — At the best focus, we obtain the maximum possible value for the local minimum fringe period of all configurations. This, then, is the optimum focus for testing the maximum degree of asphericity.
  3. *Marginal focus* — With this focus setting, the relative minimum fringe period is equal to 2.6 — better than the paraxial focus but worse than the best focus.

7.8.4.2 Spatial Linear Carrier Demodulation

These methods (described further in Chapter 8) require the introduction of a large linear carrier in the x direction. The minimum magnitude of this carrier is such that the phase increases (or decreases) in a monotonic manner with x. This condition is necessary to avoid closed loop fringes. This is possible if a tilt is introduced so that W' is always positive, as shown in the plots in Figure 7.15. In this case, the minimum slope is zero, so, ideally, a tilt larger than this must be used, but this is the minimum value. Three focus possibilities exist:

  1. *Paraxial focus* — If a tilt is introduced at the paraxial focus in order to introduce the linear carrier, the maximum local wavefront slope is increased by a factor of two, reducing the relative minimum fringe period to 0.5. A demodulation of these fringes with a spatial carrier can be performed, but this is not the

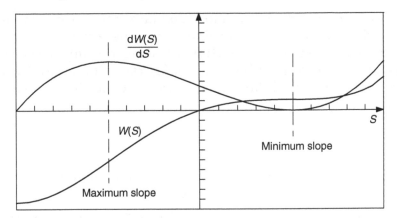

Figure 7.16 Wavefront and its radial slope at the best focus position, showing where the minimum slope occurs.

ideal amount of defocusing for achieving the maximum possible local minimum fringe period to obtain the maximum testing asphericity capacity.

2. *Best focus* — If a tilt is introduced at the best focus, we obtain the maximum possible local minimum fringe period attainable with a linear carrier, as shown in Figure 7.16. This is the ideal configuration for analyzing the fringe pattern with a modulated linear carrier.

3. *Marginal focus* — If the proper tilt is introduced at the marginal focus, a linear carrier demodulation scheme can be used; however, this is not the ideal configuration for this method. The relative fringe period is now equal to 1.2.

7.8.4.3 Spatial Circular Carrier Demodulation

(This method is described in detail in Chapter 8.) Here, no tilt is introduced, because the circular symmetry must be preserved. A focus term must be selected so the phase monotonically increases (or decreases) from the center toward the edge of the interferogram. From the three focus positions described here, only the marginal focus position is acceptable as a minimum. Ideally, a defocusing larger than this amount should be used. At the marginal focus, the wavefront radial slope does have any sign changes along the interferogram semidiameter; thus, this is the configuration to be used with radial carrier modulation. The relative minimum fringe period is equal to 2.6.

REFERENCES

Ai, C. and Wyant, J.C., Effect of piezoelectric transducer nonlinearity on phase shift interferometry, *Appl. Opt.*, 26, 1112–1116, 1987.

Ai, C. and Wyant, J.C., Effect of spurious reflection on phase shift interferometry, *Appl. Opt.*, 27, 3039–3045, 1988.

Bareket, N., Three-channel phase detector for pulsed wavefront sensing, *Proc. SPIE*, 551, 12–16, 1985.

Brophy, C.P., Effect of intensity error correlation on the computed phase of phase shifting, *J. Opt. Soc. Am. A*, 7, 537–541, 1990.

Bryngdahl, O., Polarization type interference fringe shifter, *J. Opt. Soc. Am.*, 62, 462–464, 1972.

Bryngdahl, O., Heterodyne shearing interferometers using diffractive filters with rotational symmetry, *Opt. Comm.*, 17, 43, 1976.

Bruning, J.H., Herriott, D.R., Gallagher, J.E., Rosenfeld, D.P., White, A.D., and Brangaccio, D.J., Digital wavefront measuring interferometer for testing surfaces and lenses, *Appl. Opt.*, 13, 2693–2703, 1974.

Burgwald, G.M. and Kruger, W.P., An instant-on laser for distant measurement, *Hewlett-Packard J.*, 21, 14, 1970.

Carré, P., Installation et Utilisation du Comparateur Photoelectrique et Interferentiel du Bureau International des Poids et Measures, *Metrologia*, 2, 13–23, 1966.

Chang, M., Hu, C.P., and Wyant, J.C., Phase shifting holographic interferometry, *Proc. SPIE*, 599, 149–159, 1985.

Cheng, Y.-Y. and Wyant, J.C., Multiple wavelength phase shifting interferometry, *Appl. Opt.*, 24, 804–807, 1985.

Crane, R., Interference phase measurement, *Appl. Opt.*, 8, 538–542, 1969.

Creath, K., Phase-shifting speckle interferometry, *Appl. Opt.*, 24, 3053–3058, 1985.

Creath, K., Phase-measurement interferometry techniques, in *Progress in Optics*, Vol. XXVI, Wolf, E., Ed., Elsevier Science, Amsterdam, 1988.

Creath, K. and Wyant, J.C., Aspheric measurement using phase shifting interferometry, *Proc. SPIE*, 813, 553–554, 1987.

Crescentini, L., Fringe pattern analysis in low-quality interferograms, *Appl. Opt.*, 28, 1231–1234, 1989.

de Groot, P., Vibration in phase-shifting interferometry, *J. Opt. Soc. Am. A*, 12, 354–365, 1995 (*errata*, 12, 2212, 1995).

de Groot, P. and Deck, L.L., Numerical simulations of vibration in phase-shifting interferometry, *Appl. Opt.*, 35, 2173–2181, 1996.

Deck, L., Vibration-resistant phase-shifting interferometry, *Appl. Opt.*, 34, 6555–6662, 1996.

Dorrío, B.V., Doval, A.F., López, C., Soto, R., Blanco-García, J., Fernández, J.L., and Pérez Amor, M., Fizeau phase-measuring interferometry using the moiré effect, *Appl. Opt.*, 34, 3639–3643, 1995a.

Dorrío, B.V., Blanco-García, J., Doval, A.F., López, C., Soto, R., Bugarín, J., Fernández, J.L., and Pérez Amor, M., Surface evaluation combining the moiré effect and phase-stepping techniques in Fizeau interferometry, *Proc. SPIE*, 2730, 346–349, 1995b.

Dorrío, B.V., Blanco-García, J., López, C., Doval, A.F., Soto, R., Fernández, J.L., and Pérez Amor, M., Phase error calculation in a Fizeau interferometer by Fourier expansion of the intensity profile, *Appl. Opt.*, 35, 61–64, 1996.

Freischlad, K. and Koliopoulos, C.L., Fourier description of digital phase measuring interferometry, *J. Opt. Soc. Am. A*, 7, 542–551, 1990.

Greivenkamp, J.E. and Bruning, J.H., Phase shifting interferometry, in *Optical Shop Testing*, 2nd ed., Malacara, D., Ed., John Wiley & Sons, New York, 1992.

Hariharan, P., Quasi-heterodyne hologram interferometry, *Opt. Eng.*, 24, 632–638, 1985.

Hariharan, P., Oreb, B.F., and Brown, N., Real-time holographic interferometry: a microcomputer system for the measurement of vector displacements, *Appl. Opt.*, 22, 876–880, 1983.

Hariharan, P., Oreb, B.F., and Eiju, T., Digital phase shifting interferometry: a simple error compensating phase calculator algorithm, *Appl. Opt.*, 26, 2504–2506, 1987.

Hu, H.Z., Polarization heterodyne interferometry using simple rotating analyzer. 1. Theory and error analysis, *Appl. Opt.*, 22, 2052–2056, 1983.

Ishii, Y., Chen, J., and Murata, K., Digital phase measuring interferometry with a tunable laser diode, *Opt. Lasers Eng.*, 14, 293–309, 1991.

Indebetow, G., Profile measurement using projection of running fringes, *Appl. Opt.*, 17, 2930–2933, 1978.

Johnson, G.W., Leiner, D.C., and Moore, D.T., Phase locked interferometry, *Opt. Eng.*, 18, 46–52, 1979.

Kinnstaetter, K., Lohmann, A., Schwider, W., and Streibl, J.N., Accuracy of phase shifting interferometry, *Appl. Opt.*, 27, 5082–5089, 1988.

Koliopoulos, C.L., Interferometric Optical Phase Measurement Techniques, Ph.D. dissertation, University of Arizona, Tucson, 1981.

Koliopoulos, C.L., Simultaneous phase shift interferometer, *Proc. SPIE*, 1531, 119–133, 1991.

Kothiyal, M.P. and Delisle, C., Optical frequency shifter for heterodyne interferometry using counterrotating wave plates, *Opt. Lett.*, 9, 319–321, 1984.

Kothiyal, M.P. and Delisle, C., Rotating analyzer heterodyne interferometer: error analysis, *Appl. Opt.*, 24, 2288–2290, 1985.

Kujawinska, M., Multichannel grating phase-stepped interferometers, *Optica Applicata*, 17, 313–332, 1987.

Kujawinska, M., Spatial phase measurement methods, in *Interferogram Analysis*, Robinson, D.W. and Reid, G.T., Eds., Institute of Physics, Philadelphia, PA, 1993.

Kujawinska, M. and Robinson, D.W., Multichannel phase-stepped holographic interferometry, *Appl. Opt.*, 27, 312–320, 1988.

Kujawinska, M. and Robinson, D.W., Comments on the error analysis and adjustment of the multichannel phase-stepped holographic interferometers, *Appl. Opt.*, 28, 828–829, 1989.

Kujawinska, M., Salbut, L., and Patorski, K., Three channel phase stepped system for moiré interferometry, *Appl. Opt.*, 29, 1633–1636, 1990.

Kujawinska, M., Salbut, L., and Jozwicki, R., Moiré and spatial carrier approaches to phase shifting interferometry, *Proc. SPIE*, 1553, 44–54, 1991.

Kwon, O.Y., Multichannel phase shifted interferometer, *Opt. Lett.*, 9, 59–61, 1984.

Kwon, O.Y. and Shough, D.M., Multichannel grating phase shift interferometer, *Proc. SPIE*, 599, 273–279, 1985.

Kwon, O.Y., Shough, D.M., and Williams, R.A., Stroboscopic phase-shifting interferometry, *Opt. Lett.*, 12, 855–857, 1987.

Malacara, D., Rizo, I., and Morales, A., Interferometry and the Doppler effect, *Appl. Opt.*, 8, 1746–1747, 1969.

Malacara-Hernández, D., Malacara, Z., and Servín, M., Digitization of interferograms of aspheric wavefronts, *Opt. Eng.*, 35, 2102–2105, 1996.

Massie, N.A., Heterodyne interferometry, in *Optical Interferograms: Reduction and Interpretation*, Guenther, A.H. and Liedbergh, D.H., Eds., ASTM Symp. Tech. Publ. 666, American Society for Testing and Materials, West Conshohocken, PA, 1978.

Massie, N.A., Real time digital heterodyne interferometry: a system, *Appl. Opt.*, 19, 154–160, 1980.

Massie, N.A., Digital heterodyne interferometry, *Proc. SPIE*, 816, 40–48, 1987.

Massie, N.A. and Nelson, R.D., Beam quality of acousto-optic phase shifters, *Opt. Lett.*, 3, 46–47, 1978.

Massie, N.A., Nelson, R.D., and Holly, S., High performance real-time heterodyne interferometry, *Appl. Opt.*, 18, 1797–1803, 1979.

Matthews, H.J., Hamilton, D.K., and Sheppard, C.J.R., Surface profiling by phase locked interferometry, *Appl. Opt.*, 25, 2372–2374, 1986.

Moore, D.T., Gradient Index Optics and Tolerancing, Ph.D. thesis, University of Rochester, New York, 1973.

Moore, D.T. and Truax, B.E., Phase locked moiré fringe analysis for automated contouring of diffuse surfaces, *Appl. Opt.*, 18, 91–96, 1979.

Moore, D.T., Murray, R., and Neves, F.B., Large aperture AC interferometer for optical testing, *Appl. Opt.*, 17, 3959–3963, 1978.

Moore, R.C. and Slaymaker, F.H., Direct measurement of phase in a spherical Fizeau interferometer, *Appl. Opt.*, 19, 2196–2200, 1980.

Nakadate, S. and Saito, H., Fringe scanning speckle pattern interferometry, *Appl. Opt.*, 24, 2172–2180, 1985.

Nakadate, S., Saito, H., and Nakajima, T., Vibration measurement using phase-shifting stroboscopic holographic interferometry, *Opt. Acta*, 33, 1295–1309, 1986.

Okoomian, H.J., A two beam polarization technique to measure optical phase, *Appl. Opt.*, 8, 2363–2365, 1969.

Onodera, R. and Ishii, Y., Phase-extraction analysis of laser-diode phase-shifting interferometry that is insensitive to changes in laser power, *J. Opt. Soc. Am. A*, 13, 139–146, 1996.

Robinson, D. and Williams, D., Digital phase stepping speckle interferometry, *Opt. Commun.*, 57, 26, 1986.

Salbut, L. and Patorski, K., Polarization phase shifting method for moiré interferometry and flatness testing, *Appl. Opt.*, 29, 1471–1476, 1990.

Sasaki, O. and Okasaki, H., Sinusoidal phase modulating interferometry for surface profile measurement, *Appl. Opt.*, 25, 3137–3140, 1986a.

Sasaki, O. and Okasaki, H., Analysis of measurement accuracy in sinusoidal phase modulating interferometry, *Appl. Opt.*, 25, 3152–3158, 1986b.

Sasaki, O., Okasaki, H., and Sakai, M., Sinusoidal phase modulating interferometer using the integrating-bucket method, *Appl. Opt.*, 26, 1089–1093, 1987.

Sasaki, O., Okamura, T., and Nakamura, T., Sinusoidal phase modulating Fizeau interferometer, *Appl. Opt.*, 29, 512–515, 1990a.

Sasaki, O., Takahashi, K., and Susuki, T., Sinusoidal phase modulating laser diode interferometer with a feedback control system to eliminate external disturbance, *Opt. Eng.*, 29, 1511–1515, 1990b.

Schwider, J., Burow, R., Elssner, K.-E., Grzanna, J., Spolaczyk, R., and Merkel, K., Digital wave-front measuring interferometry: some systematic error sources, *Appl. Opt.*, 22, 3421–3432, 1983.

Shagam, R.N., AC measurement technique for moiré interferograms, *Proc. SPIE*, 429, 35, 1983.

Shagam, R.N. and Wyant, J.C., Optical frequency shifter for heterodyne interferometers using multiple rotating polarization retarders, *Appl. Opt.*, 17, 3034–3035, 1978.

Slettemoen, G.Å. and Wyant, J.C., Maximal fraction of acceptable measurements in phase shifting interferometry: a theoretical study, *J. Opt. Soc. Am. A*, 3, 210–214, 1986.

Smythe, E.R. and Moore, R., Instantaneous phase measuring interferometry, *Proc. SPIE*, 429, 16–21, 1983.

Smythe, E.R. and Moore, R., Instantaneous phase measuring interferometry, *Opt. Eng.*, 23, 361–364, 1984.

Sommargren, G.E., Up–down frequency shifter for optical heterodyne interferometry, *J. Opt. Soc. Am.*, 65, 960–961, 1975.

Sommargren, G.E., Optical heterodyne profilometry, *Appl. Opt.*, 200, 610–618, 1981.

Srinivasan, V., Liu, H.C., and Halioua, M., Automatic phase-measuring profilometry: a phase measuring approach, *Appl. Opt.*, 24, 185–188, 1985.

Stetson, K.A. and Brohinsky, W.R., Phase shifting technique for numerical analysis of time average holograms of vibrating objects, *J. Opt. Soc. Am. A*, 5, 1472–1476, 1988.

Stevenson, W.H., Optical frequency shifting by means of a rotating diffraction grating, *Appl. Opt.*, 9, 649–652, 1970.

Surrel, Y., Design of phase detection algorithms insensitive to bias modulation, *Appl. Opt.*, 36, 1–3, 1997.

Susuki, T. and Hioki, R., Translation of light frequency by a moving grating, *J. Opt. Soc. Am.*, 57, 1551, 1967.

Thalmann, R. and Dändliker, R., Holographic contouring using electronic phase measurement, *Opt. Eng.*, 24, 930–935, 1985.

Wingerden van Johanes, H., Frankena, J., and Smorenburg, C., Linear approximation for measurement errors in phase shifting interferometry, *Appl. Opt.*, 30, 2718–2729, 1991.

Wyant, J.C., Use of an AC heterodyne lateral shear interferometer with real time wavefront correction systems, *Appl. Opt.*, 14, 2622–2626, 1975.

Wyant, J.C. and Shagam, R.N., Use of electronic phase measurement techniques in optical testing, *Proc. ICO-11 (Madrid)*, 659–662, 1978.

Zhao, B. and Surrel, Y., Effect of quantization error on the computed phase of phase-shifting measurements, *Appl. Opt.*, 36, 2070–2075, 1997.

Zhao, X., Susuki, T., and Sasaki, O., Sinusoidal phase-modulating laser diode interferometer capable of accelerated operations on four integrating buckets, *Opt. Eng.*, 43, 678–682, 2004.

Zhi, H., Polarization heterodyne interferometry using a simple rotating analyzer. 1. Theory and error analysis, *Appl. Opt.*, 22, 2052–2056, 1983.

8

Spatial Linear and Circular Carrier Analysis

8.1 SPATIAL LINEAR CARRIER ANALYSIS

In phase-shifting techniques several frames must be measured. This requires shifting the phase by means of piezoelectric crystals or any other equivalent device. In the spatial carrier methods described in this chapter, only a single frame is necessary to obtain the wavefront, although, if desired, several wavefronts can be averaged to improve the result. These two basic methods have several important practical differences:

1. In phase-shifting methods, at least three interferogram frames are needed. In spatial-carrier methods, only one is necessary.
2. In phase-shifting interferometry, three or more frames must be taken simultaneously to avoid the effects of vibrations. In spatial-carrier analysis, vibrations are not a problem, as only one frame is taken.
3. In phase-shifting methods, the sign of the wavefront deformations is determined. In spatial carrier methods, the sign cannot be determined, as only one frame is taken. To determine the sign it is necessary to know the sign of at least one of the aberration wavefront

components — for example, the sign of the tilt introducing the carrier.

4. In phase-shifting methods, hardware requirements are greater, as an accurately calibrated phase shifter is needed. In spatial carrier methods, more sophisticated mathematical processing by computer is necessary.

5. If a stable environment, free of vibrations and turbulence, is available (which sometimes is impossible), greater accuracy and precision are possible with phase-shifting methods than with spatial carrier methods.

8.1.1　Introduction of a Linear Carrier

A large tilt about the y-axis in an interferogram can be considered to be a linear carrier in the x direction. Interferograms with a spatial linear carrier can be analyzed to obtain the wavefront shape by processing the information in the interferogram plane (space domain) or in the Fourier plane (frequency domain). We will study both methods in this chapter. For reviews on the analysis of interferograms using a spatial carrier, see Takeda (1987), Kujawinska (1993), and Vlad and Malacara (1994).

　　The irradiance in an interferogram with a large tilt along a line parallel to the x-axis is a perfectly sinusoidal function if the two interfering wavefronts are flat. In other words, if the reference wavefront is flat and the wavefront under analysis is also flat, then the fringes are straight, parallel to the y-axis, and equidistant. If the wavefront being analyzed is not perfect, then this irradiance function is a nearly sinusoidal function with phase modulation. The phase modulation is due to the wavefront deformations, $W(x,y)$. If a tilt (θ) about the y-axis is introduced between the two wavefronts, then the signal (irradiance), $s(x,y)$, can be written from Equation 1.4 as:

$$s(x,y) = a + b\cos[2\pi fx - kW(x,y)]$$

$$= a + 0.5b\exp i[2\pi fx - kW(x,y)] \qquad (8.1)$$

$$+ 0.5b\exp{-i}[2\pi fx - kW(x,y)]$$

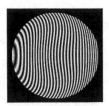

Figure 8.1 Interferogram with a linear carrier.

where the coefficients a and b can vary for different points on the interferogram; that is, they are functions of x and y, but for notational simplicity this dependence has been omitted. The carrier spatial frequency introduced by the tilt is $f = \sin\theta/\lambda$. An example of an interferogram with a linear carrier is illustrated in Figure 8.1. Here, the wavefront deformations, $W(x,y)$, are for the nontilted wavefront, before introduction of the linear carrier. To be more precise, a wavefront is said to have no tilt about the x-axis when the maximum positive or negative slopes in the x direction have the same magnitudes. The phase-modulating function $W(x,y)$ can be obtained using standard communication techniques that are quite similar to holographic techniques.

To achieve this demodulation it is necessary that, for a fixed value of y inside the aperture, the phase-modulating function $W(x,y)$ increases in a monotonic manner with the value of x. This is possible only if the tilt (θ) between the two wavefronts is chosen so that the slope of the fringes does not change sign inside the interferogram aperture. An immediate consequence of this is that no closed fringes appear in the interferogram, and no fringe in the interferogram aperture crosses any scanning line parallel to the x-axis more than once. Thus, if the tilt has a positive value, we have the following condition:

$$\frac{\partial(x\sin\theta - W(x,y))}{\partial x} > 0 \tag{8.2}$$

without any change in sign for all points inside the interferogram, or, equivalently, we have:

$$\sin\theta > \left(\frac{\partial W(x,y)}{\partial x} \right)_{\text{max}} \qquad (8.3)$$

This result can be interpreted by saying that the slope (tilt) of the reference wavefront has to be greater than the maximum (positive) slope of the wavefront under analysis in the x direction. If this wavefront is almost flat, the tilt can be almost anything between a relatively small value and the Nyquist limit (two pixels per fringe). On the other hand, Macy (1983) and Hatsuzawa (1985) showed that increasing the tilt increases the amount of measured information but reduces the precision. They found that an optimum value for the tilt is about four pixels per fringe.

An interesting point of view is to regard an interferogram with a linear carrier as an off-axis hologram. Then, Equation 8.3 is equivalent to the condition for the image spot of the first order of diffraction to be separated, without any overlap, from the zero-order point at the optical axis. A problem, when setting up the interferogram, is the selection of a tilt angle (θ) that satisfies this condition. This tilt does not have to be very precise, but it always better to be on the high side, as long as the Nyquist limit for the detector being used is not exceeded (as is described in detail later in this chapter). In the case of aspherical surfaces, it is easy to approach the Nyquist limit due to the uneven separation between the fringes. In this case, we are bounded between the lower limit for the tilt (the condition imposed by Equation 8.3) and the upper limit (imposed by the Nyquist condition). The lower limit for the tilt in Equation 8.3 was derived from purely geometrical considerations; however, in any real case the finite size or any uneven illumination of the pupil widens the diameter of the spectrum due to diffraction. The zero-order image is not a point but an Airy diffraction image (if the pupil is evenly illuminated), and the first-order image is the convolution of this Airy function with the geometrical image. This effect due to the finite size of the pupil introduces some artifacts in the results, primarily near the edge of the interferogram, but they can be minimized by any of several procedures described in Section 8.1.3.

Figure 8.2 Interferogram on which the minimum fringe slope is zero.

The approximate minimum required amount of tilt can be experimentally obtained by several different methods; for example:

1. One approach is to first adjust the interferogram tilt to obtain the maximum rotational symmetry. The tilt is then slowly introduced until the minimum local slope of a fringe in the interferogram has a value of zero (parallel to the x-axis) at the edge of the fringe, as shown in Figure 8.2. The magnitude of this tilt can be found from the interferometer adjustment.
2. Another procedure is to take the fast Fourier transform of the irradiance and to adjust the tilt in an iterative manner until the first-order lobe is clearly separated from the zero-order lobe. Then, the distance from the centroid of the first order to the zero order is the minimum amount of tilt to introduce, from a geometrical point of view. Later, we will see that a slightly greater tilt might be necessary to avoid phase errors due to diffraction effects.

8.1.2 Holographic Interpretation of the Interferogram

An interferogram with a large linear carrier is formed by interference of the wavefront to be measured with a flat wavefront forming the angle θ between them, as shown in Figure 8.3. This interferogram can be interpreted as an off-axis hologram of the wavefront $W(x,y)$. The similarity between a hologram and an interferogram has been recognized for many years (Horman,

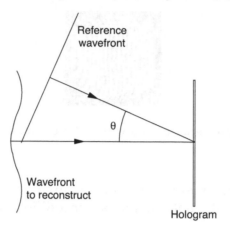

Figure 8.3   Recording of a hologram.

1965). The wavefront can be reconstructed by illumination of the hologram with a flat reference wavefront with amplitude $r(x,y)$ and tilt θ_r. This reference reconstructing wavefront does not necessarily have the same inclination (θ) as the original flat wavefront used when taking the hologram. It can be almost the same as that shown in Figure 8.4, but it can be different if desired. It will be seen later that the condition in Equation 8.3 is still valid even when these angles are very different.

The complex amplitude, $r(x,y)$, of the reconstructing reference wavefront can be written as:

$$r(x, y) = \exp i\left(2\pi f_r x\right)$$

$$= \cos\left(2\pi f_r x\right) + i \sin\left(2\pi f_r x\right) \tag{8.4}$$

where $f_r = \sin\theta_r/\lambda$. Thus, the amplitude, $e(x,y)$, in the hologram plane is given by:

$$e(x, y) = r(x, y) \cdot s(x, y) = s(x, y) \exp i\left(2\pi f_r x\right)$$

$$= a \exp i\left(2\pi f_r x\right) + 0.5 b \exp i\left[2\pi(f + f_r)x - kW(x, y)\right] \tag{8.5}$$

$$+ 0.5 b \exp - i\left[2\pi(f - f_r)x - kW(x, y)\right]$$

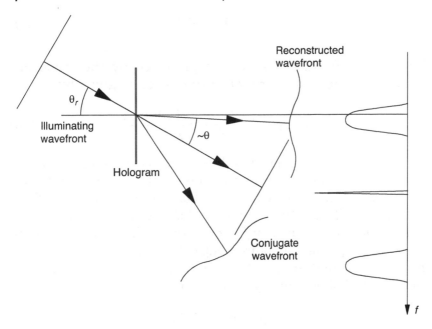

Figure 8.4 Reconstruction of a wavefront with a hologram.

These diffracted wavefronts, as expressed here, are completely general and are independent of the relative magnitude of the angles used during hologram formation and reconstruction.

These wavefronts and their frequency distribution in the Fourier plane (spectra) will now be examined. To begin, let us first remember that the phase (ϕ) of the sinusoidal function $\exp\phi$, its frequency (f), and the angular spatial frequency (ω) are related by:

$$\omega = 2\pi f = \frac{\partial \phi}{\partial x} \tag{8.6}$$

where a positive slope for the phase and hence for the wavefront is related to a positive spatial frequency. Thus, according to this sign convention, the directions of the axes on the Fourier plane must be opposite those on the interferogram. The linear carrier spatial frequency introduced by the tilt in the flat wavefront used when forming the hologram is:

$$f = \frac{\omega}{2\pi} = \frac{\sin\theta}{\lambda} \qquad (8.7)$$

The spatial frequency spectrum produced by the wavefront $W(x,y)$ in a direction parallel to the x-axis is given by:

$$f_W(x,y) = \frac{\omega_W(x,y)}{2\pi} = \frac{1}{\lambda}\frac{\partial W(x,y)}{\partial x} \qquad (8.8)$$

Thus, the spatial frequency is directly proportional to the wavefront slope in the x direction at the point (x,y). The first term in Equation 8.5 represents the flat nondiffracted wavefront with tilt θ_r. The spatial frequency of this term, with zero order, is the reference frequency f_r, and it has a delta distribution in the Fourier plane. As pointed out before, this frequency is not necessarily equal to that of the carrier, as obtained with Equation 8.6 and shown in Figure 8.4, and is given by:

$$f_r = \frac{\omega_r}{2\pi} = \frac{\sin\theta_r}{\lambda} \qquad (8.9)$$

This reference spatial frequency was defined when we determined the multiplying function $r(x,y)$ or, in other words, the angle for the reference wavefront in Equation 8.4.

The second term, with order minus one, represents a wave with deformations conjugate to those of the wavefront being reconstructed. The spatial frequency of this function in a direction parallel to the x-axis is $f_{-1}(x,y)$, given by:

$$f_{-1}(x,y) = \frac{\omega_{-1}(x,y)}{2\pi} = \frac{\sin\theta + \sin\theta_r}{\lambda} - \frac{1}{\lambda}\frac{\partial W(x,y)}{\partial x} \qquad (8.10)$$

Its deviation from this average value depends on the wavefront slope in the x direction at the point (x,y) on the interferogram — that is, in the frequency $f_W(x,y)$.

The third term, with order plus one, represents the wavefront under analysis and has a frequency of $f_{+1}(x,y)$ in the x direction, given by:

$$f_{+1}(x,y) = \frac{\omega_{+1}(x,y)}{2\pi} = \frac{\sin\theta - \sin\theta_r}{\lambda} - \frac{1}{\lambda}\frac{\partial W(x,y)}{\partial x} \qquad (8.11)$$

8.1.3 Fourier Spectrum of the Interferogram and Filtering

The expression for the spatial frequency content in the interferogram derived in the preceding section gives us the basis for an understanding of the Fourier spectrum. As pointed out before, this spectrum is geometrical; that is, this model does not take into account diffraction effects due to the pupil boundaries nor any unevenness in the pupil illumination. From Equation 8.8 we can see that the half-bandwidth f_0 along the x-axis for the first-order lobe is:

$$f_0 = \frac{1}{\lambda} \left(\frac{\partial W}{\partial x} \right)_{\max} \tag{8.12}$$

as illustrated in Figure 8.5a. Let us now assume that a spatial linear carrier with frequency f along the x-axis is introduced. The maximum and minimum frequencies, $f_{\max}$ and $f_{\min}$, along the x-axis, respectively, are:

$$f_{\max} = f + f_0 \tag{8.13}$$

and

$$f_{\min} = f - f_0 \tag{8.14}$$

When the minimum tilt required by Equation 8.3 is introduced, we obtain a spectrum like that shown in Figure 8.5b, with a minimum fringe frequency equal to zero (fringe slope zero).

It is desirable to set the linear carrier spatial frequency to its minimum allowed value if a highly aberrant wavefront is being measured in order to avoid the maximum fringe frequency and exceeding the Nyquist limit. On the other hand, if the wavefront has small deformations as compared to the wavelength, it is convenient (as is described in the next section) to select a spatial carrier with a spatial frequency much larger than the required minimum, as shown in Figure 8.5c.

The minimum allowed linear carrier spatial frequency (f) has been found with the assumption that we have a sinusoidal phase-modulated signal with no harmonic components (equivalently, we can say that the carrier is not sinusoidal, but distorted). Nevertheless, quite frequently the signal (or

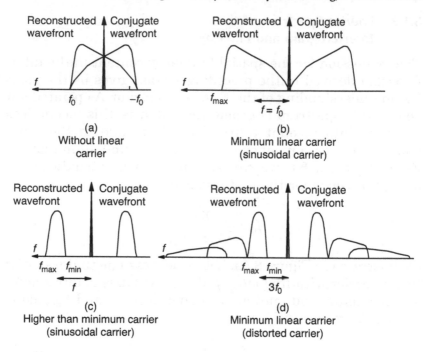

(a)
Without linear
carrier

(b)
Minimum linear carrier
(sinusoidal carrier)

(c)
Higher than minimum carrier
(sinusoidal carrier)

(d)
Minimum linear carrier
(distorted carrier)

Figure 8.5 Spatial frequency distribution along the x-axis in an interferogram with a linear carrier slightly larger than the minimum.

carrier) contains harmonics, such as when measuring Ronchi patterns, for multiple-beam interferograms, or for light detectors with nonlinear responses. In such cases, the maximum allowed linear carrier is three times the former value, as illustrated in Figure 8.5d.

It is important to remember that the finite size of the detector element acts as a low-pass filter, removing some of the harmonic frequencies before the sampling process is finished. This low-pass filtering can be quite important in preventing some high-frequency components from exceeding the Nyquist limit, thus producing aliasing noise.

If the linear carrier in the interferogram is larger than the allowed minimum, the first-order lobe can always be isolated with a suitable band-pass filter, without regard to the selected reference frequency. For practical reasons that will

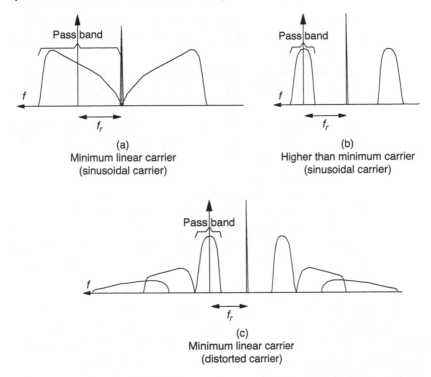

(a)
Minimum linear carrier
(sinusoidal carrier)

(b)
Higher than minimum carrier
(sinusoidal carrier)

(c)
Minimum linear carrier
(distorted carrier)

Figure 8.6 Minimum carrier frequency for three common cases.

become clear later in this chapter, it is desirable for simplicity to use a low-pass filter — in other words, a band pass centered at the origin.

Figure 8.6 shows the minimum widths of the low-pass bands that should be used when filtering three common Fourier spectrum distributions. Here, a reference frequency equal to the carrier frequency has been assumed. We can see that, in order to achieve good low-pass filtering, we must determine the values of two parameters beforehand: the carrier frequency (f) and the band half-width (f_0) of the first-order lobe. Alternatively, we must determine the maximum and minimum fringe frequencies, f_{max} and f_{min}, respectively. Several methods are available for obtaining these values (Kujawinska, 1993; Lai and Yatagai, 1994; Li and Su, 2001); for example, we can:

1. Directly set or measure these parameters when adjusting the interferometer to obtain the desired interferogram.
2. Calculate the fast Fourier transform of the interferogram and isolate the first-order lobe, either automatically or via operator intervention.
3. Automatically estimate the fringe frequencies along the x-axis with a zero crossing algorithm after high-pass filtering is used to remove constant or very low-frequency terms.
4. Calculate the wavefront using a simple rough estimation of the desired parameters, even if some errors are introduced. A better approximation for the desired parameters can be obtained from the calculated wavefront, and a new iteration will produce better results.

Let us assume that the signal is sinusoidal and phase modulated and has no harmonic components, either because they are not present in the original signal or because they have been filtered out by the sampling procedure with finite-size detectors (pixels). In this case, the reference frequency (f_r) can deviate from the carrier frequency (f) without introducing any errors if the following two conditions are met:

1. The reference frequency is within the limits:

$$\frac{f + f_0}{2} < f_r \qquad (8.15)$$

where f_0 is the band half-width of the first lobe.
2. The filtering band half-width is slightly smaller than the selected reference frequency, which can be larger than f_0.

It is interesting to note that, if the wavefront deformations are small so the carrier frequency (f) is much larger than the band half-width (f_0), this condition is transformed into:

$$\frac{f}{2} < f_r \qquad (8.16)$$

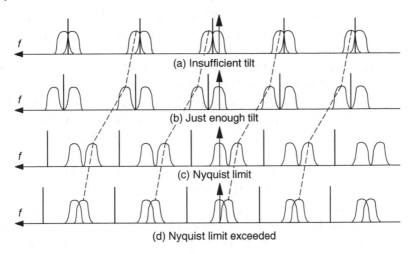

(a) Insufficient tilt

(b) Just enough tilt

(c) Nyquist limit

(d) Nyquist limit exceeded

Figure 8.7 Fourier spectrum of a sampled interferogram.

In conclusion, if the signal is not distorted and the carrier frequency is much larger than the required minimum $(f > f_0)$, then the reference frequency can have any value larger than half the signal frequency. Even in the presence of some harmonics, this criterion can help to set a good starting point in an iterative process. The discrete sampling of the interferogram, in the hologram model, can be considered as a diffraction grating superimposed on the hologram. Thus, the Fourier spectrum is split into many copies of the hologram spectrum, as shown in Figure 8.7. We can see in this figure how, by increasing the tilt between the two wavefronts, the carrier frequency is also increased, approaching the Nyquist limit.

8.1.4 Pupil Diffraction Effects

The pupil of an interferogram is not infinitely extended, but finite and most of the time circular, and its pupil illumination can be uneven; thus, our geometrical description of the Fourier spectrum of the interferogram is not complete. The correct Fourier spectrum can be obtained with the convolution of the geometrical spectrum with the Airy function, if the pupil illumination is even. This increases the width of all lobes in the spectrum, so the zero-order lobe is simply the Airy function.

The diameter of the first dark ring of the Airy function is equal to 1.22/D, where D is the diameter of the pupil. With the geometrical model, this spatial frequency corresponds to 1.22 tilt fringes. Thus, to obtain more complete separation of the first- and zero-order lobes, an additional tilt of about two to three fringes should be added to the minimum required linear carrier obtained with the geometrical model. It must be remembered, however, that the rings in the Airy diffraction pattern extend over a large area; thus, it is frequently convenient to modify the pupil boundaries in some manner so the rings are damped down, making possible good isolation of the first-order lobe. This ring damping can be achieved by one of the following two methods:

1. Extrapolation of the fringes outside the pupil boundaries; this procedure is described in detail in Chapter 3.
2. Softening the edge of the pupil with a two-dimensional Hamming filter, as proposed by Takeda et al. (1982). The Hanning or cos⁴ filter function can also be used with good results (Frankowski et al., 1989; Malcolm et al., 1989). The one-dimensional Hamming function was defined in Chapter 3, but a two-dimensional circular Hamming filter can be written as:

$$h(x,y) = 0.54 + 0.46\cos\frac{2\pi\sqrt{\left(x^2 + y^2\right)}}{D} \quad \text{for } \left(x^2 + y^2\right) < D^2 \quad (8.17)$$

$$= 0 \quad \text{elsewhere}$$

where D is the pupil diameter.

To better understand this, let us consider Figure 8.8, where we have some one-dimensional signals on the left side and their Fourier transforms on the right. In Figure 8.8a, an infinitely extended sinusoidal signal produces the Fourier transform with only delta functions; in Figure 8.8b, the signal

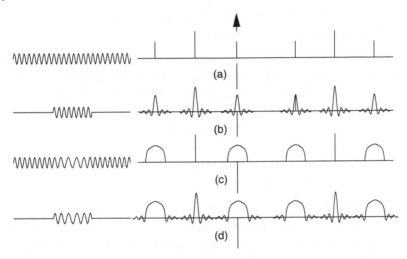

Figure 8.8 Some discretely sampled signals and their Fourier transforms: (a) infinitely extended sinusoidal signal, (b) sinusoidal signal with a finite aperture, (c) phase-modulated signal with sinusoidal signal on each side to extend it on both sides, and (d) phase-modulated signal with a finite aperture.

is limited in extension, as in any finite-size interferogram. Each of the delta functions is transformed in a sinc function for which the width is inversely proportional to the pupil size. In Figure 8.8c, the signal is no longer sinusoidal but has a phase modulation. The diffraction effects were minimized by artificially extending the pupil in both directions with sinusoidal signals. In this case, the Fourier transform terms corresponding to the orders representing the reconstructed wavefront and its conjugate wavefront are widened, as we have seen before in this chapter. Figure 8.8d shows a phase-modulated signal with a finite extension due to the pupil size.

Diffraction effects can introduce some relatively small phase errors at the edge of the pupil when the phase is calculated using phase demodulation in the space domain. These errors, however, become more important for the Fourier transform method. Both of these methods are described later in this chapter.

8.2 SPACE-DOMAIN PHASE DEMODULATION WITH A LINEAR CARRIER

The space-domain phase demodulation of interferograms with a linear carrier had its beginnings with the pioneering work by Ichioka and Inuiya (1972). Since then, several other phase demodulation methods have been developed, some of which are described in the following sections.

8.2.1 Basic Space-Domain Phase Demodulation Theory

To describe the space-domain phase demodulation method, let us follow the holographic model, where the three waves are separated by illuminating (multiplying) the hologram (interferogram) with a flat reference wave (Equation 8.4) to obtain Equation 8.5, which can be written as:

$$z(x,y) = r(x,y) \cdot s(x,y) = z_C(x,y) + iz_S(x,y) \tag{8.18}$$

where

$$z_S(x,y) = s(x,y)\sin(2\pi f_r x) \tag{8.19}$$

and

$$z_C(x,y) = s(x,y)\cos(2\pi f_r x) \tag{8.20}$$

or, using Equation 8.1, we obtain:

$$
\begin{aligned}
z_S(x,y) &= s(x,y)\sin(2\pi f_r x) \\
&= a\sin(2\pi f_r x) - \frac{b}{2}\sin\big(2\pi(f - f_r)x - kW(x,y)\big) \\
&\quad + \frac{b}{2}\sin\big(2\pi(f + f_r)x - kW(x,y)\big)
\end{aligned} \tag{8.21}
$$

and

$$
\begin{aligned}
z_C(x,y) &= s(x,y)\cos(2\pi f_r x) \\
&= a\cos(2\pi f_r x) + \frac{b}{2}\cos\big(2\pi(f - f_r)x - kW(x,y)\big) \\
&\quad + \frac{b}{2}\cos\big(2\pi(f + f_r)x - kW(x,y)\big)
\end{aligned} \tag{8.22}
$$

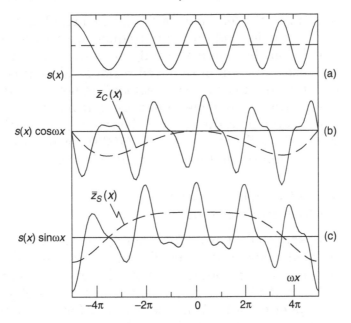

Figure 8.9 Signal along a line in an interferogram with a linear carrier (a) multiplied by a sine function (b) and cosine function (c).

These expressions are equivalent to Equations 5.27 and 5.28 in Chapter 5. An example of the functions $z_S(x,y)$ and $z_C(x,y)$ and their low-pass filtered counterparts $\bar{z}_S(x)$ and $\bar{z}_C(x)$ are illustrated in Figure 8.9. It is interesting to compare these plots with those in Figure 5.4.

With the holographic model, terms with frequency f_r and frequency $2f_r$ can be eliminated with a mask. In practice however, these two high-frequency terms are eliminated by means of a low-pass spatial filter. The filter as well as the multiplications can be implemented with analog as well as discrete sampling procedures, as described in the next few sections.

Once the high-frequency terms are filtered out, we can easily find the phase at any point x as:

$$\left[2\pi(f - f_r)x - kW(x, y)\right] = -\tan^{-1}\left[\frac{\bar{z}_S(x, y)}{\bar{z}_C(x, y)}\right] \qquad (8.23)$$

The first term on the left side, $2\pi(f - f_r)x$, is a residual tilt that appears if the carrier and reference frequencies are not exactly equal, but it can be removed easily, if desired, in the final result. The exact amount of removed residual tilt (a procedure sometimes referred to as *carrier removal*) is not important in most cases; however, in some applications it might be important, and several procedures have been designed with this purpose in mind. Fernández et al. (1998) have provided a review of this subject and a comparison of several methods.

8.2.2 Phase Demodulation with an Aspherical Reference

If the ideal shape of the wavefront being measured is aspherical, this ideal shape is subtracted from the calculated wavefront deformations to obtain the final wavefront error. A slightly different alternative procedure can be employed by using an aspherical wavefront instead of a flat wavefront as a reference. Let us now study this method to assess its relative advantages or disadvantages. Because the interferogram can be interpreted as a hologram of the wavefront $W(x,y)$, with a reference wavefront with an inclination θ, the flat reference wavefront can be reconstructed if we illuminate this interferogram with the wavefront $W(x,y)$. Hence, a null test can be obtained if we illuminate (reconstruct) with the ideal aspherical wavefront (W_r) as follows:

$$r(x,y) = \exp i[2\pi f_r x - kW_r(x,y)] \qquad (8.24)$$

Thus, we obtain:

$$\begin{aligned}
s(x,y) \cdot r(x,y) &= s(x,y)\exp[2\pi f_r x - kW_r(x,y)] \\
&= a\exp i[2\pi f_r x - kW_r(x,y)] \\
&\quad + \frac{b}{2}\exp i[2\pi(f + f_r)x - k(W(x,y) + W_r(x,y))] \\
&\quad + \frac{b}{2}\exp - i[2\pi(f - f_r)x - k(W(x,y) - W_r(x,y))]
\end{aligned} \qquad (8.25)$$

The first term after the equal sign represents the tilted ideal aspherical wavefront, with a frequency equal to that of the carrier. The second term represents a wavefront with a large asphericity and a frequency equal to about twice the carrier frequency. The last term represents a wavefront with a shape equal to the difference between the actual measured wavefront and the ideal aspherical wavefront. If all terms in these signals with frequencies equal to or greater than the carrier frequency are removed by means of a low-pass filter, only the last term remains, with real and imaginary components given by the signals $z_S(x,y)$ and $z_C(x,y)$ of an ideal aspherical wavefront with tilt (shown in Figure 8.2), as follows:

$$\bar{z}_S(x,y) = -\frac{b}{2}\sin\left[2\pi(f - f_r)x - k(W(x,y) - W_r(x,y))\right] \quad (8.26)$$

and

$$\bar{z}_C(x,y) = \frac{b}{2}\cos\left[2\pi(f - f_r)x - k(W(x,y) - W_r(x,y))\right] \quad (8.27)$$

Then, the wavefront deformations $W(x,y) - W_r(x,y)$ are given by:

$$\left[2\pi(f - f_r)x - k(W(x,y) - W_r(x,y))\right] = -\tan^{-1}\left[\frac{\bar{z}_S(x,y)}{\bar{z}_C(x,y)}\right] \quad (8.28)$$

which are the wavefront deviations with respect to the ideal aspherical wavefront.

We can see in Figure 8.10 that the width of the spectrum of the reconstructed wavefront (under test) is much narrower when an aspherical wavefront is used as a reference. On the other hand, the width of the spectrum of the conjugate wavefront is duplicated, because its asphericity is duplicated. The Nyquist limit is reached with the same sampling frequency as in the normal case, thus no improvement is obtained in this respect; however, because the width of the spectrum of the reconstructed wavefront is much narrower, the low-pass filter has to be narrower in this case.

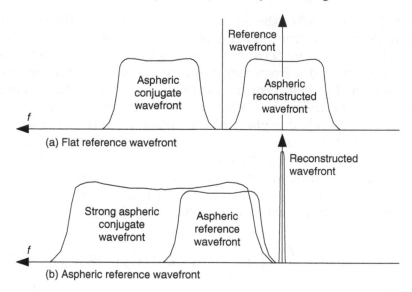

Figure 8.10 Spectra when reconstructing with (a) a flat wavefront and (b) an aspherical wavefront.

8.2.3 Analog and Digital Implementations of Phase Demodulation

As mentioned before, Ichioka and Inuiya (1972) used analog electronics to implement a simple phase-demodulation procedure. Several years later, another, slightly different phase demodulation method was described by Mertz (1983) that still utilized electronics hardware. He made three measurements in a small interval where the phase could be considered to change linearly with the distance. The measurements were separated 120° in their phase. Macy (1983) studied Mertz's method but utilized software calculations instead of hardware.

    Commercial interferometers have been constructed that evaluate two-dimensional wavefront deformations by direct digital phase demodulation (Dörband et al., 1990; Freischlad et al., 1990a,b; Küchel, 1990). The multiplications and spatial filtering are implemented through the use of dedicated digital electronics hardware, and the image is captured via a two-dimensional array of 480 × 480 pixels. Many image frames were obtained at a rate of 30 per second, and then a wavefront averaging technique was used to reduce the effects of atmospheric turbulence. The random wavefront measurement error is inversely proportional to the square root of the number of averaged wavefronts.

Another practical implementation of the digital demodulation of interferograms with a linear carrier has been described by Womack (1984). The interferogram is digitized with a two-dimensional array of light detectors (for example, with a charge-coupled device [CCD] television camera), and the irradiance values are sampled at every pixel in the detector. All operations are performed numerically, instead of using illumination with a real hologram. The sampled signal values are multiplied by the reference functions $\sin(2\pi f_r x)$ and $\cos(2\pi f_r x)$ to obtain the values of the functions $z_S(x,y)$ and $z_C(x,y)$, respectively. Thus, we can write:

$$z_S(x,y) = \sum_{i=1}^{M} s(\alpha_i,y)\sin(2\pi f_r \alpha_i) \cdot \delta(x - \alpha_i) \qquad (8.29)$$

and

$$z_C(x,y) = \sum_{i=1}^{M} s(\alpha_i,y)\cos(2\pi f_r \alpha_i) \cdot \delta(x - \alpha_i) \qquad (8.30)$$

where M is the number of pixels in a horizontal line to be scanned and sampled.

8.2.4 Spatial Low-Pass Filtering

The Fourier theory developed in Chapter 5 is not directly applicable here because we need to calculate the phase for all values of x, not only at the origin; thus, the complete low-pass filtering convolution for all values of x must be performed. As we have seen in Section 8.1.3, we require the elimination of undesired spatial frequencies at all values of x along the interferogram measured line. Thus, a common filtering function, $h(x)$, can be used for $z_S(x)$ and $z_C(x)$. This low-pass filter transforms $z_S(x,y)$ and $z_C(x,y)$ into the functions $\bar{z}_S(x)$ and $\bar{z}_C(x)$, respectively, as follows:

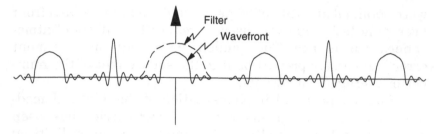

Figure 8.11   Filtering with a low-pass filter.

$$\bar{z}_S(x,y) = \sum_{i=-N}^{N} s(\alpha_i,y)\sin(2\pi f_r\alpha_i)h(x-\alpha_i) \qquad (8.31)$$

and

$$\bar{z}_C(x,y) = \sum_{i=-N}^{N} s(\alpha_i,y)\cos(2\pi f_r\alpha_i)h(x-\alpha_i) \qquad (8.32)$$

where N is the number of pixels taken before and after the point (x) being considered. We have assumed a finite spatial filter extent of $2N + 1$ pixels for the filtering function ($i = -N$ to $+N$).

These two functions are evaluated in two steps. First, the interferogram signal values on every pixel are multiplied by the reference functions sine and cosine to obtain $z_S(x,y)$ and $z_C(x,y)$. Then, the spatial low filtering process with the filtering function $h(x)$ is performed. As shown in Figure 8.11, the purpose of the low-pass filter is to filter out all undesired high frequencies in order to isolate the desired first-order lobe in the Fourier spectrum.

The low-pass filter can be any symmetric filter — for example, the two-dimensional Hanning, Hamming, $\cos^2$, or any other kernel filter described earlier. In Equations 8.31 and 8.32, a kernel with $2N + 1$ elements is assumed.

Because none of the spectral responses of the usual low-pass filters has a sharp edge, some attenuation of the high spatial frequencies in the wavefront can occur, as illustrated

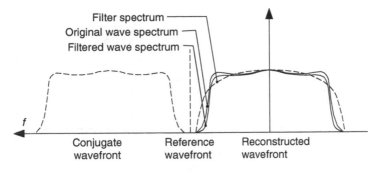

Filter spectrum
Original wave spectrum
Filtered wave spectrum

f

Conjugate wavefront

Reference wavefront

Reconstructed wavefront

Figure 8.12 Attenuation of high spatial frequencies in the measured wavefront with a low-pass filter.

in Figure 8.12. This attenuation is the same in the real part as well as in the imaginary part of the Fourier transform of the filtered wavefront, as the same filter is used for both $z_S(x,y)$ and $z_C(x,y)$; thus, no phase error is introduced. Figure 8.13 shows an example of phase demodulation using a linear carrier and discrete sampling of the interferogram.

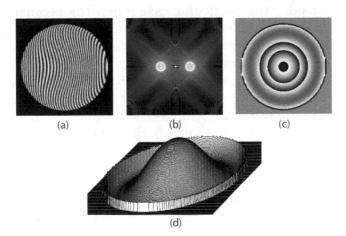

(a)   (b)   (c)

(d)

Figure 8.13 Phase demodulation with a linear carrier: (a) interferogram, (b) Fourier transform of interferogram, (c) wrapped phase, and (d) unwrapped phase.

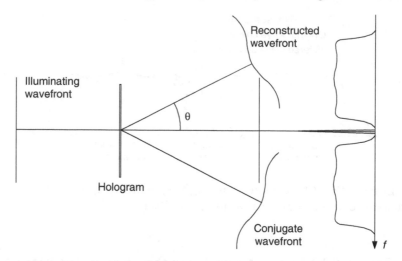

Figure 8.14   Reconstruction with a hologram using a normal reference wavefront.

8.2.5   Sinusoidal Window Filter Demodulation

We will now describe another space-domain demodulation method using a sinusoidal filtering window (Womack, 1984). Let us consider the particular case when the reconstruction frequency is quite different from the carrier frequency and is equal to zero. It this case, reconstruction in the hologram is achieved using a flat wavefront impinging perpendicularly on the hologram, as shown in Figure 8.14. In this case, the spectra for the wavefront being reconstructed and the wavefront being analyzed are symmetrically placed with respect to the origin, as shown in Figure 8.15. Under these conditions, a low-pass filter does not allow us to isolate the spectrum of the desired wavefront from the rest. Only the zero-order beam can be isolated with a low-pass filter.

A sinusoidal filter, $h_S(x)$, as described in a previous chapter, allows for beam separation. On the other hand, a cosinusoidal filter, $h_C(x)$, can be used to eliminate the zero-order beam; that is, we need a set of two filters in quadrature, acting as a band-pass filter, to isolate the first-order beam. The band-pass filtering can then be performed using the relations:

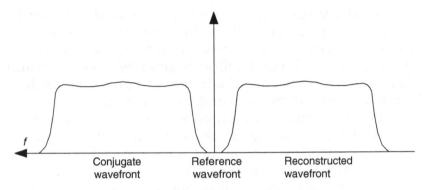

Figure 8.15 Spectrum from a hologram using a normal reference wavefront.

$$\bar{z}_S(x,y) = \sum_{i=-N}^{N} s(\alpha_i, y)h_S(x - \alpha_i) \qquad (8.33)$$

and

$$\bar{z}_C(x,y) = \sum_{i=-N}^{N} s(\alpha_i, y)h_C(x - \alpha_i) \qquad (8.34)$$

as shown in Figure 8.16.

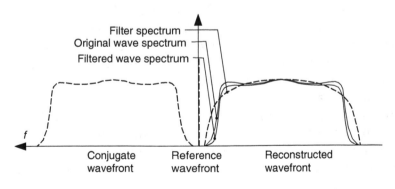

Figure 8.16 Filtering with a sinusoidal window band-pass filter. Notice that the origin is not at the same location as in Figure 8.12.

An advantage of this method is that multiplication by the reference functions and the filtering operations are performed in a single step by means of the appropriate kernel. The frequency width of the filter is given by the space width of the square function and the frequency position of the filter by the frequency of the sine and cosine functions.

Once the proper convolution kernels for $h_S(x)$ and $h_C(x)$ have been found, the signal phase at the first pixel in the interval is calculated. The kernel is then moved one pixel to the right, and the signal phase is again calculated for this new pixel until a whole line is scanned. The wavefront shape can be expressed as:

$$W(x,y) = -\frac{1}{k}\tan^{-1}\left[\frac{\bar{z}_S(x,y)}{\bar{z}_C(x,y)}\right] \qquad (8.35)$$

8.2.6 Spatial Carrier Phase-Shifting Method

The spatial carrier phase-shifting method introduced by Shough et al. (1990) is a spatial application of the temporal phase-shifting techniques. The basic assumption is that in a relatively small window the wavefront can be considered flat, so, in a small interval, the phase varies linearly and the phase difference between adjacent pixels is constant. The interval length is chosen so that the number of pixels it contains is equal to the number of sampling points. The signal phase is calculated, using a phase-shifting sampling algorithm, at some point in the first interval on a line being scanned, then the interval is moved one pixel to the right and the signal phase is again calculated. In this manner, the procedure continues until an entire line is scanned.

We can see that this method is equivalent to the sinusoidal window filter demodulation method described earlier. Here, the chosen phase-shifting sampling algorithm defines the filtering functions used. The Fourier theory developed in Chapter 5 is directly applicable, as the phase is to be determined at the local origin of each interval.

Many different phase-shifting sampling algorithms can be used. A frequent important requirement is that asynchronous or detuning-insensitive algorithms must be used, as the frequency in the interval is not always well known, mainly if the wavefront is aspherical or has strong deformations. A second useful requirement is low sensitivity to harmonics.

The simplest approach when the spatial carrier frequency is well known and the wavefront deviations from sphericity are small is to use the three-step algorithms — for example, three 120° equally spaced points or Wyant's three-step algorithm, as described by Kujawinska and Wójciak (1991a,b), using a phase step of $\pi/2$ between any two consecutive pixels. As pointed out before, when the wavefront is defocused or aspherical the spacing between the fringes is not constant and significant detuning errors are likely to appear, because the fringe spacing is quite variable inside the aperture. To solve this problem, Kujawinska and Wójciak (1991a,b) used the Schwider and Hariharan self-calibrating, five-sampling-point approach. Frankowski et al. (1989) published a report on their efforts to experimentally determine the degree of correction obtained with the asynchronous approach originally proposed by Toyoka and Tominaga (1984) and described in Chapter 6.

To test strongly aspherical surfaces it is better to assume that the phase step between adjacent pixels is not constant and has to be determined. The phase can then be found using an asynchronous algorithm — for example, the Carré algorithm, as proposed by Melozzi et al. (1995), although almost any other asynchronous detection algorithm, such as those described in Chapter 6, can be used.

A practical way to obtain the signal phase at all points in the pupil is to calculate the two functions $\bar{z}_S(x)$ and $\bar{z}_C(x)$ by means of a convolution of the signal with two one-dimensional kernels, $h_S(x)$ and $h_C(x)$, and then use Equation 8.35. The two kernels are defined by the chosen phase-shifting algorithm. Figure 8.17 shows the one-dimensional kernels for three common phase-shifting algorithms with phase equations:

$$\tan\phi = \frac{-s_1 + s_3}{s_1 - 2s_2 + s_3} \tag{8.36}$$

with shifts of –90°, 0°, and +90°, and

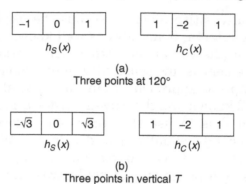

(a)
Three points at 120°

(b)
Three points in vertical T

Figure 8.17 Two one-dimensional kernels for phase-shifting algorithms with three sampling points.

$$\tan\phi = \sqrt{3}\,\frac{-s_1 + s_3}{s_1 - 2s_2 + s_3} \qquad (8.37)$$

with shifts of −120°, 0°, and +120°.

In the Zeiss Direct 100 interferometer, Küchel (1997) used a linear carrier with an angular orientation at 45° and a magnitude such that two consecutive horizontal or vertical pixels had a phase difference of 90°. As pointed out by Küchel (1994), the advantages of a linear carrier with this orientation include the following:

1. A 3 × 3 convolution kernel measures five steps in the perpendicular direction to the fringes.
2. The distance between pixels in the perpendicular

   direction to the fringes is $1/\sqrt{2}$ smaller than the distance in a horizontal or vertical direction, thus enhancing spatial resolution.

Figure 8.18a shows a 3 × 3 kernel suggested by Küchel (1994). This kernel is obtained by a combination of three inverted T algorithms shifted 90°, the second with respect to the first and the third with respect to the second. This kernel is symmetrical about its diagonal at −45°, due to the inclination of the carrier fringes at 45°. Unfortunately, complete detuning insensitivity is not obtained as in the Schwider algorithm,

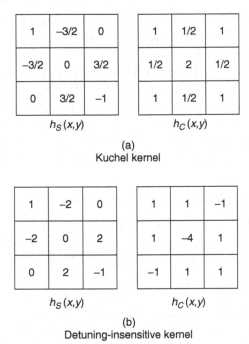

(a)
Kuchel kernel

(b)
Detuning-insensitive kernel

Figure 8.18 Two 3 × 3 kernels for spatial phase-shifting phase demodulation.

because the three algorithms have the same weights when linearly combined. Nevertheless, this kernel has a relatively low sensitivity to detuning. Its phase equation is:

$$\tan\phi = \frac{s_1 - 3s_2 + 3s_4 - s_5}{s_1 + s_2 - 4s_3 + s_4 + s_5} \tag{8.38}$$

Better results can be obtained if detuning-insensitive algorithms are used. A similar algorithm, but one that is detuning insensitive, is obtained if the second algorithm of the combination is given a weight of two (in the numerator as well as in the denominator of its phase equation), thus obtaining:

$$\tan\phi = \frac{s_1 - 4s_2 + 4s_4 - s_5}{s_1 + 2s_2 - 6s_3 + 2s_4 + s_5} \tag{8.39}$$

The kernel for this algorithm is shown in Figure 8.18b. Greater flexibility and thus better results can be obtained with a properly designed 5×5 kernel.

It is important to notice that the function $\tan^{-1}$ gives the result modulo 2π. This means that in all of these phase demodulation methods the wavefront $W(x,y)$ is calculated modulo λ. This is what is referred to as a *wrapped phase*. Unwrapping is a general problem in interferogram analysis, and methods to unwrap the phase are studied in detail in Chapter 11.

8.2.7 Phase-Locked Loop Demodulation

Phase-locked loop (PLL) demodulation, another method for interferogram analysis with a linear carrier, is based on the phase-locked loop method used in electrical communications. The PLL technique has been used since 1950 in electronic communications to demodulate electrical signals; however, its use in interferometry occurred later (Servín and Rodríguez-Vera, 1993; Servín et al., 1995). A PLL can be considered a narrow band-pass adapting filter the central frequency of which tracks the instantaneous fringe pattern frequency along the scanning line. Figure 8.19 shows the building blocks of a typical electronic PLL with its basic components.

The basic principle of this phase-tracking loop is the following: The phase changes of a phase-modulated input signal are compared with the output of a voltage-controlled oscillator (VCO) by means of a multiplier (see Figure 8.19). The PLL works in such a way that the phase difference between the modulated input signal and the output signal of the VCO eventually vanishes. This phase tracking is achieved by means of a closed loop and feeding the input of the VCO with the output signal, which is proportional to the modulating signal. When evaluating an interferogram, this VCO is not actually a piece of hardware but rather is simulated by computer software. For convenience, the term "VCO" will be used here, even though the signals are not voltage signals but are numbers. Let us assume that the input phase-modulated signal with amplitude $s(x)$ has a carrier angular frequency of ω and a phase modulation of $\phi(x)$ given by:

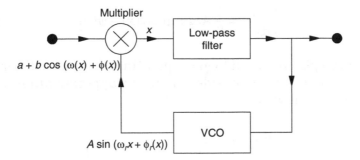

Figure 8.19 Building blocks for an electronic phase-locked loop.

$$s(x) = a + b\cos\psi(x) = a + b\cos(\omega x + \phi(x)) \qquad (8.40)$$

The VCO is an oscillator tuned to produce a sinusoidal reference signal with angular frequency ω_r in the absence of a control voltage. When a control voltage is applied to the VCO, its frequency output changes to a new value. The low-pass filter shown in Figure 8.19 is a one-pole filter that can be represented by the following first-order differential equation:

$$\frac{d\phi_r(x)}{dx} = Ag\big[a + b\cos(\omega x + \phi(x))\big]\sin(\omega_r x + \phi_r(x)) \quad (8.41)$$

where g is the gain of the low-pass filter of the PLL. This equation can also be rewritten as:

$$\frac{d\phi_r(x)}{dx} = Ag\big[a + b\cos(\psi(x))\big]\sin(\psi_r(x)) \qquad (8.42)$$

The right-hand term of Equation 8.42 can be rewritten as:

$$\frac{d\phi_r(x)}{dx} = Aag\sin\psi_r(x) + \frac{1}{2}Abg\sin(\psi_r(x) + \psi(x)) +$$

$$+ \frac{1}{2}Abg\sin(\psi_r(x) - \psi(x)) \qquad (8.43)$$

The first-order differential equation filters out all high frequencies. This eliminates the first and second terms, leaving only the last term with the lowest frequency:

$$\frac{d\phi_r(x)}{dx} = \frac{1}{2} Abg \sin(\psi_r(x) - \psi(x)) \qquad (8.44)$$

When the phase-locked loop is operating, the phase difference is small enough to consider a linear approximation valid. Hence, we can write:

$$\frac{d\phi_r(x)}{dx} = \frac{1}{2} Abg(\psi_r(x) - \psi(x)) \qquad (8.45)$$

To understand how this loop works, let us consider a system initially in equilibrium, where $\omega_r = \omega$. Then, due to the phase modulation on the input signal, its frequency changes momentarily, producing a change in its phase. This change produces a change in the input of the low-pass filter that acts on the VCO, increasing its frequency of oscillation. A new equilibrium point is found when the phase of the oscillator matches that of the input. Of course, the change in the phase of the input signal is reflected in an increase in the input of the VCO; thus, the low-pass filter output is the demodulated signal.

Normalizing the gain of the VCO ($A = 1$), we can write:

$$\frac{d\phi_r(x)}{dx} = \frac{1}{2} b\tau(\psi_r(x) - \psi(x)) \qquad (8.46)$$

where τ is the closed-loop gain. This differential equation tells us that the rate of change of the phase of the VCO is directly proportional to the demodulated signal. The output phase of the VCO will follow the input phase continuously as long as the input signal does not have any large discontinuities.

If the product of the closed-loop gain (τ) multiplied by the signal amplitude (b) is less than one, we can compute the modulation signal by the more precise expression:

$$\frac{d\phi_r(x)}{dx} = b\tau \cos \psi(x) \sin \psi_r(x) \qquad (8.47)$$

because a first-order system with a small closed-loop gain (τ) behaves as a low-pass filter; that is, due to the low τ value, no explicit low-pass filtering is required.

This theory can be applied to interferogram fringe analysis if the input signal is replaced by signal values along a horizontal scanning line in the interferogram. The variations in the illumination can be filtered out using a high-pass filter. High-pass filtering is also convenient because the phase-locked loop low-pass filter rejects only an unwanted signal with twice the carrier frequency of the interferogram. As pointed out in Chapter 3, a very simple high-pass filter is achieved simply by substituting the signal function with its derivative with respect to x. Thus, Equation 8.47 can be written as:

$$\frac{d\phi_r(x)}{dx} = -b\tau \frac{ds(x)}{dx} \cos \psi_r(x) \tag{8.48}$$

One possible way to scan a two-dimensional fringe pattern using a PLL can be found in Servín and Rodríguez-Vera (1993). Figure 8.20 shows an example of phase demodulation using the phase-locked loop method and the two-dimensional scanning strategy proposed in Servín and Rodríguez-Vera (1993). This demodulation method has been applied to aspherical wavefront measurement and also to demodulating Ronchi patterns (Servín et al., 1994).

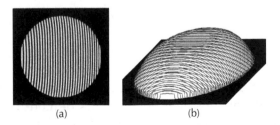

(a)   (b)

Figure 8.20 Example of phase demodulation using the phase-locked loop method: (a) interferogram to be demodulated, and (b) two-dimensional demodulated phase.

8.3 CIRCULAR SPATIAL CARRIER ANALYSIS

For some systems of closed fringes, the introduction of a linear carrier is not practical for some reason — for example, because the minimum needed carrier is of such a high spatial frequency that the Nyquist limit is exceeded. This situation can arise when the wavefront being measured is highly aspherical or aberrant; in this case, demodulation must be performed without a linear carrier. One alternative to a linear carrier is a circular carrier that introduces large defocusing, as shown in the interferogram in Figure 8.21. The irradiance function in the interferogram produced by the interference between a reference spherical wavefront and the wavefront under consideration is:

$$s(x,y) = a + b\cos k\left[D\left(x^2 + y^2\right) - W(x,y)\right]$$

$$= a + b\cos k\left[DS^2 - W(x,y)\right]$$

$$= a + \frac{b}{2}\exp + ik\left[DS^2 - W(x,y)\right] + \qquad (8.49)$$

$$+ \frac{b}{2}\exp - ik\left[DS^2 - W(x,y)\right]$$

where $S^2 = x^2 + y^2$. The radial carrier spatial frequency is:

$$f(x,y) = \frac{2DS}{\lambda} \qquad (8.50)$$

Again using the holographic analogy, we can interpret the interferogram as an on-axis or Gabor hologram. This hologram can be demodulated by illuminating it with a reference wavefront, either spherical or flat. This demodulation can be achieved only if the phase in the irradiance function increases or decreases in a monotonic manner from the center toward the edge of the pupil. Thus, if the defocusing term is positive, we require that

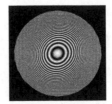

Figure 8.21 Interferogram with a circular carrier.

$$\frac{\partial\left[DS^2 - W(x,y)\right]}{\partial S} > 0 \qquad (8.51)$$

or

$$D > \frac{1}{2S}\frac{\partial W(x,y)}{\partial S} \qquad (8.52)$$

This condition assures us that two fringes in the interferogram aperture do not have the same order of interference. In other words, no fringe crosses more than once any line traced from the center of the interferogram to its edge. In the vicinity of the center of the interferogram the carrier frequency is so small that the demodulated phase in this region is not reliable. This is a disadvantage of this method. To reduce this problem, the circular carrier frequency should be as large as possible, provided the Nyquist limit is not exceeded.

8.4 PHASE DEMODULATION WITH A CIRCULAR CARRIER

Phase demodulation of an interferogram (hologram reconstruction) can be performed using an on-axis spherical or tilted spherical wavefront. These two methods, although quite similar, have some small but important differences.

8.4.1 Phase Demodulation with a Spherical Reference Wavefront

Demodulation using an on-axis spherical wavefront with almost the same curvature used to introduce the circular

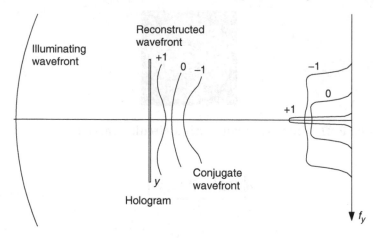

Figure 8.22   Phase demodulation in an interferogram with a circular carrier using a spherical reference wavefront.

carrier is illustrated in Figure 8.22 (Garcia-Marquez et al., 1998). This spherical reference wavefront can be written as:

$$r(x,y) = \exp i\left[kD_r\left(x^2 + y^2\right)\right] = \exp i\left[kD_r S^2\right] \qquad (8.53)$$

where $S^2 = x^2 + y^2$, and the curvature of this wavefront is close to that of the original spherical wavefront that produced the hologram (circular carrier). In other words, the value of coefficient D_r for the reference beam must be as close as possible to the value of coefficient D for the spherical beam introducing the circular carrier.

The product between the interferogram irradiance, $s(x,y)$, in Equation 8.51 and the illuminating wavefront amplitude, $r(x,y)$, is:

$$s(x,y) \cdot r(x,y) = a \exp i\left[kD_r S^2\right] +$$

$$+ \frac{b}{2}\exp ik\left[(D + D_r)S^2 - W(x,y)\right] + \qquad (8.54)$$

$$+ \frac{b}{2}\exp - ik\left[(D - D_r)S^2 - W(x,y)\right]$$

The first term is the zero-order beam corresponding to the illuminating spherical wavefront. Its spatial frequency is zero at the center, and it increases with the square of S toward the edge of the pupil:

$$f_r(x,y) = \frac{2D_r S}{\lambda} \qquad (8.55)$$

The second term is the minus first order. It is the conjugate wavefront with deformations opposite those of the wavefront being analyzed. Its curvature is about twice the reference wavefront curvature, and its spatial frequency is:

$$f_{-1}(x,y) = \frac{2(D+D_r)S}{\lambda} - \frac{1}{\lambda}\frac{\partial W(x,y)}{\partial S} \qquad (8.56)$$

The third term is the first order of diffraction and represents the reconstructed wavefront, with only a slight difference in curvature, and its spatial frequency is:

$$f_{+1}(x,y) = \frac{2(D-D_r)S}{\lambda} - \frac{1}{\lambda}\frac{\partial W(x,y)}{\partial S} \qquad (8.57)$$

The Fourier spectra of these three beams are concentric and overlap each other; however, the wavefront to be measured can still be isolated due to the different diameters of these spectra. Equation 8.54 can be rewritten as:

$$s(x,y) \cdot r(x,y) = z_C(x,y) + iz_S(x,y)$$
$$= s(x,y)\cos\left[kD_r S^2\right] + is(x,y)\sin\left[kD_r S^2\right] \qquad (8.58)$$

We see that phase demodulation of an interferogram with a circular carrier can be achieved by multiplying the signal by the functions cosine and sine with a quadratic phase, close to that used to introduce the circular carrier.

Using a two-dimensional, digital, low-pass filter, we can eliminate the first two terms in Equation 8.54 to obtain:

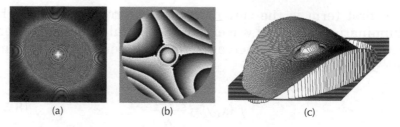

(a)   (b)   (c)

Figure 8.23 Phase demodulation of the interferogram with a circular carrier (see Figure 5.19): (a) spectrum, (b) phase map, and (c) unwrapped phase.

$$\bar{z}_C(x,y) + i\bar{z}_S(x,y) = \frac{b}{2}\exp{-ik\left[(D-D_r)S^2 - W(x,y)\right]}$$

$$= \frac{b}{2}\cos k\left[(D-D_r)S^2 - W(x,y)\right] - \quad (8.59)$$

$$-i\frac{b}{2}\sin k\left[(D-D_r)S^2 - W(x,y)\right]$$

Thus, the wavefront being reconstructed is given by:

$$k\left[(D-D_r)S^2 - W(x,y)\right] = -\tan^{-1}\left[\frac{\bar{z}_S(x,y)}{\bar{z}_C(x,y)}\right] \quad (8.60)$$

An example of phase demodulation using a circular carrier is provided in Figure 8.23.

8.4.2 Phase Demodulation with a Tilted-Plane Reference Wavefront

This method, described by Moore and Mendoza-Santoyo (1995), is basically a modification of that of Kreis (1986a,b) for the Fourier method. Here, we consider a circular carrier, but we will see that this method is more general and also applies to interferograms with systems of closed fringes. To understand how demodulation can be achieved with closed fringes, let us consider the interference along one diameter in an interferogram with a circular carrier. Figure 8.24a shows a flat wavefront interfering with a spherical wavefront.

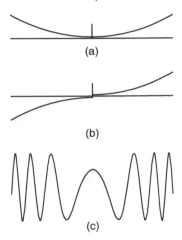

Figure 8.24 Interfering wavefronts: (a) flat wavefront and spherical wavefront, (b) flat wavefront and discontinuous wavefront with two spherical portions, and (c) signal for both cases.

In Figure 8.24b, the spherical wavefront has been replaced by a discontinuous wavefront in which the sign of the left side has been reversed. Both pairs of wavefronts produce the same interferogram with the same signal, as shown in Figure 8.24c.

In the first case, the phase increases monotonically from the center to the edges. In the second case, the phase increases monotonically from the left to the right. If we assume that what we have is the second case, we can perform phase demodulation in the standard manner, multiplying by the functions sine and cosine and then low-pass filtering these two functions; however, to obtain the correct result we must reverse the sign of the left half of the wavefront.

Now, using the holographic analogy, let us consider an interferogram with a circular carrier and illuminated with a tilted-plane wavefront, as illustrated in Figure 8.25. This illuminating tilted-plane reference wavefront can be written as:

$$r(x,y) = \exp i(2\pi f_r x) = \cos(2\pi f_r x) + i\sin(2\pi f_r x) \quad (8.61)$$

where this reference tilt has to be larger than half the maximum tilt in the wavefront along the x-axis.

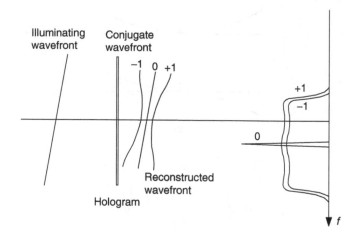

Figure 8.25 Phase demodulation in an interferogram with a circular carrier using a tilted-plane reference wavefront.

The product of the interferogram irradiance, $s(x,y)$, in Equation 8.49 and the illuminating wavefront amplitude, $r(x,y)$, gives us:

$$s(x, y) \cdot r(x, y) = a \exp i\left[2\pi f_r x\right] +$$

$$+ \frac{b}{2} \exp i\left[(2\pi f_r x) + k\left(DS^2 - W(x, y)\right)\right] + \quad (8.62)$$

$$+ \frac{b}{2} \exp - i\left[(-2\pi f_r x) + k\left(DS^2 - W(x, y)\right)\right]$$

The first term is the tilted, flat wavefront (zero order), the second term is the conjugate wavefront, and the last term is the reconstructed wavefront to be measured. The wavefront to be measured and the conjugated wavefront differ only in the sign of the deformations with respect to the reference plane.

The Fourier spectrum of Equation 8.62 is illustrated in Figure 8.26. We see that these three spots are concentric but shifted laterally with respect to the axis. If we use a rectangular low-pass filter as shown on the right side of Figure 8.26, we can see that we are isolating the reconstructed wavefront

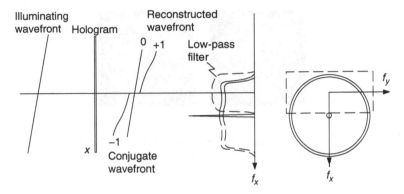

Figure 8.26 Fourier spectrum produced by an interferogram with a circular carrier (Gabor hologram) when illuminated with a tilted, flat reference wavefront.

for the $+y$ half-plane and the conjugate wavefront for the $-y$ half-plane. The conjugate wavefront is equal in magnitude to the reconstructed wavefront but has the opposite sign. Thus, we obtain the wavefront being measured simply by changing the sign of the retrieved wavefront deformations for the negative half-plane. It is easy to understand that singularities are present in the vicinity of the points where the slope of the fringes is zero.

We can also write Equation 8.62 as:

$$s(x,y) \cdot r(x,y) = z_C(x,y) + iz_S(x,y)$$
$$= s(x,y)\cos(2\pi f_r x) + is(x,y)\sin(2\pi f_r x) \tag{8.63}$$

Again, we see that the phase demodulation of an interferogram with a circular carrier can be achieved by multiplying the signal by the functions cosine and sine with a reference frequency. This reference frequency has to be larger than half the maximum spatial frequency in the interferogram, and the filter edge in the Fourier domain has to be sharp enough.

Using two-dimensional, digital, low-pass filtering, the first two terms in Equation 8.62 are eliminated, so we obtain:

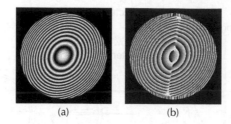

(a)   (b)

Figure 8.27 Phase map of demodulated interferogram with a circular carrier: (a) interferogram, and (b) retrieved phase. A reference frequency near the highest value in the interferogram was used.

$$\bar{z}_C(x,y) + i\bar{z}_S(x,y) = \frac{b}{2}\exp - i\left[(-2\pi f_r x) + k\left(DS^2 - W(x,y)\right)\right]$$

$$= \frac{b}{2}\cos\left[(-2\pi f_r x) + k\left(DS^2 - W(x,y)\right)\right] - \quad (8.64)$$

$$- i\frac{b}{2}\sin\left[(-2\pi f_r x) + k\left(DS^2 - W(x,y)\right)\right]$$

Thus, the retrieved wavefront is given by:

$$\left[(-2\pi f_r x) + k\left(DS^2 - W(x,y)\right)\right] = -\tan^{-1}\left[\frac{\bar{z}_S(x,y)}{\bar{z}_C(x,y)}\right] \quad (8.65)$$

which, as we know, gives us the wavefront to be measured by changing the sign of the phase for negative values of *y*. Examples of phase demodulation using a circular carrier and a tilted-plane reconstruction wavefront are shown in Figure 8.27.

8.5 FOURIER TRANSFORM PHASE DEMODULATION WITH A LINEAR CARRIER

Wavefront deformations in an interferogram with a linear carrier can also be calculated with a procedure using Fourier transforms. This method was originally proposed by Takeda et al. (1982) using one-dimensional Fourier transforms along one scanning line. Later, Macy (1983) applied Takeda's method to

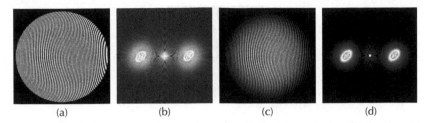

Figure 8.28 Interferogram and its Fourier transform, before and after applying the Hamming filter: (a) interferogram, (b) its Fourier transform, (c) same interferogram after applying Hamming function, and (d) its Fourier transform.

extend the Fourier transform to two dimensions by adding the information from many scanning lines and obtaining slices of the two-dimensional phase. Bone et al. (1986) extended Macy's work by using two-dimensional Fourier transforms and suggested techniques to reduce phase errors introduced by the finite boundaries.

Let us assume that we are calculating the Fourier transform of an interferogram with a large tilt. The minimum magnitude of this tilt from a geometrical point of view is the same as that used in direct interferometry; however, even if this tilt is increased, the images with orders of minus one and plus one still partially overlap the light with the zero order of diffracation. The reason is that diffraction effects due to the finite size of the aperture produce rings around the three Fourier images. The presence of these rings makes it impossible to completely separate the three images so the zero-order image can be isolated. These diffraction rings due to the finite boundary of the interferogram can be substantially reduced by any of two mechanisms, as described in Section 8.1.4. Figure 8.28 shows the result of applying a two-dimensional Hamming window to an interferogram and its effect on the Fourier transform.

Another important precaution for avoiding the presence of high spatial frequency noise in the Fourier images is to subtract irradiance irregularities in the continuum. These can be easily subtracted by measuring the irradiance in a pupil without interference fringes and then subtracting the irregularities from the interference pattern. This continuum can

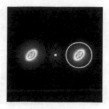

Figure 8.29 Isolating desired spectrum spot in interferogram using the Fourier method.

be measured in many ways, as described by Roddier and Roddier (1987), who also described several ways to eliminate the effects of turbulence in the interferogram.

Once the interference pattern has been cleaned up and the fringes extended outside of the pupil or the Hamming filter has been applied, a fast Fourier transform (see Chapter 2) is used to obtain the Fourier space images. When the three Fourier spots are clear and separated from each other, a circular boundary is selected around one of the first-order images (Figure 8.29). All irradiance values outside this circular boundary are multiplied by zero to isolate only the selected image. After the desired image is isolated, its center is shifted to the origin and its Fourier transform is obtained. The result is the wavefront under test.

To describe this procedure mathematically, let us write the expression for the signal in the form:

$$s(x,y) = g(x,y) + h(x,y)\exp i(2\pi fx) + h^*(x,y)\exp - i(2\pi fx) \quad (8.66)$$

where * denotes a complex conjugate and f_c is the carrier spatial frequency. The variable $s(x,y)$ is the signal in the interferogram after subtracting the irradiance irregularities and the Hamming filter has been applied or the fringes have been extrapolated outside of the pupil. We have written all variables with lower-case letters, so the Fourier transforms are represented with upper-case letters, and $h(x,y)$ is defined by:

$$h(x,y) = 0.5b(x,y)\exp - ik(W(x,y)) \quad (8.67)$$

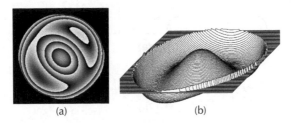

(a)   (b)

Figure 8.30 Phase demodulation of interferogram shown in Figure 8.26 using the Fourier transform method: (a) phase map, and (b) wavefront deformations after phase unwrapping.

If we take the Fourier transform of the signal $s(x,y)$ using some Fourier transform properties, we can write:

$$S(f_x, f_y) = G(f_x, f_y) + H(f_x - f_0, f_y) + H^*(f_x + f_0, f_y) \quad (8.68)$$

where the coordinates in the Fourier plane are f_x and f_y.

A low-pass filter function can be used to isolate the desired term (for example, the Hamming filter), thus obtaining:

$$S(f_x, f_y) = H(f_x - f_0, f_y) \quad (8.69)$$

Shifting this function to the origin in the Fourier plane we have:

$$S(f_x, f_y) = H(f_x, f_y) \quad (8.70)$$

Now, taking the inverse Fourier transform of this term we obtain:

$$h(x, y) = 0.5b(x, y) \exp - ik(W(x, y)) \quad (8.71)$$

Hence, the wavefront deformation is given by:

$$W(x, y) = -\frac{1}{k} \tan^{-1} \frac{\text{Im}\{h(x, y)\}}{\text{Re}\{h(x, y)\}} \quad (8.72)$$

As an example, the wavefront obtained from the interferogram in Figure 8.21 is shown in Figure 8.30.

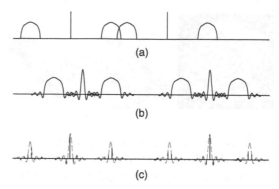

Figure 8.31 Graphical illustration of errors due to the discrete nature of the fast Fourier transform: (a) aliasing, (b) energy leakage, and (c) picket fence.

Reviews on the Fourier method have been published by Takeda (1989) and Kujawinska et al. (1989). Kujawinska and Wójciak (1991a,c) have described practical details for the implementation of Fourier demodulation, and Simova and Stoev (1993) have applied this technique to holographic moiré fringe patterns.

8.5.1 Sources of Error in the Fourier Transform Method

The Fourier transform method has some advantages but also some important limitations compared to other phase-modulation methods for analyzing interferograms with linear carriers. Several factors can introduce errors into phases calculated by the Fourier transform method, as pointed out in detail by, for example, Nugent (1985), Takeda (1987, 1989), Green et al. (1988), Frankowski et al. (1989), Malcolm et al. (1989), Kujawinska and Wójciak (1991a,c), and Schmit et al. (1992). The main errors are inherent to the discrete nature of the fast Fourier transform. The continuous Fourier transform cannot be evaluated; instead, the discrete fast Fourier transform is used. The following are some of the possible sources of phase errors:

1. *Aliasing* — If the sampling frequency is not high enough, as in Figure 8.31a, the Nyquist limit is exceeded and some nonexistent spatial frequencies can appear in the computed wavefront.

2. *Picket fence* — This error is produced by discrete calculation of the fast Fourier transform. We see in Figure 8.31c that not all frequency components appear in the calculated discrete Fourier transform. It is easy to see that after filtering and taking the inverse Fourier transform some wavefront spatial frequencies can disappear in the calculated wavefront.

3. *Energy leakage* — This is the most important source of phase errors in the Fourier method. As we pointed out before, if the tilt is not high enough and the pupil is finite, the side ripples of the Fourier transforms of each order interfere with each other, as in Figure 8.31b. This effect can cause serious phase errors in the retrieved wavefront due to leakage of the energy of some spatial frequencies into adjacent spatial frequencies. Increasing the tilt, using window functions such as the Hamming filter, or extrapolating fringes outside of the pupil limits can reduce this error.

4. *Multiple reflection or spurious fringes in the interferogram* — Multiple reflection or spurious fringe inside the interferogram pupil as well as outside can produce phase errors. These fringes distort the signal, introducing harmonic components. In this case, the minimum frequency of the linear carrier is three times that required by Equation 8.3, as discussed in Section 8.1.3. The reason is that the harmonic components cannot be filtered out if their spatial frequency is lower than the maximum fringe frequency in the interferogram. The proper low-pass filtering should then be performed.

5. *Light detector nonlinearity* — Nugent (1985) showed that if the light detector has a nonlinear response to the light irradiance then the harmonics due to this nonlinearity produce phase errors.

6. *Random noise* — Bone et al. (1986) showed that the expected root mean square (*rms*) phase error is:

$$\delta\phi_{rms} = \pi\sqrt{\frac{\alpha}{2}}\sigma m \tag{8.73}$$

where $\alpha = n/N$ is the ratio of the number of spectral sample points (n) in the filter band pass to the number of sample points (N), σ is the *rms* value of the noise, and m is the mean modulation amplitude.

7. *Quantization errors* — Frankowski et al. (1989) proved that quantization noise cannot contribute to phase errors. The error for 6 bits is smaller than 1/1000 of a wavelength.

A comparison of phase-shifting interferometry and the Fourier transform method from the viewpoint of their noise characteristics has been published by Takeda (1987).

8.5.2 Spatial Carrier Frequency, Spectrum Width, and Interferogram Domain Determination

The magnitude of the spatial carrier frequency, the filter width, and the interferogram domain limits are three important parameters that must be determined with the highest possible precision. They can be obtained automatically, as described by Kujawinska (1993), but they can also be obtained using operator-assisted methods. As pointed out before, to measure and then to remove the spatial carrier (tilt) from the interferogram, Takeda et al. (1982), Macy (1983), and Lai and Yatagai (1994) performed a lateral translation of the Fourier transform of the interferogram. However, the magnitude of the translation must be determined beforehand but it cannot be figured exactly, as the Fourier transform is calculated at discrete spatial frequency values. As a result, we are bound to obtain a residual tilt in the calculated interferogram, but this linear term can then be removed in the final result.

Filter width determination is another problem that must be solved. Takeda and Mutoh (1983) suggested that the limits of the Fourier band to be filtered and preserved are the maximum and minimum local fringe spatial frequencies. This is true for large wavefront deformations, where we can neglect diffraction effects. Kujawinska et al. (1990) suggested another method to determine both the carrier frequency and the spectrum width. The carrier frequency is determined by locating the maximum value of the Fourier transform, and the filter width is determined by isolating the area in the frequency space where Fourier transform values above a certain threshold are found.

The simplest (but not most precise) way to determine the filter width and location is through operator intervention, by observing on the computer screen the image of the two-dimensional Fourier transform and manually selecting a circle around the first order of a visually estimated location and size.

8.6 FOURIER TRANSFORM PHASE DEMODULATION WITH A CIRCULAR CARRIER

We have seen in Section 8.4.1 that an interferogram with a circular carrier can be demodulated, following the holographic analogy, using a tilted, flat reconstruction wavefront without a linear carrier. This method can also be used for demodulation using the Fourier transform. In this case, the flat reconstructing wavefront does not need to be tilted, as illustrated in Figure 8.32. This method of demodulating with closed fringes was described by Kreis (1986a,b). If all frequencies greater than or equal to zero are filtered out, as shown in Figure 8.33, then we can isolate the reconstructed wavefront for the $+y$ half-plane and the conjugate wavefront for the $-y$ half-plane. The wavefront to be measured is obtained if the sign of the phase for positive values of y is changed. Kreis (1986a,b) showed that this method can be extended to demodulation of fringe patterns with closed fringes, not necessarily with a circular carrier. The fringe pattern has to be processed

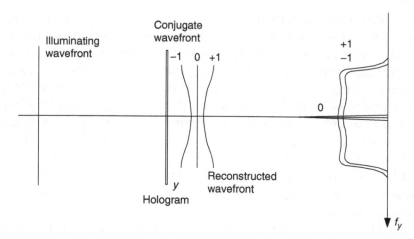

Figure 8.32 Demodulation of an interferogram with a circular carrier (Gabor hologram) with a flat reference wavefront.

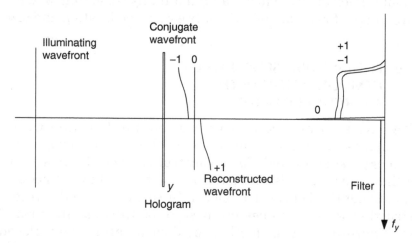

Figure 8.33 Spatial frequencies in an interferogram with a circular carrier (Gabor hologram) when illuminated with a flat reference wavefront, after filtering out all positive spatial frequencies (f_y).

with two orthogonal rectangular filters as shown in Figure 8.34. The problem of analyzing an interferogram with closed fringes, as well as the problem of recording in a single interferogram information about two events using crossed fringes, has been studied by Pirga and Kujawinska (1995, 1996).

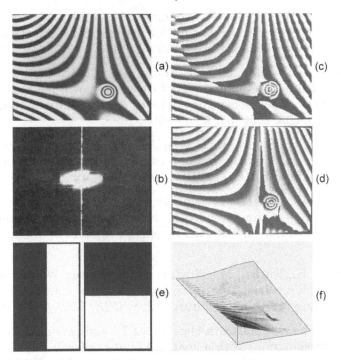

Figure 8.34 Demodulation of an interferogram with closed fringes with a flat reference wavefront. (a) Interferogram, (b) spectrum, and (c) phase maps; (e) filters; (f) calculated phase. (From Kreis, T., *J. Opt. Soc. Am. A*, 3, 847–855, 1986. With permission.)

REFERENCES

Bone, D.J., Bachor, H.-A., and Sandeman, R.J., Fringe-pattern analysis using a 2D Fourier transform, *Appl. Opt.*, 25, 1653–1660, 1986.

Burton, D.R. and Lalor, M.J., Managing some of the problems of Fourier fringe analysis, *Proc. SPIE*, 1163, 149–160, 1989.

Chan, P.H., Bryanston-Cross, P.J., and Parker, S.C., Spatial phase stepping method of fringe pattern analysis, *Opt. Lasers Eng.*, 23, 343–356, 1995.

Choudry, A. and Kujawinska, M., Fourier transform method for the automated analysis of fringe pattern, *Proc. SPIE*, 1135, 113–118, 1989.

Dörband, B., Wiedmann, W., Wegmann, U., Kübler, W., and Freis-chlad, K.R., Software concept for the new Zeiss interferometer, *Proc. SPIE*, 1332, 664–672, 1990.

Fernández, A., Kaufmann, G.H., Doval, A.F., Blanco-García, J., and Fernández, J.L., Comparison of carrier removal methods in the analysis of TV holography fringes by the Fourier transform method, *Opt. Eng.*, 37, 2899–2905, 1998.

Frankowski, G., Stobbe, I., Tischer, W., and Schillke, F., Investigation of surface shapes using a carrier frequency based analysis system, *Proc. SPIE*, 1121, 89–100, 1989.

Freischlad, K., Küchel, M., Schuster, K.H., Wegmann, U., and Kaiser, W., Real-time wavefront measurement with lambda/10 fringe spacing for the optical shop, *Proc. SPIE*, 1332, 18–24, 1990a.

Freischlad, K., Küchel, M., Wiedmann, W., Kaiser, W., and Mayer, M., High precision interferometric testing of spherical mirrors with long radius of curvature, *Proc. SPIE*, 1332, 8–17, 1990b.

Garcia-Marquez, J., Malacara-Hernandez, D., and Servín, M., Analysis of interferograms with a spatial radial carrier or closed fringes and its holographic analysis, *Appl. Opt.*, 37, 7977–7982, 1998.

Green, R.J., Walker, J.G., and Robinson, D.W., Investigation of the Fourier transform method of fringe pattern analysis, *Opt. Lasers Eng.*, 8, 29–44, 1988.

Hatsuzawa, T., Optimization of fringe spacing in a digital flatness test, *Appl. Opt.*, 24 2456–2459, 1985.

Horman, M.H., An application of wavefront reconstruction to interferometry, *Appl. Opt.*, 4, 333–336, 1965.

Ichioka, Y., and Inuiya, M., Direct phase detecting system, *Appl. Opt.*, 11, 1507–1514, 1972.

Kreis, T., Digital holographic interference-phase measurement using the Fourier transform method, *J. Opt. Soc. Am. A*, 3, 847–855, 1986a.

Kreis, T., Fourier transform evaluation of holographic interference patterns, *Proc. SPIE*, 814, 365–371, 1986b.

Küchel, M., The new Zeiss interferometer, *Proc. SPIE*, 1332, 655–663, 1990.

Küchel, M., Methods and Apparatus for Phase Evaluation of Pattern Images Used in Optical Measurement, U.S. Patent Number 5361-312, 1994.

Küchel, M., Personal communication, 1997.

Kujawinska, M., Spatial phase measurement methods, in *Interferogram Analysis*, Robinson, D.W. and Reid, G.T., Eds., Institute of Physics, Philadelphia, PA, 1993.

Kujawinska, M. and Wójciak, J., High accuracy Fourier transform fringe pattern analysis, *Opt. and Lasers in Eng.*, 14, 325–339, 1991a.

Kujawinska, M. and Wójciak, J., Spatial-carrier phase shifting technique of fringe pattern analysis, *Proc. SPIE*, 1508, 61–67, 1991b.

Kujawinska, M. and Wójciak, J., Spatial phase shifting techniques of fringe pattern analysis in photomechanics, *Proc. SPIE*, 1554, 503–513, 1991c.

Kujawinska, M., Spik, A., and Wójciak, J., Fringe pattern analysis using Fourier transform techniques, *Proc. SPIE*, 1121, 130–135, 1989.

Kujawinska, M., Salbut, M., and Patorski, K., Three channel phase stepped system for moiré interferometry, *Appl. Opt.*, 29, 1633–1636, 1990.

Lai, G. and Yatagai, T., Use of the fast Fourier transform method for analyzing linear and equispaced Fizeau fringes, 33, 5935–5940, 1994.

Li, W. and Su, X., Real-time calibration algorithm for phase shifting in phase-measuring profilometry, 40, 761–766, 2001.

Macy, W.W., Jr., Two-dimensional fringe pattern analysis, *Appl. Opt.*, 22, 3898–3901, 1983.

Malcolm, A., Burton, D.R., and Lalor, M.J., A study of the effects of windowing on the accuracy of surface measurements obtained from the Fourier analysis of fringe patterns, in *Proc. FASIG Fringe Analysis 1989*, Loughborough, UK, 1989.

Melozzi, M., Pezzati, L., and Mazzoni, A., Vibration-insensitive interferometer for on-line measurements, *Appl. Opt.*, 34, 5595–5601, 1995.

Mertz, L., Real time fringe pattern analysis, *Appl. Opt.*, 22, 1535–1539, 1983.

Moore, A.J. and Mendoza-Santoyo, F., Phase demodulation in the space domain without a fringe carrier, *Opt. Lasers Eng.*, 23, 319–330, 1995.

Nugent, K.A., Interferogram analysis using an accurate fully automatic algorithm, *Appl. Opt.*, 24, 3101–3105, 1985.

Peng, X., Shou, S.M., and Gao, Z., An automatic demodulation technique for a non-linear carrier fringe pattern, *Optik*, 100, 11–14, 1995.

Pirga, M. and Kujawinska, M., Two directional spatial-carrier phase-shifting method for analysis of crossed and closed fringe patterns, *Opt. Eng.*, 34, 2459–2466, 1995.

Pirga, M. and Kujawinska, M., Errors in two directional spatial-carrier phase-shifting method, *Proc. SPIE*, 2544, 112–121, 1996.

Ransom, P.L. and Kokal, J.V., Interferogram analysis by a modified sinusoid fitting technique, *Appl. Opt.*, 25, 4199, 1986.

Roddier, C. and Roddier, F., Interferogram analysis using Fourier transform techniques, *Appl. Opt.*, 26, 1668–1673, 1987.

Roddier, C. and Roddier, F., Wavefront reconstruction using iterative Fourier transforms, *Appl. Opt.*, 30, 1325–1327, 1991.

Schmit, J., Creath, K., and Kujawinska, M., Spatial and temporal phase-measurement techniques: a comparison of major error sources in one dimension, *Proc. SPIE*, 1755, 202–211, 1992.

Servín, M. and Cuevas, F.J., A novel technique for spatial phase-shifting interferometry, *J. Mod. Opt.*, 42, 1853–1862, 1995.

Servín, M. and Rodríguez-Vera, R., Two-dimensional phase locked loop demodulation of interferogram, *J. Mod. Opt.*, 40, 2087–2094, 1993.

Servín, M., Malacara, D., and Cuevas, F.J., Direct phase detection of modulated Ronchi rulings using a phase locked loop, *Opt. Eng.*, 33, 1193–1199, 1994.

Servín, M., Rodríguez-Vera, R., and Malacara, D., Noisy fringe pattern demodulation by an iterative phase locked loop, *Opt. Lasers Eng.*, 23, 355–366, 1995.

Shough, D.M., Kwon, O.Y., and Leary, D.F., High speed interferometric measurement of aerodynamic phenomena, *Proc. SPIE*, 1221, 394–403, 1990.

Simova, E.S. and Stoev, K.N., Automated Fourier transform fringe-pattern analysis in holographic moiré, *Opt. Eng.*, 32, 2286–2294, 1993.

Takeda, M., Temporal versus spatial carrier techniques for heterodyne interferometry, *Proc. SPIE*, 813, 329–330, 1987.

Takeda, M., Spatial carrier heterodyne techniques for precision interferometry and profilometry: an overview, *Proc. SPIE*, 1121, 73–88, 1989.

Takeda, M. and Mutoh, K., Fourier transform profilometry for the automatic measurement of 3D object shapes, *Appl. Opt.*, 22, 3977–3982, 1983.

Takeda, M. and Ru, Q.-S., Computer-based highly sensitive electron-wave interferometry, *Appl. Opt.*, 24, 3068–3071, 1985.

Takeda, M. and Tung, Z., Subfringe holographic interferometry by computer-based spatial-carrier fringe-pattern analysis, *J. Optics (Paris)*, 16, 127–131, 1985.

Takeda, M., Ina, H., and Kobayashi, S., Fourier-transform method of fringe-pattern analysis for computer-based topography and interferometry, *J. Opt. Soc. Am.*, 72, 156–160, 1982.

Toyooka, S., Phase demodulation of interference fringes with spatial carrier, *Proc. SPIE*, 1121, 162–165, 1990.

Toyooka, S. and Iwaasa, Y., Automatic profilometry of 3D diffuse objects by spatial phase detection, *Appl. Opt.*, 25, 1630–1633, 1986.

Toyooka, S. and Tominaga, M., Spatial fringe scanning for optical phase measurement, *Opt. Commun.*, 51, 68–70, 1984.

Toyooka, S., Ohashi, K., Yamada, K., and Kobayashi, K., Real-time fringe processing by hybrid analog–digital system, *Proc. SPIE*, 813, 33–35, 1987.

Vlad, V.I. and Malacara, D., Direct spatial reconstruction of optical phase from phase-modulated images, in *Progress in Optics*, Vol. XXXIII, Wolf, E., Ed., Elsevier, Amsterdam, 1994.

Womack, K.H., Interferometric phase measurement using spatial synchronous detection, *Opt. Eng.*, 23, 391–395, 1984.

9

Interferogram Analysis
with Moiré Methods

9.1 MOIRÉ TECHNIQUES

When two slightly different periodic structures are superimposed, a moiré fringe pattern appears (Sciammarella, 1982; Reid, 1984; Patorski, 1988). Traditionally, moiré patterns have been analyzed from a geometrical point of view, but alternative approaches have also been used. Chapter 1 described some of the typical applications for moiré techniques, the use of which is explored in this chapter as a tool for the analysis of interferograms. The superposition of periodic structures to form moiré patterns can be performed in two different ways:

1. *Multiplication of the irradiances of the two images —* This process can be implemented by, for example, superimposing the slides of two images, which is the most common method. The irradiance transmission of the combination is equal to the product of the two transmittances; thus, the contrast in the moiré is smaller than the contrast in each of the two images. An interesting holographic interpretation of the multiplicative moiré is described later in this chapter.

2. *Addition or subtraction of the irradiances of the two images* — This method is less commonly used than the multiplicative method because it is more difficult to implement in practice (Rosenblum et al., 1992). The advantage of this method is that, because the two images (irradiances) are additively superimposed, the contrast in the moiré image is higher than in the multiplicatively superimposed images.

9.2 MOIRÉ FORMED BY TWO INTERFEROGRAMS WITH A LINEAR CARRIER

To analyze the moiré fringes from a geometrical point of view, using the multiplicative method, let us consider a photographic slide with a phase-modulated structure, such as an interferogram with a linear carrier (tilt), for which the transmittance (assuming maximum contrast) can be described by:

$$T(x,y) = 1 + \cos(kx\sin\theta - kW(x,y)) \tag{9.1}$$

where $W(x,y)$ represents the wavefront deformations with respect to a close reference sphere (frequently a plane), and the angle θ introduces the linear carrier by means of a wavefront tilt about the x-axis.

Let us now superimpose this interferogram to be evaluated on another reference interferogram with an irradiance transmittance given by:

$$T_r(x,y) = 1 + \cos\left(\frac{2\pi}{d_r}x - kW_r(x,y) + \phi\right) \tag{9.2}$$

where $W_r(x,y)$ is any possible aspherical deformation of the wavefront producing this interferogram, with respect to the same reference sphere used to measure $W(x,y)$, d_r is the vertex spatial period of the reference linear carrier, and ϕ is its phase at the origin. The transmittance of the combination is the product of these two individual transmittances. Thus, if the moiré pattern is produced by the multiplicative method, the transmitted signal $s(x,y)$ is:

$$s(x,y) = \left[1 + \cos k(x\sin\theta - W(x,y))\right] \times$$

$$\times \left[1 + \cos\left(\frac{2\pi}{d_r}x - kW_r(x,y) + \phi\right)\right] \tag{9.3}$$

from which we obtain:

$$s(x,y) = 1 + \cos k(x\sin\theta - W(x,y))\cos\left(\frac{2\pi}{d_r}x - kW_r(x,y) + \phi\right) +$$

$$+ \cos k(x\sin\theta - W(x,y)) + \cos\left(\frac{2\pi}{d_r}x - kW_r(x,y) + \phi\right) \tag{9.4}$$

Let us now use the following trigonometrical identity:

$$\cos\alpha\cos\beta = \frac{1}{2}\cos(\alpha + \beta) + \frac{1}{2}\cos(\alpha - \beta) \tag{9.5}$$

to obtain:

$$s(x,y) = 1 + \frac{1}{2}\cos\left[\left(k\sin\theta - \frac{2\pi}{d_r}\right)x - \phi - k(W(x,y) - W_r(x,y))\right] +$$

$$+ \frac{1}{2}\cos\left[\left(k\sin\theta + \frac{2\pi}{d_r}\right)x + \phi - k(W(x,y) - W_r(x,y))\right] + \tag{9.6}$$

$$+ \cos k[x\sin\theta - W(x,y)] + \cos\left(\frac{\lambda}{d_r}x - kW_r(x,y) + \phi\right)$$

It is important to note that, although each of the cosine functions can have a positive or negative value, the total signal function has only positive values.

This result applies to spherical as well as aspherical wavefronts. The following sections consider a reference interferogram with tilt fringes and a reference aspherical interferogram.

9.2.1 Moiré with Interferograms of Spherical Wavefronts

When the wavefront that produced the interferogram to be evaluated is nearly spherical the reference interferogram must be ideally perfect, which, as pointed out before, means that it is formed by straight, parallel, equidistant fringes. If we assume that the reference wavefront is spherical and $W_r(x,y)$ is equal to zero, then Equation 9.6 becomes:

$$s(x,y) = 1 + \frac{1}{2}\cos\left[\left(k\sin\theta - \frac{2\pi}{d_r}\right)x - \phi - k(W(x,y))\right] +$$

$$+ \frac{1}{2}\cos\left[\left(k\sin\theta + \frac{2\pi}{d_r}\right)x + \phi - k(W(x,y))\right] + \qquad (9.7)$$

$$+ \cos k[x\sin\theta - W(x,y)] + \cos\left(\frac{\lambda}{d_r}x + \phi\right)$$

The first term on the right side of Equation 9.7 is a constant, so it has zero spatial frequency. Because ϕ is a constant, we see that the spatial frequency along the x coordinate of the second term is $f_2(x,y)$, written as:

$$f_2(x,y) = \pm\left(f - f_r - \frac{1}{\lambda}\frac{\partial W(x,y)}{\partial x}\right) \qquad (9.8)$$

the spatial frequency along the x coordinate of the third term is $f_3(x,y)$, written as:

$$f_3(x,y) = \pm\left(f + f_r - \frac{1}{\lambda}\frac{\partial W(x,y)}{\partial x}\right) \qquad (9.9)$$

and the spatial frequency along the x coordinate of the fourth term is $f_4(x,y)$, written as:

$$f_4(x,y) = \pm\left(f - \frac{1}{\lambda}\frac{\partial W(x,y)}{\partial x}\right) \qquad (9.10)$$

where the interferogram carrier frequency (f) and the reference carrier frequency (f_r) are given by:

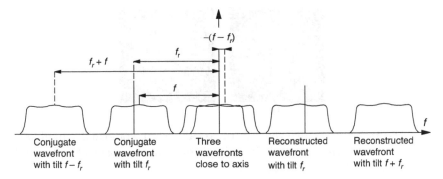

Figure 9.1 Fourier spectrum with the spatial frequencies of the moiré pattern.

$$f = \frac{\sin \theta}{\lambda} \quad \text{and} \quad f_r = \frac{1}{d_r} \tag{9.11}$$

Finally, the frequency of the fifth term is the reference frequency. Figure 9.1 shows the Fourier spectrum with the spatial frequency distribution of this moiré pattern.

Equation 9.7 represents the resulting irradiance pattern, but when observing moiré patterns the high-frequency components must be filtered out by any of several possible methods — for example, by defocusing or digital filtering. It is important to notice that the low-pass filtering reduces the contrast of the pattern.

Let us assume that the carrier frequencies f and f_r are close to each other. We also impose the condition that the central frequency lobes in Figure 9.1 are sufficiently separated from their neighbors so they can be isolated. Thus, the carrier spatial frequency of the interferogram, along the x coordinate, must have a value such that:

$$f > \frac{2}{\lambda} \left(\frac{\partial W(x, y)}{\partial x} \right)_{\max} \tag{9.12}$$

for all points inside the pattern.

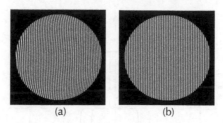

(a)   (b)

Figure 9.2 (a) Interferogram of an aberrant spherical wavefront with a linear carrier, and (b) interferogram of a perfect spherical wavefront with a linear carrier.

If we use a low-pass filter that cuts out all spatial frequencies higher than $f/2$, leaving only the central lobes in Figure 9.1, then we get:

$$s(x,y) = 1 + \frac{1}{2}\cos\left[\left(k\sin\theta - \frac{2\pi}{d_r}\right)x + kW(x,y) + \phi\right] \quad (9.13)$$

which is the signal or irradiance of the interferogram, without any tilt (if $f = f_r$). From this result, we can derive two important conclusions:

1. The moiré between the interferogram with a large tilt and the linear ruling modifies the carrier frequency or removes it if $f = f_r$. It is interesting to note that, to remove this carrier with the moiré effect, the minimum allowed linear carrier is twice the value required to phase demodulate the interferogram with a linear carrier using the methods in Chapter 7.

2. The phase of the final interferogram after the low-pass filter can be changed if the constant phase (ϕ) of the linear ruling is changed. This effect has been utilized in some phase-shifting schemes (Dorrio et al., 1995a,b; 1996).

Figure 9.2a shows an example of an aberrant spherical interferogram. The reference interferogram has a perfect wavefront with tilt, as shown in Figure 9.2b. The resulting moiré pattern is provided in Figure 9.3a, and Figure 9.3b shows the moiré image after low-pass filtering.

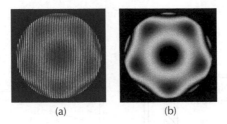

(a)   (b)

Figure 9.3 (a) Moiré formed by interferograms (one aberrant) of spherical wavefront with a linear carrier, and (b) moiré image after low-pass filtering. The histogram has been adjusted to compensate for the reduction in the contrast due to the low-pass filtering.

The magnification or minification and, hence, the spatial frequency of the reference ruling can be modified to change the appearance of the moiré pattern. Two possible ways are illustrated in Figure 9.4. In Figure 9.4a, the two slides are placed one over the other, with a short distance between them. The apparent magnification is changed by moving the reference ruling a small distance along the optical axis to change the separation between the two slides. In Figure 9.4b, the interferogram is placed at an integer multiple of the Rayleigh magnification of the reference ruling so an autoimage of the ruling is located close to the interferogram. Then, the magnification is modified by moving the collimator along the optical axis to make the light beam slightly convergent or divergent.

When a ruling with a linear carrier is used as a reference, the magnification change can be a useful tool to visually remove the linear carrier or to change its magnitude. If the interferogram has a high-frequency linear carrier, the spatial carrier (tilt) of the observed interferogram can be modified at will by moving the collimator along the axis. If the linear ruling is rotated, a spatial carrier (tilt) component in the y direction as well as in the x direction is introduced. We pointed out before that a lateral movement of the reference linear ruling introduces a constant phase shift (piston term). These effects can be used for teaching or demonstration purposes.

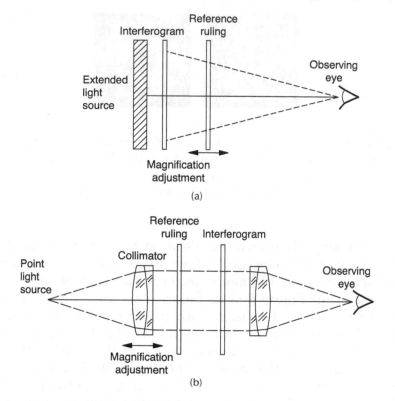

Figure 9.4 Optical arrangement to observe the moiré between an interferogram with a linear carrier and a linear ruling, with adjustable linear carrier frequency.

9.2.2 Moiré with Interferograms of Aspherical Wavefronts

When two perfect aspherical interferograms are superimposed, a moiré pattern formed by straight and parallel lines is observed. If the two interferograms are slightly different, the moiré fringes represent the difference between the two wavefronts, producing a null test. The general Equation 9.5 must now be used. The first term on the right-hand side of Equation 9.6 has zero spatial frequency. The spatial frequency in the x direction of the second term is $f_2(x,y)$, written as:

$$f_2(x,y) = \pm\left(f - f_r - \frac{1}{\lambda}\frac{\partial(W(x,y) - W_r(x,y))}{\partial x}\right) \qquad (9.14)$$

the spatial frequency in the x direction of the third term is $f_3(x,y)$, written as:

$$f_3(x,y) = \pm\left(f + f_r - \frac{1}{\lambda}\frac{\partial(W(x,y) + W_r(x,y))}{\partial x}\right) \qquad (9.15)$$

the spatial frequency in the x direction of the fourth term is $f_4(x,y)$, written as:

$$f_4(x,y) = \pm\left(f - \frac{1}{\lambda}\frac{\partial(W(x,y))}{\partial x}\right) \qquad (9.16)$$

and, finally, the frequency of the fifth term is f_5, written as:

$$f_5 = f_r - \frac{1}{\lambda}\frac{\partial(W_r(x,y))}{\partial x} \qquad (9.17)$$

The Fourier spectrum for this case, when an aspherical interferogram forms the moiré with a reference aspherical interferogram, is shown in Figure 9.5. As pointed out before, when we observe moiré patterns the high-frequency components are filtered out.

Let us now assume that the frequencies f and f_r are close to each other. We use a low-pass filter that cuts out all spatial frequencies equal to or higher than the width of the central lobes. To be able to isolate the lowest frequency terms, we impose the condition that

$$f > \frac{1}{\lambda}\left(\frac{\partial(2W(x,y) - W_r(x,y))}{\partial x}\right)_{max} \qquad (9.18)$$

and we find:

$$s(x,y) = 1 + \frac{1}{2}\cos\left[\left(k\sin\theta - \frac{2\pi}{d_r}\right)x - k(W(x,y) - W_r(x,y))\right] \qquad (9.19)$$

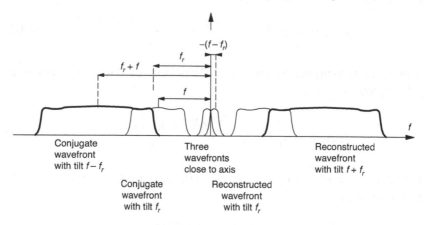

Figure 9.5 Fourier spectrum with the spatial frequencies of the moiré pattern when an aspherical reference is used.

Figure 9.6a shows an interferogram with spherical aberration plus some other high-order aberrations. Figure 9.6b shows an interferogram with pure spherical aberration, to be used as a reference. The transmittance of the combination is shown in Figure 9.7a, and Figure 9.7b shows the low-pass filtered moiré for two aspherical wavefronts. If the wavefront under consideration is equal to the reference wavefront, we obtain a pattern of straight, parallel, equidistant lines; if the linear carriers of both interferograms are different, the result is like that found in any null test.

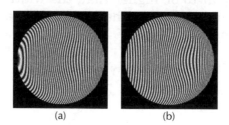

Figure 9.6 (a) Interferogram of an aberrant aspherical wavefront with a linear carrier, and (b) interferogram of a perfect aspherical wavefront with a linear carrier.

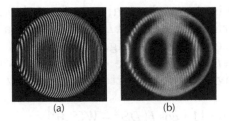

(a)   (b)

Figure 9.7 (a) Moiré produced by the superposition of two aspherical interferograms (one aberrant), and (b) low-pass filtered moiré after contrast enhancement.

9.3 MOIRÉ FORMED BY TWO INTERFEROGRAMS WITH A CIRCULAR CARRIER

Let us now study the moiré patterns between an interferogram with a circular carrier (defocusing) and an interferogram of a perfect wavefront with defocusing (circular ruling). All equations are now written in polar coordinates (S,θ), as defined in Chapter 4, Section 4.3.1. The first image is an aberrant interferogram with a circular carrier (defocusing), for which the transmittance can be written as:

$$T(S,\theta) = 1 + \cos k\left(DS^2 - W(S,\theta)\right) \tag{9.20}$$

where $W(S,\theta)$ is the wavefront deformation, and kDS^2 is the radial spatial phase of the circular carrier.

Let us now superimpose on this interferogram another reference interferogram of a nonaberrant, aspherical interferogram. This interferogram has perfect circular symmetry, but it can be decentered in the positive direction of x a small distance a with an irradiance transmittance given by:

$$\begin{aligned}
T_r(S,\theta) &= 1 + \cos k\left(D_r\left((x-a)^2 + y^2\right) - W_r(S,\theta)\right) \\
&= 1 + \cos k\left(D_r S^2 + a^2 - 2ax - W_r(S,\theta)\right)
\end{aligned} \tag{9.21}$$

where $W_r(S,\theta)$ is the aspherical wavefront deformation of the reference interferogram, and kD_rS^2 is the radial spatial phase of the reference circular ruling.

The transmittance of the combination is the product of these two individual transmittances, given by $s(S,\theta)$ as:

$$s(S,\theta) = \left[1 + \cos k\left(DS^2 - W(S,\theta)\right)\right] \times$$

$$\times \left[1 + \cos k\left(D_rS^2 + a^2 - 2ax - W_r(S,\theta)\right)\right] \quad (9.22)$$

from which we obtain:

$$s(S,\theta) = 1 + \cos k\left[DS^2 - W(S,\theta)\right] \times$$

$$\times \cos k\left[D_rS^2 + a^2 - 2ax - W_r(S,\theta)\right] +$$

$$+ \cos k\left[DS^2 - W(S,\theta)\right] + \quad (9.23)$$

$$+ \cos k\left[D_rS^2 + a^2 - 2ax - W_r(S,\theta)\right]$$

Using Equation 9.5, we obtain:

$$s(S,\theta) = 1 + \frac{1}{2}\cos k\begin{bmatrix}(D - D_r)S^2 - a^2 + 2ax \\ -(W(S,\theta) - W_r(S,\theta))\end{bmatrix} +$$

$$+ \frac{1}{2}\cos k\begin{bmatrix}(D + D_r)S^2 + a^2 - 2ax \\ -(W(S,\theta) - W_r(S,\theta))\end{bmatrix} + \quad (9.24)$$

$$+ \cos k\left(DS^2 - W(S,\theta)\right) +$$

$$+ \cos k\left(D_rS^2 + a^2 - 2ax - W_r(S,\theta)\right)$$

This result is valid for a spherical as well as aspherical reference interferogram.

9.3.1 Moiré with Interferograms of Spherical Wavefronts

If the wavefront that produced the interferogram to be evaluated is nearly spherical, the reference interferogram must have a spherical wavefront with defocusing, similar to a Fresnel zone plate or Gabor plate. If the reference wavefront is spherical and $W_r(x,y)$ is equal to zero, then Equation 9.24 becomes:

$$s(S,\theta) = 1 + \frac{1}{2}\cos k\left[(D - D_r)S^2 - a^2 + 2ax - W(S,\theta)\right] +$$

$$+ \frac{1}{2}\cos k\left[(D + D_r)S^2 + a^2 - 2axW(S,\theta)\right] + \tag{9.25}$$

$$+ \cos k\left(DS^2 - W(S,\theta)\right) + \cos k\left(D_rS^2 + a^2 - 2ax\right)$$

Because the reference pattern is centered ($a = 0$), the first term in the right-hand side of Equation 9.24 has zero spatial frequency. The radial spatial frequency of the second term, $f_2(S,\theta)$, is:

$$f_2(S,\theta) = f(S) - f_r(S) - \frac{1}{\lambda}\frac{\partial W(S,\theta)}{\partial S} \tag{9.26}$$

the radial spatial frequency of the third term, $f_3(S,\theta)$, is:

$$f_3(S,\theta) = f(S) + f_r(S) - \frac{1}{\lambda}\frac{\partial W(S,\theta)}{\partial S} \tag{9.27}$$

and the radial spatial frequency of the fourth term, $f_4(x,y)$, is:

$$f_4(S,\theta) = f(S) - \frac{1}{\lambda}\frac{\partial W(S,\theta)}{\partial S} \tag{9.28}$$

where

$$f(S) = 2kDS \quad \text{and} \quad f_r(S) = 2kD_rS \tag{9.29}$$

Finally, the frequency of the fifth term is the reference frequency $f_r(S)$. Equation 9.17 represents the resulting irradiance pattern, but when we observe moiré patterns the high-frequency components are filtered out by any of many possible methods (for example, by defocusing).

Let us assume that the values of the linear carriers of both interferograms are close to each other. We also assume that the lowest frequency terms can be isolated by requiring that the minimum radial frequency in the interferogram is such that

$$f > \frac{2}{\lambda}\left(\frac{\partial W(S,\theta)}{\partial S}\right)_{max} \tag{9.30}$$

for all points inside the moiré pattern.

If we use a low-pass filter that cuts out all spatial frequencies equal to or greater than the reference frequency $f_r(S)$, then the second term is eliminated because its frequency is more than twice the carrier frequency. After the low-pass filtering process we have:

$$s(S,\theta) = 1 + \frac{1}{2}\cos k\left[(D - D_r)S^2 - a^2 + 2ax - W(S,\theta)\right] \tag{9.31}$$

which is an interferogram with a spherical reference wavefront (defocus magnitude changed) that is modified or made flat (defocus removed) when $D = D_r$. Also, a tilt is added with a value of a. An example of an interferogram of this type is shown in Figure 9.8a, and Figure 9.8b illustrates the reference interferogram with a perfect wavefront and circular carrier. The moiré pattern obtained by the superposition of these two structures is illustrated in Figure 9.9a, and Figure 9.9b shows the low-pass filtered moiré.

9.3.2   Moiré with Interferograms of Aspherical Wavefronts

If the wavefront to be evaluated is aspherical (see Figure 9.10), the reference interferogram can also be aspherical. In this case, $W_r(x,y)$ is not equal to zero, and general Equation 9.24 must

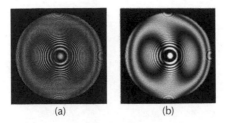

Figure 9.8 (a) Interferogram of an aberrant spherical wavefront with a circular carrier, and (b) reference interferogram of a perfect spherical wavefront with a circular carrier.

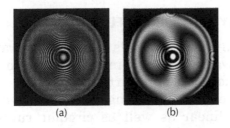

Figure 9.9 (a) Moiré produced by interferograms with spherical wavefronts (one aberrant) with a circular carrier, and (b) filtered moiré after contrast enhancement.

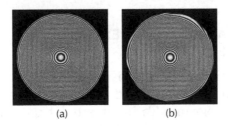

Figure 9.10 (a) Interferogram of an aberrant aspherical wavefront with a circular carrier, and (b) interferogram of a perfect aspherical wavefront with a circular carrier.

be used. We now have a null test for aspherical surfaces. The moiré pattern produced by these two interferograms is shown in Figure 9.11a, the low-pass filtered moiré in Figure 9.11b.

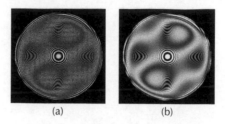

(a)   (b)

Figure 9.11 (a) Moiré produced by interferograms of aspherical wavefronts (one aberrant) with a circular carrier, and (b) filtered moiré after contrast enhancement.

9.4 SUMMARY OF MOIRÉ EFFECTS

Moiré methods are useful tools to detect aberrations in interferograms as well as for teaching demonstrations of the effect of tilts and defocusing on interferograms. The apparent magnification of the reference ruling can be changed. These effects are useful in linear as well as circular rulings. Table 9.1 summarizes the main operations that can be performed with moiré patterns of interferograms by modifying the axial position (magnification) or the lateral position of the reference ruling.

9.5 HOLOGRAPHIC INTERPRETATION OF MOIRÉ PATTERNS

The holographic approach to studying interferograms (see Chapter 8) can also be applied to interpreting the moiré patterns of interferograms. To illustrate, let us consider the case of a linear reference ruling. Let us assume that the linear ruling is illuminated with a plane wavefront perpendicularly impinging on this ruling (Figure 9.12a). Three diffracted beams will now illuminate the hologram. After passing through the hologram, each of these flat wavefronts will generate its own three wavefronts: the zero-order wavefront, the wavefront under reconstruction, and the conjugate wavefront. So, on the other side of the hologram we will have a total of nine wavefronts, as illustrated in Figure 9.13. The lowest and uppermost wavefronts in this figure are the wavefront under

TABLE 9.1 Effect Produced by Displacement of the Reference Pattern

Reference Ruling	Reference Ruling Displacement	
	Lateral Displacement	Axial Displacement (Magnification)
Linear	Piston term (phase)	Tilt (linear carrier)
Circular	Tilt (linear carrier)	Focus (circular carrier)

reconstruction and the conjugate wavefront, which correspond respectively to the exp{−iz} and exp{+iz} components of the cos function in the fourth term in Equation 9.7. We now have a reconstructed image of the interferogram and a reconstructed

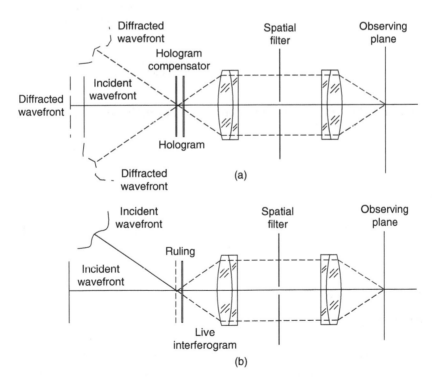

Figure 9.12 Moiré patterns between an interferogram and a ruling: (a) with a recorded interferogram, and (b) with a live interferogram.

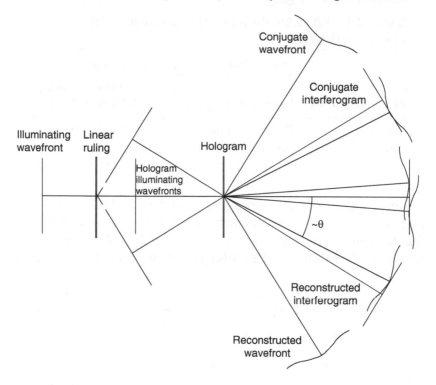

Figure 9.13 Holographic interpretation of moiré patterns; generation of nine wavefronts.

image of the conjugate interferogram corresponding respectively to the second and last terms in Equation 9.7. Near the optical axis, almost overlapping, are the reconstructed wavefront, its conjugate, and a flat wavefront, which come from the third term and the constant term.

9.6 CONCLUSION

We must point out an important conclusion that can be derived from the theory just described, particularly from Equation 9.24. If two interferograms are formed by the interference between a flat reference wavefront and a distorted wavefront, different in each case, then the moiré pattern formed by these two interferograms is identical to the interferogram that would be obtained by the interference of the two distorted wavefronts. In other words, the moiré pattern of two interferograms represents the difference between the wavefront distortions (aberrations) in these two interferograms; thus, any aberration common to both interferograms is canceled out.

REFERENCES

Dorrío, B.V., Doval, A.F., López, C., Soto, R., Blanco-García, J., Fernández, J.L., and Pérez Amor, M., Fizeau phase-measuring interferometry using the moiré effect, *Appl. Opt.*, 34, 3639–3643, 1995a.

Dorrío, B.V., Blanco-García, J., Doval, A.F., López, C., Soto, R., Bugarín, J., Fernández, J.L., and Pérez Amor, M., Surface evaluation combining the moiré effect and phase-stepping techniques in Fizeau interferometry, *Proc. SPIE*, 2730, 346–349, 1995b.

Dorrío, B.V., Blanco-García, J., López, C., Doval, A.F., Soto, R., Fernández, J.L., and Pérez-Amor, M., Phase error calculation in a Fizeau interferometer by Fourier expansion of the intensity profile, *Appl. Opt.*, 35, 61–64, 1996.

Patorski, K., Moiré methods in interferometry, *Opt. Lasers Eng.*, 8, 147–170, 1988.

Reid, G.T., Moiré fringes in metrology, *Opt. Lasers Eng.*, 5, 63–93, 1984.

Rosenblum, W.M., O'Leary, D.K., and Blaker, W.J., Computerised moiré analysis of progressive addition lenses, *Optom. Vis. Sci.*, 69, 936–940, 1992.

Sciammarella, C.A., The moiré method: a review, *Exp. Mech.*, 22, 418–433, 1982.

10

Interferogram Analysis without a Carrier

10.1 INTRODUCTION

In this chapter, we analyze interferometric techniques to demodulate a single fringe pattern containing closed fringes. Elsewhere in this book we have addressed the problem of analyzing a single interferogram when a spatial carrier is introduced (Takeda et al., 1982) — that is, whenever the modulating phase of the interferogram contains a linear component large enough to guarantee that the total modulating phase would remain an increasing function in a given direction of the two-dimensional space. Why is it interesting to demodulate a single interferogram or a series of interferograms having no spatial or temporal carriers, knowing that it is substantially more difficult? The answer is that, although we always try to obtain a single interferogram or a series of interferograms with spatial and/or temporal carriers (Malacara et al., 1998), sometimes the very nature of the experimental setup does not allow us to obtain them. One reason could be that we are studying fast transient phenomena and lack the time necessary to introduce a carrier. In these cases, though, we still want to demodulate the interferograms to evaluate quantitatively the physical variable under study.

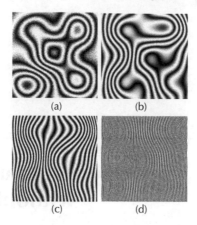

(a)   (b)

(c)   (d)

Figure 10.1 Process of spatial carrier introduction: (a) fringe pattern without carrier; (b) fringe image with a small carrier; (c) fringe image with the minimum amount of carrier, which permits its demodulation using standard phase demodulation techniques; and (d) maximum carrier that can be introduced.

10.2 MATHEMATICAL MODEL OF THE FRINGES

A mathematical model for the measured signal, $s(x,y)$, from a single interferogram without a carrier is:

$$s(x, y) = a(x, y) + b(x, y)\cos[\phi(x, y)] \qquad (10.1)$$

An example of such an interferogram can be seen in Figure 10.1a. It is convenient at this point to remind the reader that, when a spatial carrier is introduced, the usual mathematical model of the fringe pattern can be written as:

$$s(x, y) = a(x, y) + b(x, y)\cos[\omega_0 + \phi(x, y)] \qquad (10.2)$$

and the carrier frequency ω_0 must be large enough to guarantee that the total phase will be a monotonic increasing function of the x coordinate in this case. This last condition is equivalent to opening all the fringes of the interferogram, as shown in Figure 10.1d, where the phase $\phi(x,y)$ is the same

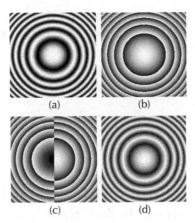

(a) (b)

(c) (d)

Figure 10.2 A simple closed-fringe interferogram: (a) fringe pattern of a defocused wavefront; (b) desired demodulated phase; (c) wrong phase, which produces the same fringes; and (d) yet another phase that produces the same fringes.

except for the linear carrier term, which in this case is large enough to open all the fringes. As we increase the linear carrier, we can see that the central closed fringe moves away from the center of the interferogram in the x direction until this closed fringe moves outside the pupil of the interferogram, as seen in Figure 10.1. If we continue to increase the carrier frequency (tilting the reference mirror in the interferometer), we will observe that the open fringes straighten and approach the maximum resolution of the digital camera used to grab the interferogram.

In Figure 10.2a, the modulating phase of the interferogram is:

$$\phi(x,y) = 4\lambda\left(x^2 + y^2\right), \quad \left(x^2 + y^2\right) < 1 \qquad (10.3)$$

where λ is the wavelength of the laser used in the interferometer. Figure 10.2b shows the wrapped phase of this interferogram. This radially symmetric phase corresponds to a defocused wavefront. The main problem with closed fringes is that the demodulated wavefront is not unique; that is, we can have many wavefronts for which the cosines are identical. For example, the following two wavefronts would give the same fringe pattern:

$$\phi_1(x,y) = |\phi(x,y)|$$

$$\phi_2(x,y) = \phi(x,y), \quad x \le 0 \tag{10.4}$$

$$- \phi(x,y), \quad x > 0$$

These two phases are shown in Figures 10.2c and 10.2d. Even some spatial combination of these two phases can also give the same fringe pattern. In fact, these two "wrong" solutions can be obtained from Equation 10.1 relatively easily, as we will see later in this chapter. Unfortunately, however, we are not interested in either of these phases. The main feature that distinguishes the phases in Equation 10.4 from the desired one (Equation 10.3) is the smoothness of the desired solution. The expected solution (Equation 10.3, Figure 10.2b) is smoother than the competing ones (Equation 10.4, Figures 10.2c,d). So, the algorithms that have been designed to deal with this problem in some form must introduce the fact that the smoothest solution among the infinitely many competing ones is the desired one.

The first attempt to demodulate a single interferogram with closed fringes was made by Kreis (1986). In this first attempt a unidimensional Hilbert transform was used. The problem with this approach is that the recovered phase is always a monotonically increasing function of a space coordinate, so in some way we must change the sign of the recovered phase. This has been done quite often by an expert viewing the interferogram on a computer screen.

One might wonder what would happen if we used some of the phase determination formulas studied in this book to find the modulating phase of an interferogram without a carrier. Probably the simplest demodulating formula that can be used for this task is the three-step phase-shifting formula applied along the x spatial coordinate. For convenience, we reproduce this simple three-step algorithm here:

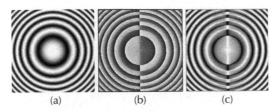

(a)　　　　　　(b)　　　　　　(c)

Figure 10.3 Demodulation of a single interferogram with closed fringes using a three-step phase-shifting algorithm: (a) fringe pattern of a defocused wavefront; (b) incorrectly demodulated phase, observing its monoticity; and (c) cosine of the incorrectly demodulated phase in (b).

$$\phi_3(x,y) = \tan^{-1}\left(\frac{(1-\cos\alpha)[s(x-1,y)-s(x+1,y)]}{\sin\alpha[2s(x,y)-s(x-1,y)-s(x+1,y)]} \right) \quad (10.5)$$

The parameter α is the phase step between the samples. Because we have no spatial carrier, parameter α is undefined; nevertheless, we can set a low value (e.g., $\alpha = 0.1$; see Figure 10.3a) with the poor but sometimes useful result shown in Figure 10.3b. The cosine of the demodulated phase is shown in Figure 10.3c, where the phase distortion obtained is more clear. We then encounter two problems with using the phase-shifting formulas presented in this book: (1) phase distortion due to the absence of a carrier, and (2) a monotonic demodulated phase regardless of the real modulating phase. The phase shown in Figure 10.3b was obtained using Equation 10.5 but is not what we would like to have as a demodulated phase. What we expect as the demodulated phase is shown in Figure 10.2b. Using any phase demodulation formula given earlier in this book will give us slightly better or similar results. To summarize, the difficulty when dealing with a single, closed-fringe interferogram resides in the fact that the fringe patterns given by:

$$\cos\phi = \cos\phi_1 = \cos\phi_2 = \cos\phi_3 \quad (10.6)$$

all look alike, so even when these phases are clearly very different they all give the same observed fringe pattern. In

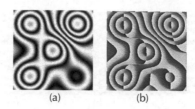

(a)   (b)

Figure 10.4 A more complicated fringe pattern demodulated using a simple phase-shifting algorithm: (a) fringe pattern, and (b) incorrectly demodulated phase.

the past, some researchers tried to automatically set the sign of the demodulated phase as the one given by Equation 10.5. This automatic sign correction turned out to be a very difficult thing to achieve (as can be seen in Figure 10.4), and this approach never gained wide acceptance.

In the following paragraphs we will analyze two recent approaches to dealing with a single interferogram that contains closed fringes. One approach is a generalization of the phase-locked loop (PLL) interferometry that was analyzed in Chapter 8. The PLL has been generalized by Servín et al. (2001, 2004) to two dimensions, a procedure we refer to as the *regularized quadrature and phase tracker* (RPT), or simply the *phase tracker*, which involves interferogram demodulation by sequentially tracking the local phase of the interferogram. The other approach was first proposed by Larkin et al. (2001), who used an isotropic Hilbert transform to avoid the distortion found in the one-dimensional Hilbert transform used by Kreis et al. (1986). Servín et al. (2003) proposed another fringe analysis technique based on and closely related to that proposed by Larkin et al. (2001). This technique is, among other things, an n-dimensional generalization of the work by Larkin et al. (2001). In the work by Servín et al. (2003) and Larkin et al. (2001), we must unwrap the orientation of the fringes using an approach based on the works by Quiroga et al. (2002), Ghiglia and Pritt (1998), and Servín et al. (1999).

10.3 THE PHASE TRACKER

A very simple yet useful way to demodulate closed fringe interferograms is a system we refer to as the regularized phase tracker. Suppose that we have a small neighborhood N within an interferometer (for example, a 7×7 pixel region) around the data pixel located at (x_1, y_1) of an interferogram. Additionally, assume that such a neighborhood is so small that within N the modulating phase may be considered linear. That is, within N we assume that the following phase plane well represents the local modulating phase:

$$p(x, y) = \phi_0 + \omega_x (x - x_1) + \omega_y (y - y_1) \qquad (10.7)$$

Now we want to find the triad $(\phi_0, \omega_x, \omega_y)$ that minimizes the following quadratic cost functional:

$$U_{(x,y)}(\phi_0, \omega_x, \omega_y) = \sum_{(x,y) \in N} \left(s'(x,y) - \cos \begin{bmatrix} \phi_0 + \omega_x (x - x_1) \\ + \omega_y (y - y_1) \end{bmatrix} \right)^2 \qquad (10.8)$$

where $s'(x,y)$ is the high-pass filtered version of $s(x,y)$ in Equation 10.1, used to remove the background term $a(x,y)$. We can find this minimum using a fixed-step gradient descent:

$$\phi_0^{k+1} = \phi_0^k - \tau \frac{\partial U}{\partial \phi_0}$$

$$\omega_x^{k+1} = \omega_x^k - \tau \frac{\partial U}{\partial \omega_x} \qquad (10.9)$$

$$\omega_y^{k+1} = \omega_y^k - \tau \frac{\partial U}{\partial \omega_v}$$

where the initial condition is equal to zero:

$$\phi_0^0 = 0, \quad \omega_z^0 = 0, \quad \omega_y^0 = 0$$

When the optimum values for the phase plane parameters have been found, we obtain a very good estimation of not only the modulating phase ϕ_0 but also the spatial frequencies (ω_x, ω_y) at point (x_1, y_1). Now, let's move one pixel away from (x_1, y_1). We want to determine the phase plane parameters at the neighborhood point $(x_1 + 1, y_1)$. Assuming that the modulating phase is a smooth continuous function, we can expect that the phase plane given by the triad $(\phi_0, \omega_x, \omega_y)$ at the neighborhood pixel $(x_1 + 1, y_1)$ would be very close to the triad previously found at (x_1, y_1); therefore, we can use the previously found parameters for the phase plane (instead of zero) as our starting point in the gradient descent formula. We have moved only slightly toward minimizing the cost functional, given that we are already very close to the sought minimum. By applying this algorithm throughout the entire fringe pattern image we can determine its modulating phase.

This simple RPT can be improved in several ways (e.g., Servín et al., 2004), but one immediate way of improving the cost functional given by Equation 10.8 is to add the derivatives of the fringe data. The new cost functional then reads:

$$U = \sum_{(x,y)\in N} \left[\left(s' - \cos p\right)^2 + \eta\left(s'_x + \omega_x \sin p\right)^2 + \eta\left(s'_y + \omega_y \sin p\right)^2 \right] \quad (10.10)$$

where for clarity the (x,y) dependence has been omitted. The parameter η can be greater than 1 (usually 10) because, normally, at low frequencies the derivative terms will make a smaller contribution to the cost functional U. The phase plane $p(x,y)$ is as given before in Equation 10.7.

Another way to improve the RPT is by using a scanning strategy. If the scanning strategy is conducted on a row-by-row basis (as in a television set), then the RPT will not work properly, particularly when it passes through local extrema of the modulating phase $\phi(x,y)$, as shown in Figure 10.5. This is because the RPT does not know how to handle the different kinds of stable points, such as minima, maxima, or saddle points, when the phase plane, $p(x,y)$, of the RPT has no information regarding the local curvature. A better way of dealing

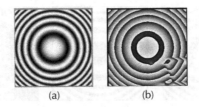

(a)   (b)

Figure 10.5 Phase demodulation of a simple closed fringe interferogram using the phase tracker along with a demodulation
scanning strategy based on row-by-row, television-like scanning: (a)
fringe pattern of a defocused interferogram, and (b) incorrectly
demodulated phase.

with this problem is to follow the scanning path traced by the
fringes of the interferogram. By scanning the interferogram
with this fringe-following strategy, we can eliminate crossing
through these extrema points. A consequence of this is that
the RPT will only "see" N open fringes within its small neighborhood. To develop this scanning strategy, we can use an
algorithm published by Ströbel (1996), where the image is
scanned according to the quality of the different regions of
the image, beginning with regions having higher signal-to-
noise ratios. In our case, however the scanning strategy has
nothing to do with the local signal-to-noise ratio but will be
assigned arbitrarily as follows:

If $s'(x,y) \geq 0$, we have "good" data.

If $s'(x,y) < 0$, we have "bad" data.

As mentioned, $s'(x,y)$ is the high-pass filtered version of $s(x,y)$
in Equation 10.1. The opposite of these criteria can also be
used. In this case, the algorithm proposed by Ströbel (1996)
will drive the RPT system along the fringes as shown in
Figures 10.6 and 10.7. With this scanning strategy, the local
phase along the fringes will have an almost constant phase
value and only the local frequencies will smoothly change,
thus improving the demodulation of the fringe pattern.

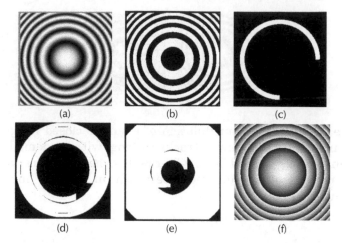

Figure 10.6 Demodulated fringe pattern using the phase tracker and scanning strategy following the fringes of the interferogram: (a) fringe pattern; (b) path suggested by the interferogram; (c), (d), (e) path actually followed by the RPT during its demodulation process; and (f) demodulated phase.

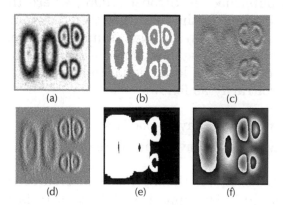

Figure 10.7 Demodulation process using the phase tracker following the path of the fringes: (a) experimentally obtained fringe pattern; (b) demodulation path derived from the fringes; (c), (d) derivative of the fringe pattern along the x and y directions; (e) snapshot of the demodulation sequence where the white zone is the demodulated zone; and (f) correctly demodulated phase.

10.4 THE *N*-DIMENSIONAL QUADRATURE TRANSFORM

Now we will analyze another way to find the modulating phase of a single closed-fringe interferogram which is based on a quadrature filter. The aim of a quadrature transform can be stated mathematically as:

$$Q\{b(\vec{r})\cos[\phi(\vec{r})]\} = -b(\vec{r})\sin[\phi(\vec{r})] \qquad (10.11)$$

where $\vec{r} = (x, y)$ is the two-dimensional vector position. As seen in this equation, a cosinusoidal signal must be transformed into a sinusoidal signal, which in turn it is useful to determine the modulating phase of the interferogram by:

$$\phi(\vec{r}) = -\frac{b(\vec{r})\sin[\phi(\vec{r})]}{b(\vec{r})\cos[\phi(\vec{r})]} \qquad (10.12)$$

Therefore, as we have seen in the previous chapters, the quadrature of a signal is of outmost importance when determining the modulating phase of an interferogram. In previous chapters, having three or more phase-shifted interferograms allowed us to obtain the modulating phase, but, in the case considered here, in which just a single interferogram (without spatial carrier) is available, we cannot apply these techniques. In the last section, we discussed how the regularized phase tracker can be used to demodulate a single interferogram, but now we will examine a different method, which was proposed by Larkin et al. (2001) and uses complex signal representation. This method was extended using vectorial calculus to n dimensions by Servín et al. (2003), an approach discussed here.

The first step toward obtaining the quadrature signal is calculating the gradient of the (high-pass filtered) fringe pattern:

$$\nabla s(\vec{r}) = \cos[\phi(\vec{r})]\nabla b(\vec{r}) + b(\vec{r})\nabla\big[\cos[\phi(\vec{r})]\big] \qquad (10.13)$$

Because in most practical situations the contrast $b(\vec{r})$ is a low-frequency signal, the first term of this last equation can be neglected with respect to the second one to obtain:

$$\nabla s(\bar{r}) \approx b(\bar{r}) \nabla \big[\cos[\phi(\bar{r})]\big] \qquad (10.14)$$

Hereafter, we will assume this approximation to be valid so the approximation sign will be replaced by an equal sign. Of course, for the special case of a constant contrast, $b(\bar{r}) = b_0$, the above mathematical relation is exact. Applying the chain rule for differentiation, we obtain:

$$\nabla s(\bar{r}) \approx -b(\bar{r}) \sin[\phi(\bar{r})] \nabla \phi(\bar{r}) \qquad (10.15)$$

If it were possible to know the real sign and magnitude of the local frequency $\nabla \phi(\bar{r})$, we could use this information as follows:

$$\nabla s(\bar{r}) \cdot \nabla \phi(\bar{r}) \approx -b(\bar{r}) \sin[\phi(\bar{r})] |\nabla \phi(\bar{r})|^2 \qquad (10.16)$$

and the quadrature of the interferogram can be obtained by dividing both sides of this equation by the squared magnitude of the local frequency $|\phi(\bar{r})|^2$:

$$Q\big\{b(\bar{r})\cos[\phi(\bar{r})]\big\} = \frac{\nabla \phi(\bar{r})}{|\nabla \phi(\bar{r})|^2} \cdot \nabla s(\bar{r}) = -b(\bar{r})\sin[\phi(\bar{r})] \qquad (10.17)$$

We now have the result we were looking for, but this result is a little misleading because, as far as we know, no linear system applied to our fringe pattern $I(\bar{r})$ gives us $\nabla \phi(\bar{r})$ in a direct way. We can rewrite the above equation in a slightly different way as:

$$Q\{s(\bar{r})\} = \frac{\nabla \phi(\bar{r})}{|\nabla \phi(\bar{r})|} \cdot \frac{\nabla s(\bar{r})}{|\nabla \phi(\bar{r})|} = \vec{H}\{s(\bar{r})\} \cdot \vec{n}\{s(\bar{r})\} \qquad (10.18)$$

Although it may seem superfluous, this rearrangement nevertheless separates the problem into two complementary and independent problems — namely, an isotropic two-dimensional Hilbert transform given by:

$$\vec{H}\{s(\bar{r})\} = \frac{\nabla s(\bar{r})}{|\nabla \phi(\bar{r})|} \qquad (10.19)$$

which is a vector field, and another two-dimensional vector field given by:

$$\vec{n}[s(\vec{r})] = \frac{\nabla\phi(\vec{r})}{|\nabla\phi(\vec{r})|} \qquad (10.20)$$

which is the orientation vector field of the fringes. Therefore, the quadrature of the signal is the scalar product of two vector fields.

10.4.1 Using the Fourier Transform To Calculate the Isotropic Hilbert Transform

Servín et al. (2003) demonstrated that the two-dimensional vector field $\vec{H}\{I(\vec{r})\}$ can also be calculated in the frequency domain as:

$$F\{\vec{H}[s(\vec{r})]\} = \left[\frac{-iu}{\sqrt{u^2 + v^2}}\hat{i} + \frac{-iv}{\sqrt{u^2 + v^2}}\hat{j}\right]F\{s(\vec{r})\} \qquad (10.21)$$

where $F\{\cdot\}$ is the Fourier transform of a signal, and we define:

$$F\{a\hat{i} + b\hat{j}\} = F\{a\}\hat{i} + F\{b\}\hat{j}$$

As can be seen from this equation the transform $\vec{H}\{\cdot\}$ is easily computed in the frequency domain using a technique first proposed by Larkin (2001) for use with complex numbers. The filter within the square brackets can be put in complex notation given that the complex plane is homeomorphic with the Euclidian plane. By doing this, we can rewrite Equation 10.21 as (Larkin 2001):

$$\vec{H}[s(\vec{r})] = F^{-1}\{e^{i\arctan(u/v)}F\{s(\vec{r})\}\} \qquad (10.22)$$

the filter $e^{i\arctan(u/v)}$ was given the name *vortex* by Larkin et al. (2001), and it is easy to see that it is equivalent in two dimensions to the filter in Equation 10.21, provided the vectors $\hat{i}$ and $\hat{j}$ are replaced by the real 1 and the imaginary

$i = \sqrt{-1}$, respectively. Equation 10.22 is a good practical way to calculate the vector field $\vec{H}\{\cdot\}$.

10.4.2 The Fringe Orientation Term

The other factor in Equation 10.12 is the fringe orientation term $\vec{n}\{s(\vec{r})\}$. This term is by far more difficult to calculate than $\vec{H}\{s(\vec{r})\}$. The reason is that the orientation in an interferogram is a wrapped signal. The orientation term has an associate fringe orientation angle given by:

$$\arctan[\theta_{2\pi}(x,y)] = \frac{\bar{n}(x,y) \cdot \bar{j}}{\bar{n}(x,y) \cdot \bar{i}} = \frac{\left(\dfrac{\partial \phi(x,y)}{\partial y}\right)}{\left(\dfrac{\partial \phi(x,y)}{\partial x}\right)} \qquad (10.23)$$

As can be seen from this equation, the fringe orientation can be readily known once the modulating phase is known, but this seems to be a vicious circle. For starters, we do not know the modulating phase of the interferogram. What *is* knowable from the fringe irradiance is the fringe orientation angle modulo π, which is:

$$\tan[\theta_{\pi}(x,y)] = \frac{\left(\dfrac{\partial s(x,y)}{\partial y}\right)}{\left(\dfrac{\partial s(x,y)}{\partial x}\right)} \qquad (10.24)$$

This formula is valid provided the fringe pattern $s(x,y)$ has been previously normalized. The orientation modulo π corresponding to the computer-generated noiseless fringe patterns in Figures 10.8a and 10.9a are shown in Figures 10.8b and 10.09b, respectively. To obtain the orientation modulo 2π (shown in Figures 10.8c and 10.9c), we will need an unwrapping process. This unwrapping process is not like the ones seen before in this book, as this unwrapping must be performed along the direction of the fringes, following the fringe

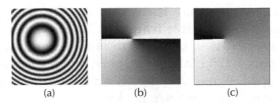

(a)   (b)   (c)

Figure 10.8   Fringe orientation of a simple closed-fringe interferogram: (a) interferogram of a defocused wavefront; (b) orientation
of the fringes modulo π (θ_π) obtained from the irradiance using
Equation 10.22; and (c) orientation of the fringes modulo 2π ($\theta_{2\pi}$)
obtained from (b) by the process of unwrapping the orientation θ_π
along the path of the fringes.

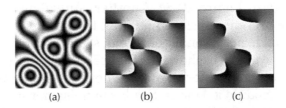

(a)   (b)   (c)

Figure 10.9   Fringe orientation unwrapping of a more complicated
interferogram: (a) interferogram; (b) fringe orientation modulo π
(θ_π), and (c) unwrapped fringe orientation modulo 2π ($\theta_{2\pi}$).

path, which can be easily seen by comparing Figures 10.8b
and 10.8c. Here, we will outline the main ideas behind a
technique proposed by Quiroga et al. (2002) to unwrap the
fringe orientation angle modulo π to obtain the required orientation angle modulo 2π. As a consequence, the relation
between the fringe orientation angle $\theta\pi$ modulo π with the
modulo 2π orientation angle $\theta_{2\pi}$ is:

$$\theta_\pi = \theta_{2\pi} + k\pi \qquad (10.25)$$

where k is an integer. Using this relation, we can multiply
both sides by 2 and write the wrapped $W[\cdot]$ orientation formula as:

$$W[2\theta_\pi] = W[2\theta_{2\pi} + 2k\pi] = W[2\theta_{2\pi}] \qquad (10.26)$$

This relation states that the value for the wrapped angle, $W[2\theta_\pi]$, is indistinguishable from that for the wrapped version, $W[2\theta_{2\pi}]$; therefore, it is possible to obtain the unwrapped $\theta_{2\pi}$ by unwrapping $W[2\theta_\pi]$ (along the path of the fringes), dividing the unwrapped signal $2\theta_{2\pi}$ by 2, and finally obtaining $\theta_2\pi$, which is the quantity we are seeking.

Unwrapping $W[2\theta_\pi]$, however, cannot be carried out by standard path-independent techniques — for example, least squares (Ghiglia and Pritt, 1998), where the modulating phase of the interferogram is wrapped perpendicular to the fringe direction. The fringe orientation modulo π must be unwrapped along the fringe direction to obtain the desired fringe orientation modulo 2π to move from the image shown in Figure 10.8b to the image in Figure 10.8c. Another equivalent condition is that, in the presence of closed fringes, the wrapped orientation phase $W[2\theta\pi]$ is not a consistent field, so path-dependent strategies must be used. As shown in Figures 10.8b and 10.9b, along the fringes of the interferogram is where the fringe orientation is wrapped modulo π. Due to the large noise normally encountered in practice for $W[2\theta_\pi]$ (due to the ratio of two derivatives in Equation 10.23), again, we must use robust path-dependent strategies. The algorithm that best fits these requirements is the unwrapping algorithm based on the RPT (Servín et al., 1999). A more detailed account of unwrapping the fringe orientation angle and some interesting examples are provided by Quiroga et al. (2002).

10.5 CONCLUSION

In this chapter, we reviewed two techniques to demodulate a single fringe pattern having closed fringes. The first reviewed technique, the regularized phase tracker (RPT), was initially proposed by Servín et al. (2001, 2004). In this approach, the fringe pattern can be considered as having a single spatial frequency in a small neighborhood around the pixel being demodulated. Within this neighborhood, the local phase can be modeled by a plane. The optimum phase plane is built using the optimum phase and optimum spatial frequencies. Another approach was proposed by Larkin et al. (2001) and extended to n dimensions by Servín et al. (2003). In this method, the demodulating problem is split into two separate problems — namely, an isotropic Hilbert transform multiplied by the fringe orientation. These two methods allow us to demodulate a single-image interferometer when the modulating phase is not monotonical. Before concluding, we should mention yet another fully automatic technique that was proposed by Marroquín et al. (1997, 1998) in which the modulating phase is considered a smooth Markovian field.

REFERENCES

Ghiglia, D.C. and Pritt, M.D., *Two-Dimensional Phase Unwrapping: Theory, Algorithms, and Software*, John Wiley & Sons, New York, 1998.

Kreis, T., Digital holographic interference-phase measurement using the Fourier transform method, *J. Opt. Soc. Am. A*, 3, 847–855, 1986.

Larkin, K.G., Bone, D.J., and Oldfield, M.A., Natural demodulation of two dimensional fringe patterns. I. General background of the spiral phase quadrature transform, *J. Opt. Soc. Am. A*, 18, 1862–1870, 2001.

Malacara, D., Servín, M., and Malacara, Z., *Interferogram Analysis for Optical Testing*, Marcel Dekker, New York, 1998.

Marroquín, J.L., Servín, M., and Rodriguez-Vera, R., Adaptive quadrature filters and the recovery of phase from fringe pattern images, *J. Opt. Soc. Am. A*, 14, 1742–1753, 1997.

Marroquín, J.L., Rodriguez-Vera, R., and Servín, M., Local phase from local orientation by solution of a sequence of linear systems, *J. Opt. Soc. Am. A*, 15, 1536–1543, 1998.

Quiroga, J.A., Servín, M., and Cuevas, F.J., Modulo 2π fringe-orientation angle estimation by phase unwrapping with a regularized phase tracking algorithm, *J. Opt. Soc. Am. A*, 19, 1524–1531, 2002.

Servín, M., Cuevas, F.J., Malacara, D., and Marroquín, J.L., Phase unwrapping through demodulation using the RPT technique, *Appl. Opt.*, 38, 1934–1940, 1999.

Servín, M., Marroquín, J.L., and Cuevas, F.J., Fringe-following regularized phase tracker for demodulation of closed-fringe interferogram, *J. Opt. Soc. Am. A*, 18, 689–695, 2001.

Servín, M., Quiroga, J.A., and Marroquín, J.L., General n-dimensional quadrature transform and its application to interferogram demodulation, *J. Opt. Soc. A*, 20, 925–934, 2003.

Servín, M., Marroquín, J.L., and Quiroga, J.A., Regularized quadrature and phase tracking from a single closed-fringe interferogram, *J. Opt. Soc. Am.*, 21, 411–419, 2004.

Ströbel, B., Processing of interferometric phase maps as complex value phasor images, *Appl. Opt.*, 35, 2192–2198, 1996.

Takeda, M., Ina, H., and Kobayashi, S., Fourier transform method for fringe pattern analysis, *J. Opt. Soc. Am.*, 72, 156–160, 1982.

11

Phase Unwrapping

11.1 THE PHASE UNWRAPPING PROBLEM

Optical interferometers can be used to measure a wide range of physical quantities. Among the interesting data supplied by the interferometer is the fringe pattern, which is a cosinusoidal function phase modulated by the wavefront distortions being measured. As shown in Chapter 1, a fringe pattern or interferogram can be modeled by the expression:

$$s(x, y) = a(x, y) + b(x, y) \cos \phi(x, y) \qquad (11.1)$$

where $a(x,y)$ is a slowly varying background illumination; $b(x,y)$ is the amplitude modulation, which also is a low-frequency signal; and $\phi(x,y)$ is the phase being measured. The purpose of computer-aided fringe analysis is automatic detection of the two-dimensional phase variation, $\phi(x,y)$, that occurs over the interferogram due to the spatial change of the corresponding physical variable. The continuous interferogram is then imaged over a charge-coupled device (CCD) video camera and digitized using a video frame grabber for further analysis in a digital computer.

Several techniques can be used to measure the desired spatial phase variation of $\phi(x,y)$, including phase-shifting interferometry, which requires at least three phase-shifted interferograms. The phase shift among the interferograms must be known over the entire interferogram. In this case, we can estimate the modulating phase at each resolvable image pixel. Phase-shifting interferometry is the technique chosen first whenever atmospheric turbulence and mechanical conditions of the interferometer remain constant over the time required to obtain the three phase-shifted interferograms. When these requirements are not met, we can analyze just one interferogram, if carrier fringes are introduced to the fringe pattern, to obtain a spatial carrier frequency interferogram. We can then analyze this interferogram using such well-known techniques as the Fourier transform, spatial carrier demodulation, spatial phase shifting, and phase-locked loop (PLL), among others. Except for the PLL technique, which does not introduce any phase wrapping, in all other methods the detected phase is wrapped. Carré's method wraps the phase modulo π, but all other methods wrap the phase modulo 2π, due to the arc tangent function involved in the phase estimation process.

Ideally, the functions that calculate the arc tangent must have as input parameters not the final value of the tangent but the values of the numerator ($\sin\phi$) and the denominator ($\cos\phi$) to avoid losing useful information. This pair of values allows calculation of the angle in the entire circle from $0°$ to 2π or from $-\pi$ to $+\pi$. After we calculate the angle ϕ in the interval from $-\pi/2$ to $+\pi/2$, a correction is made as shown in Tables 11.1 and 11.2 to obtain the angle in the entire circle. For this purpose, the signs of $\sin\phi$ and $\cos\phi$ are used. If the range from $-\pi$ to $+\pi$ is desired, Table 11.1 is used. If the range from $0°$ to $+2\pi$ is desired, Table 11.2 is used.

An example of a phase map is given in Figure 11.1, where we have represented the 2π dynamic range in gray levels. Black represents the phase value of $-\pi$, and white the value of π. All other gray levels represent intermediate and linearly mapped phase values. The relationship between the wrapped phase and the unwrapped phase can be stated as:

$$\phi(x_i, y_j) = \phi_W(x_i, y_j) + 2\pi m(x_i, y_j); \quad 1 \le i \le N; 1 \le j \le M \quad (11.2)$$

TABLE 11.1 Phase and Range of Values According to the Signs in the Numerator (sinφ) and Denominator (cosφ) in the Expression for tanφ

sinφ	cosφ	Adjusted Phase
sinφ > 0	cosφ > 0	φ
sinφ > 0	cosφ < 0	φ + π
sinφ < 0	cosφ < 0	φ − π
sinφ < 0	cosφ > 0	φ
sinφ > 0	cosφ = 0	π/2
sinφ = 0	cosφ < 0	π
sinφ < 0	cosφ = 0	3π/2
sinφ = 0	cosφ > 0	0

Note: The final range of phases is from −π to +π.

TABLE 11.2 Phase and Range of Values According to the Signs in the Numerator (sinφ) and Denominator (cosφ) in the Expression for tanφ

sinφ	cosφ	Adjusted Phase
sinφ > 0	cosφ > 0	φ
sinφ > 0	cosφ < 0	φ + π
sinφ < 0	cosφ < 0	φ + π
sinφ < 0	cosφ > 0	φ + 2π
sinφ > 0	cosφ = 0	π/2
sinφ = 0	cosφ < 0	π
sinφ < 0	cosφ = 0	3π/2
sinφ = 0	cosφ > 0	0

Note: The final range of phases is from 0° to +2π.

where $\phi_W(x,y)$ is the wrapped phase, $\phi(x,y)$ is the unwrapped phase, and $m(x,y)$ is an integer-valued number known as the field number.

Figure 11.1   Wrapped phase data mapped to gray levels for display purposes.

The unwrapping problem is trivial for phase maps calculated from good-quality fringe data for which both of the following conditions are satisfied:

1. The signal is free of noise.
2. The Nyquist condition is not violated, which means that the absolute value of the phase difference between any two consecutive phase samples (pixels) is less than π.

The Nyquist condition can be expressed mathematically by:

$$\frac{\partial W(x,y)}{\partial x} < \frac{\lambda}{2(\Delta x)} \qquad (11.3)$$

where Δx is the distance between the two consecutive pixels. In other words, the wavefront slope has a maximum value that cannot be exceeded.

Figure 11.2 illustrates the phase wrapping of a one-dimensional function. The lower zigzag curve is the wrapped function and the upper curve, passing through the small circles, is the unwrapped function. To unwrap, several of the phase values should be shifted by an integer multiple of 2π to any of the small circles. The vertical distance between the circles is 2π. The phase step from pixel 2 to pixel 3 is smaller than π if the phase goes from point A to point B, which is the correct point; however, the phase step from point A to point C, which is the incorrect point, is larger than π. This is because the Nyquist condition is fulfilled. The phase step (pixel 3 to pixel 4) going to the correct point, D, is larger than π, and the

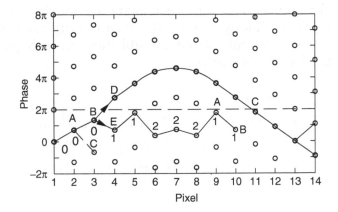

Figure 11.2 Phase unwrapping in one direction, without noise, and the appropriate Nyquist-limited sampling frequency.

phase step going to the incorrect point, E, is smaller than π. In this case, the correct and incorrect phase steps are reversed because the Nyquist condition is not fulfilled. Thus, we can also write the Nyquist condition as:

$$|\Delta\Phi(x, y)| < \pi \qquad (11.4)$$

where $\Delta\phi(x,y)$ is the correct phase step between two consecutive pixels. The problem here is that once the phase has been calculated it is frequently difficult to determine if the Nyquist condition has been violated or not. This uncertainty is because we do not know which of the two possible phase jumps is the correct one. Ideally, it is better to ensure that we have fringe separation everywhere in the x and y directions larger than half the pixel separation.

Assuming that the Nyquist condition is fulfilled at all points, unwrapping is thus a simple matter of adding or subtracting 2π offsets at each discontinuity encountered in the phase data (Macy, 1983; Bone, 1991) or integrating the wrapped phase differences along a given coordinate (Itoh, 1982; Ghiglia et al., 1987; Ghiglia and Romero, 1994).

The unwrapping procedure consists of finding the correct field number for each phase measurement. In Figure 11.2, the field numbers, $m(x)$, for each pixel are marked near the wrapped value. Taking $m(x_1) = 0$, we can easily see that this field number has only three possibilities at each pixel, as expressed by (Kreis, 1986):

$$
\begin{aligned}
m(x_1) &= 0 \\
m(x_i) &= m(x_{i-1}) && \text{if} \quad |\phi(x_i) - \phi(x_{i-1})| < \pi \\
m(x_i) &= m(x_{i-1}) + 1 && \text{if} \quad \phi(x_i) - \phi(x_{i-1}) \le -\pi \\
m(x_i) &= m(x_{i-1}) - 1 && \text{if} \quad \phi(x_i) - \phi(x_{i-1}) \ge \pi
\end{aligned}
\tag{11.5}
$$

$$
i = 1, 2, \ldots, N
$$

Kreis (1986) has also described a method for unwrapping in two dimensions. Unwrapping becomes more difficult when the absolute phase differences between adjacent pixels at points other than discontinuities in the arctan function are greater than π. These discontinuities can be introduced by (Figure 11.3):

1. High-frequency, high-amplitude noise
2. Discontinuous phase jumps
3. Regional undersampling in the fringe pattern

Ghiglia et al. (1987) considered unwrapping the phase by isolating these erroneous discontinuities before beginning the unwrapping process. Erroneous discontinuities or phase inconsistencies can be detected when the sum of the wrapped-phase differences around a square path of size L is not zero. Inconsistencies generate phase errors (unexpected phase jumps) which propagate along the unwrapping direction. As a consequence, the unwrapping process becomes path dependent; that is, we can obtain different unwrapped phase fields depending on the unwrapping direction chosen.

An important step toward obtaining a robust path-independent phase unwrapper was made by Ghiglia and Romero (1994), who applied the ideas of Fried (1977) and Hudgin (1977) regarding least-squares integration of phase gradients (Noll, 1978; Hunt, 1979; Takajo and Takahashi, 1988) to the

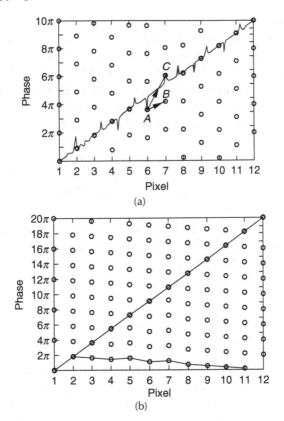

Figure 11.3 Phase unwrapping (a) in the presence of noise and (b) with oversampling.

unwrapping problem. The phase gradient required by Ghiglia and Romero (1994) is obtained as wrapped-phase differences along the x and y directions. This wrapped-gradient field is then least-squares integrated to obtain the continuous phase. More recently, Marroquín and Rivera (1995) extended the technique of least-squares integration of wrapped-phase gradients by adding a regularization term in the form of a norm of potentials. Using this technique, it is possible to filter out some noise in the unwrapped phase as well as interpolate the solution over regions of invalid phase data (such as holes) with a well-defined behavior.

One drawback of the least-squares integration or its regularized extension stems from the assumption that the phase difference between adjacent pixels is less than π in absolute value. That is, these techniques take the wrapped differences of the wrapped phase as if it were a true gradient field; unfortunately, however, this is not the case when severely noisy phase maps are being unwrapped. The phase gradient obtained here is actually wrapped in regions of high phase noise and high phase gradients. Using the least-squares unwrapping technique in very noisy phase maps leads to unwrapping errors due to a reduction of the dynamic range in the unwrapped phase.

In areas in an interferogram where the spatial frequency is low, phase unwrapping is relatively easy. Su and Xue (2001) pointed out that, by filtering the interferogram with a Hanning filter, phase unwrapping becomes more reliable in some cases.

11.2   UNWRAPPING CONSISTENT PHASE MAPS

In this section, we analyze two simple unwrapping techniques that apply to consistent phase maps. The first one unwraps full-field wrapped phase data. The second one deals with the unwrapping problem of consistent data within an arbitrary simple connected region.

11.2.1   Unwrapping Full-Field Consistent Phase Maps

The phase unwrapping technique shown in this section is one of the simplest methods for unwrapping a good or nearly consistent (small phase noise) smooth phase map. The technique consists of integrating phase differences along a scanning path (Figure 11.4). Let us assume that the full-field phase map is given by $\phi_W(x,y)$ in a regular two-dimensional lattice L of size $N \times N$ pixels. We can unwrap this phase map by unwrapping the first row ($y = 0$) of it and afterwards taking the last value of it as our initial condition to unwrap along

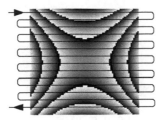

Figure 11.4 Scanning path followed by the proposed full-field phase unwrapper.

the following row of the phase map in a positive direction. We can do this along the first row by using the following formula:

$$\phi(x_{i+1},y_0) = \phi(x_i,y_0) + V[\phi_w(x_{i+1},y_0) - \phi(x_i,y_0)]; \quad 1 \le i \le N \quad (11.6)$$

where the wrapping function is $V(x) = [x - 2\pi\,\text{int}(x/\pi)]^2$, valid in the interval $(-\pi, +\pi)$. This function is equal to $V(x) = \tan^{-1}(\sin(x)/\cos(x))$ in the same range. In Equation 11.6, we can use as our initial condition:

$$\phi(x_0,y_0) = \phi_0 \quad (11.7)$$

Having unwrapped along the first row, we can use the last unwrapped phase value as our initial condition to unwrap the following row ($j = 1$) in the backward direction; that is:

$$\phi(x_{i-1},y_1) = \phi(x_i,y_1) + V[\phi_w(x_{i-1},y_1) - \phi(x_i,y_1)]; \quad 1 \le i \le N \quad (11.8)$$

For the backward unwrapping direction (Equation 11.8), we must use as our initial condition:

$$\phi(x_{N-1},y_1) = \phi(x_{N-1},y_0) + V[\phi_w(x_{N-1},y_1) - \phi(x_{N-1},y_0)] \quad (11.9)$$

The unwrapping then proceeds to the next row ($j = 2$) in the forward direction as:

$$\phi(x_{i+1},y_2) = \phi(x_i,y_2) + V[\phi_w(x_{i+1},y_2) - \phi(x_i,y_2)]; \quad 1 \le i \le N \quad (11.10)$$

Figure 11.5   Unwrapped full-field phase data using the sequential technique.

and our initial condition is:

$$\phi(x_0, y_2) = \phi(x_0, y_1) + V\left[\phi_w(x_0, y_2) - \phi(x_0, y_1)\right] \quad (11.11)$$

The scanning procedure just described is followed until the full-field phase map is unwrapped. The phase surface obtained using this sequential procedure is shown in Figure 11.5.

11.2.2 Unwrapping Consistent Phase Maps within a Simple Connected Region

On the other hand, what if we do not have a full-field phase map? If the shape of the consistent phase map is bounded by an arbitrary, simply connected region, such as the one shown in Figure 11.6, then the previous algorithm (Equations 11.6 to 11.11) cannot be used. For this situation, we can apply the following algorithm to unwrap a consistent phase map. To start, define and set to zero an indicator function, $\sigma(x,y)$, inside the domain (D) of valid phase data (as shown in Figure 11.6).

Figure 11.6   An example of a simple connected region containing valid phase data.

Then, choose a seed or starting point inside D and assign to it an arbitrary phase value of $\phi(x,y) = \phi_0$. Mark the visited site as unwrapped; that is, set $\sigma(x,y) = 1$. Now that the seed pixel phase is defined, we can carry out the unwrapping process:

1. Choose a pixel, (x,y), inside D (at random or in any prescribed order).
2. Test if the visited site, (x,y), inside D is already unwrapped.
   - If the selected site is marked as unwrapped ($\sigma(x,y) = 1$), then return to the first statement.
   - If the visited site is wrapped ($\sigma(x,y) = 0$), then test for any adjacent unwrapped pixel, (x',y').
   - If no adjacent pixel has already been unwrapped, then return to the first statement.
   - If an adjacent pixel, (x',y'), is found to be unwrapped, then take its phase value, $\phi(x',y')$, and use it to unwrap the current site, (x,y), as:

$$\phi(x,y) = \phi(x',y') + V\left[\phi_w(x,y) - \phi(x',y')\right] \quad (11.12)$$

   where $V(.)$ is the wrapping function defined before.
3. Mark the selected site as unwrapped ($\sigma(x,y) = 1$).
4. Return to the first statement until all the pixels in D are unwrapped.

The algorithm just described will unwrap any simply connected bounded region D having valid and consistent wrapped phase data, as shown in Figure 11.7.

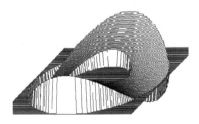

Figure 11.7   Noise-free phase unwrapped using the algorithm given in Section 11.2.2.

11.3 UNWRAPPING NOISY PHASE MAPS

We can still use the above-described algorithm to unwrap inconsistent phase maps corrupted by a small amount of noise. This can be done by marking the inconsistent wrapped phase pixels and excluding them from the unwrapping process as forbidden regions. Inconsistencies occur when multiples of 2π rad cannot be added to each wrapped phase sample over a two-dimensional grid to eliminate all adjacent phase differences greater than π rad in magnitude. Marking the inconsistent pixels is not practical as the noise increases greatly given that the number of inconsistent marked pixels can grow very quickly. For that reason, we will not provide the details of such techniques here.

Although many algorithms have been proposed for phase unwrapping in the presence of noise, we will limit our discussion here to the two algorithms that we feel are the most important for unwrapping inconsistent phase maps of smooth continuous functions. These algorithms are *least-squares integration of wrapped phase differences* (Ghiglia 1994) and the *regularized phase tracking* (RPT) *unwrapper*. Our discussion will not address the algorithms and techniques that can handle phase maps of noisy or discontinuous functions (Huntley, 1989, 1994; Huntley and Saldner, 1993; Buckland et al., 1995; Ströbel, 1996), because we feel that these techniques fall outside the scope of this book.

11.3.1 Unwrapping Using Least-Squares Integration

The least-squares technique was first introduced by Ghiglia et al. (1994) to unwrap inconsistent phase maps. To apply this method, begin by estimating the wrapped phase gradient along the x and y direction; that is,

$$\phi_y(x_i, y_j) = V\big[\phi_w(x_i, y_j) - \phi_w(x_i, y_{j-1})\big]$$

$$\phi_x(x_i, y_j) = V\big[\phi_w(x_i, y_j) - \phi_w(x_{i-1}, y_j)\big]$$

(11.13)

Because we have an oversampled phase map, the phase differences in Equation 11.13 will be everywhere in the range $(-\pi, +\pi)$; in other words, the estimated gradient will be unwrapped. Now we can integrate the phase gradient in a consistent way by means of a least-squares integration. The integrated or continuous phase we are seeking will be the one that minimizes the following cost function:

$$U(\phi) = \sum_{i=2}^{N} \sum_{j=2}^{M} \left[\phi(x_i, y_i) - \phi(x_{i-1}, y_j) - \phi_x(x_i, y_j)\right]^2 +$$

$$+ \sum_{i=2}^{N} \sum_{j=2}^{M} \left[\phi(x_i, y_j) - \phi(x_i, y_{j-1}) - \phi_y(x_i, y_j)\right]^2 \qquad (11.14)$$

This expression applies whenever we have a full-field wrapped phase. Let us assume that we have valid phase data only inside a two-dimensional region marked by an indicator function, $\sigma(x,y)$; that is, we will have valid phase data for $\sigma(x,y) = 1$ and invalid phase data for $\sigma(x,y) = 0$. We then can modify our cost function to include the indicator function as follows:

$$U = \sum_{i=2}^{N} \sum_{j=2}^{M} \left[\phi(x_i, y_j) - \phi(x_{i-1}, y_j) - \phi_x(x_i, y_j)\right]^2 \sigma(x_i, y_j)\sigma(x_{i-1}, y_j) +$$

$$+ \sum_{i=2}^{N} \sum_{j=2}^{M} \left[\phi(x_i, y_j) - \phi(x_i, y_{j-1}) - \phi_y(x_i, y_j)\right]^2 \sigma(x_i, y_j)\sigma(x_i, y_{j-1}) \qquad (11.15)$$

The estimated unwrapped phase $\phi(x,y)$ can be found, for example, by using a simple gradient descent at all pixels:

$$\phi^{k+1}(x, y) = \phi^k(x, y) - \tau \frac{\partial U}{\partial \phi(x, y)} \qquad (11.16)$$

where k is the iteration number and τ is the convergence rate of the gradient search system (typically around $\tau = 0.1$). Among the faster algorithms for obtaining the unwrapped phase are the techniques of conjugate gradient or the transform methods (Ghiglia and Romero, 1994).

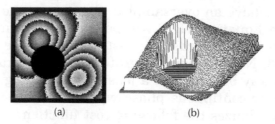

(a)   (b)

Figure 11.8 (a) Computer-generated noisy phase map; (b) unwrapped phase using least-squares integration of wrapped differences.

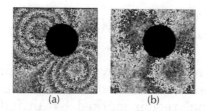

(a)   (b)

Figure 11.9 (a) Highly noisy phase map; (b) phase map obtained after unwrapping and then wrapping again for comparison purposes. We can see that the technique fails to recover the full dynamic range of the modulating phase because the wrapped first-order difference is a bad estimator of the true phase gradient in such a noisy phase map.

Consider the noisy phase map of Figure 11.8a. In this map, the wrapped phase, $\phi_w(x,y)$, is obtained as the sum of two Gaussians with different signs. Figure 11.8b shows the unwrapped phase map obtained using the least-squares integration technique developed by Ghiglia and Romero (1994). Figure 11.9b shows the phase after unwrapping and then wrapping again for comparison purposes. This phase, again, was obtained using the least-squares integration technique of wrapped differences applied to the same phase map (Ghiglia and Romero, 1994), but with more noise added. Note that the method is not as successful as with less noise, and a substantial decrease in the phase dynamic range can be observed.

11.3.2 The Regularized Phase Tracking Unwrapper

From Equation 11.2 we can see that the unwrapping inverse problem is ill posed; that is, the $m(x,y)$ field is not uniquely determined by the observations. This means that the unwrapping problem cannot be solved unless additional (prior) information about the expected unwrapped phase, $\phi(x,y)$, is provided. Smoothness is a typical piece of prior information that constrains the search space of unwrapped functions, and this information can be incorporated into the unwrapping algorithm using regularization theory (Marroquín and Rivera, 1995).

To regularize the phase unwrapping problem, it is necessary to find a suitable merit function that uses at least two terms that contribute to constraining the unwrapped field we are seeking. These terms are related by the following factors:

1. Fidelity between the estimated function and the observations
2. Prior knowledge about the spatial behavior of the unwrapped phase

It is then assumed that the phase function we seek is the one that minimizes this merit function.

In classical regularization we use a pixel-wise error between the sought function and the observed data and the norm of a differential operator over the this function as regularizer. In the proposed RPT technique however, we assume that in a small region of the image one can consider the data smooth enough so it can be modeled by a plane. This plane must be close to the observed phase map in the wrapped space (statement 1, above). A phase plane such as this must adapt itself to every region in the phase map so its local slope changes continuously in the two-dimensional space. We postulate that the phase of the estimated fringe pattern, $\phi(x,y)$, must minimize the following merit function at each site (x,y) containing valid phase data:

$$U_{x,y}(\phi,\omega_x,\omega_y) = \sum_{(\varepsilon,\eta)\in(N_{x,y}\cap L)} \left\{ \begin{array}{l} V\big[\phi_w(\varepsilon,\eta) - \phi_e(x,y,\varepsilon,\eta)\big] \\ + \lambda\big[\phi(\varepsilon,\eta) - \phi_e(x,y,\varepsilon,\eta)\big]^2 \sigma(\varepsilon,\eta) \end{array} \right\} \quad (11.17)$$

and

$$\phi_e(x,y,\varepsilon,\eta) = \phi(x,y) + \omega_x(x,y)(x-\varepsilon) + \omega_y(x,y)(y-\eta) \quad (11.18)$$

The functions $\phi_W(x,y)$, and $\phi(x,y)$ are the wrapped and unwrapped phases, respectively, estimated at pixel (x,y); L is the two-dimensional domain having valid wrapped phase data; and $N_{x,y}$ is a small neighborhood around the coordinate (x,y). As explained below, the function $\sigma(\varepsilon,\eta)$ is an indicator field that equals one if the site (ε,η) has already been unwrapped and zero otherwise. We can see from Equation 11.18 that we are approximating the local behavior of the unwrapped phase by a plane for which the parameters $\phi(x,y)$, $\omega_x(x,y)$, and $\omega_y(x,y)$ are determined in such a way that the merit function $U_{x,y}(\phi,\omega_x,\omega_y)$ at each site (x,y) in L is minimized.

The first term in Equation 11.17 attempts to keep the local phase model close to the observed phase map in a least-squares sense within the neighborhood $N_{x,y}$ (statement 1, above). The second term enforces our assumption of smoothness and continuity of the unwrapped phase (statement 2, above) using only previously unwrapped pixels marked by $\sigma(x,y)$. We can see that the second term will contribute a small amount to the value of the merit function $U_{x,y}(\phi,\omega_x,\omega_y)$ only for smooth unwrapped phase functions. Note also that the local phase plane is adapted simultaneously to the observed data (in the wrapped space using the wrapping operator $V[x]$) and to the continuous unwrapped phase marked by $\sigma(x,y)$.

To unwrap the phase map $\phi_W(x,y)$ we need to find the minimum of the merit function $U_{x,y}(\phi,\omega_x,\omega_y)$ (Equation 11.17) with respect to the fields $\phi(x,y)$, $\omega_x(x,y)$, and $\omega_y(x,y)$. To this end, we propose to find a minimum of $U_{x,y}(\phi,\omega_x,\omega_y)$ according to the sequential unwrapping algorithm described next.

The proposed unwrapping strategy in L is calculated as follows. To begin, we set the indicator function to zero $(m(x,y) = 0$ in $L)$ and choose a seed or starting point inside L to begin the unwrapping process. We then optimize the chosen site for $U_{x,y}(\phi,\omega_x,\omega_y)$ by adapting the triad $\phi_0(x,y)$, $\omega_x(x,y)$, $\omega_y(x,y)$ until a minimum is reached and mark the visited site as unwrapped; that is, we set $\sigma(x,y) = 1$. Now that the seed pixel is unwrapped, we can begin the unwrapping process as follows:

1. Choose a pixel inside L (at random or in any prescribed order).
2. Test whether or not the visited site is unwrapped:
   - If the selected site is marked as unwrapped (i.e., $\sigma(x,y) = 1$), then return to the first statement.
   - If the visited site is wrapped (i.e., $\sigma(x,y) = 0$), then test for any adjacent unwrapped pixel (x',y').
   - If no adjacent pixel (x',y') has already been unwrapped, then return to the first statement.
   - If an adjacent pixel (x',y') is found to be unwrapped, then take its optimized triad (ϕ,ω_x,ω_y) and use it as the initial condition to minimize the merit function $U_{x,y}(\phi,\omega_x,\omega_y)$ (Equation 11.18) at the chosen site (x,y).
3. When the minimum for $U_{x,y}(\phi,\omega_x,\omega_y)$ in (x,y) is reached, mark the selected site as unwrapped (i.e., $\sigma(x,y) = 1$).
4. Return to the first statement until all the pixels in L are unwrapped.

An intuitive way of regarding this iteration is as a "crystal growing" (CG) process in which new molecules (planes) are added to the bulk in that particular orientation (slope) to minimize the local crystal energy given the geometric orientation of the adjacent and previously positioned molecules.

We can use simple gradient descent to optimize $U_{x,y}$ by moving the triad (ϕ,ω_x,ω_y) as follows:

$$\phi^{k+1}(x,y) = \phi^k(x,y) - \tau\frac{\partial U_{x,y}(\phi,\omega_x,\omega_y)}{\partial\phi(x,y)}$$

$$\omega_x^{k+1}(x,y) = \omega_x^k(x,y) - \tau\frac{\partial U_{x,y}(\phi,\omega_x,\omega_y)}{\partial\omega_x(x,y)} \qquad (11.19)$$

$$\omega_y^{k+1}(x,y) = \omega_y^k(x,y) - \tau\frac{\partial U_{x,y}(\phi,\omega_x,\omega_y)}{\partial\omega_y(x,y)}$$

where τ is the convergence rate of the gradient search system. As mentioned before, the initial condition for Equation 11.19 is chosen from any adjacent unwrapped pixel. In practice, the τ parameter in the first relation in Equation 11.19 can be multiplied by about 10 to accelerate the convergence rate of the gradient search.

The first global phase estimation just described is usually very close to the actual unwrapped phase; if needed, one can perform additional global iterations to improve the phase estimation process. The additional iterations can be performed using Equation 11.19, but we now take as our initial condition the last estimated values at the same site (x,y) (not the ones at a neighborhood site, (x',y'), as done in the first global CG iteration). Note that for the additional global phase estimations, the indicator function $\sigma(x,y)$ in Equation 11.17 is now everywhere equal to one; therefore, we can scan the lattice in any desired order whenever all the sites are visited at each global iteration. In practice, only one or two additional global iterations are needed to reach a stable minimum of $U_{x,y}(\phi,\omega_x,\omega_y)$ at each site (x,y) in the two-dimensional lattice L.

One can argue that only the first term in Equation 11.17 can suffice to unwrap the observed phase map, but the simplified system was found to give good results only for small phase noise (between -0.2π and 0.2π). For higher amounts of phase noise (between -0.7π and 0.7π), the second term (the regularizing plane over the unwrapped phase) makes a substantial improvement in the noise robustness of the RPT system.

The parameter λ and the size of the neighborhood $(N_{x,y})$ are related to the unwrapped phase bandwidth and to the robustness of the RPT algorithm. For example, a very low-frequency, highly inconsistent phase map the size of $N_{x,y}$ should be large so the RPT system can properly track the smooth unwrapped phase in such a noisy field. When the size of $N_{x,y}$ has been chosen, the value of the λ parameter in Equation 11.7 is not very critical. A value of $\lambda = 2$ was used all over the results herein presented. The computational speed of the RPT technique is related to the size of the neighborhood $(N_{x,y})$ as well as the size of the lattice (L). In the literature, the size of $N_{x,y}$ has ranged from 5×5 pixels to 11

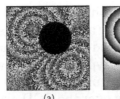

(a) (b)

Figure 11.10 (a) Highly noisy phase map (also shown in Figure 11.9a). (b) Phase obtained using the regularized phase tracking (RPT) technique and shown after unwrapping and then wrapping again for comparison purposes. We can see that the RPT technique works better than the least-squares technique (Figure 11.9b) for severe phase noise.

$\times$ 11 pixels. Given reasonably good phase maps, a neighborhood $N_{x,y}$ of 3 × 3 pixels can be sufficient, and the RPT system will give very quick and reliable results.

As in a crystal growing process, the size of the neighborhood $(N_{x,y})$ in the RPT technique is very critical. If it succeeds, the RPT system will move the entire unwrapping system to the correct attractor. If the crystal growing algorithm reaches a wrong attractor, the RPT system will give a wrong result. In these cases, we must try another neighborhood $(N_{x,y})$ for the RPT system and compute the solution again.

Figure 11.10b shows the phase obtained from the noisy phase map of Figure 11.10a after unwrapping using the RPT unwrapper and then wrapping again for comparison purposes. Inspecting this figure, we can appreciate the capacity of the RPT system to remove noise while preserving, almost unchanged, the original phase dynamic range. The noise introduced in Figure 11.10a can roughly be considered to be the maximum noise tolerated by the proposed RPT unwrapper. Notice how the unwrapped phase is almost unaffected near the image boundaries despite the large amount of noise.

11.4 UNWRAPPING SUBSAMPLED PHASE MAPS

Testing of aspherical wavefronts is routinely achieved in the optical shop by the use of commercial interferometers. The testing of deep aspheres is limited by aberrations of the imaging optics of the interferometer as well as the spatial resolution of the CCD video camera used to gather the interferometric data. The CCD video arrays typically come with 256×256 or 512×512 image pixels. The number of CCD pixels limits the highest recordable frequency over the CCD array to π rad/pixel. As seen in Chapter 2, this maximum recordable frequency is known as the Nyquist limit of the sampling system. The detected phase map of an interferogram having frequencies higher than the Nyquist limit contains false fringes and is said to be aliased. Another factor to take into account is the fact that CCD detector elements have a finite size, which can be almost as large as the pixel separation. In this case, the contrast of the sub-Nyquist sampled image is strongly reduced, as described in Chapter 2 and illustrated in Figure 2.12. Thus, aliasing fringes cannot be observed with these kind of detectors, unless a CCD detector is used that has detector elements of a size much smaller than their separation. Unfortunately, aliasing fringes can be recorded only if the size of each individual detector is smaller than half the maximum spatial frequency contained in the interferogram (the separation between the detector can be larger).

A specially constructed sparse array detector that has detector elements much smaller than their separation (Greivenkamp, 1987) is quite expensive and must be specially manufactured. This kind of detector can be simulated if some elements are eliminated in an image obtained with a normal detector for which the size of the elements is equal to their separations. The undesired elements can be eliminated before detection by means of placing a mask with holes over the desired detector elements or after detection when digitally processing the image. Of course, this simulation is not a real practical advantage and only serves the purpose of testing the unwrapping procedure. Aliasing fringes are quite useful for unwrapping sub-Nyquist sampled phase maps when utilizing any of the several methods described in the following sections.

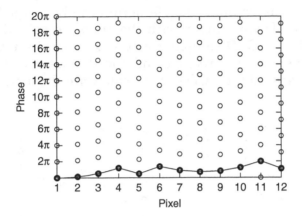

Figure 11.11 Wrapped phase for a wavefront with spherical aberration, with sub-Nyquist sampling.

11.4.1 Greivenkamp's Method

Subsampled phase maps cannot be unwrapped using standard techniques such as those presented so far; nevertheless, we can still unwrap an undersampled phase map if aliasing fringes are obtained and:

1. We have enough knowledge about the wavefront being tested to null test the wavefront under analysis (Greivenkamp, 1987; Servín and Malacara, 1996a).
2. The expected wavefront is smooth, in which case we can introduce this prior knowledge into the unwrapping process (Greivenkamp, 1987; Servín and Malacara, 1996b).

To illustrate the principle of Greivenkamp's sub-Nyquist phase unwrapping in one dimension, Figure 11.11 shows the unwrapped phase in a wavefront produced by an optical system with spherical aberration. The correct unwrapping result is shown in Figure 11.12a; however, if no previous knowledge about the wavefront shape is available, the result in Figure 11.12b would be obtained.

The undersampled interferogram can be imaged directly over the CCD video array with the aid of an optical interferometer. If the CCD sampling rate is x_s over the x direction, and y_s over the y direction and the diameter of the light-sensitive area of the CCD is d, we can write the mathematical

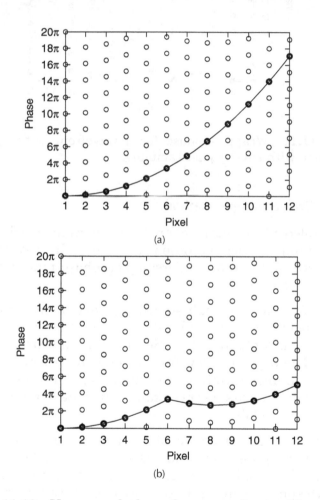

(a)

(b)

Figure 11.12 Unwrapped phase for a wavefront with spherical aberration, with sub-Nyquist sampling: (a) correct phase, and (b) phase obtained if no previous knowledge is available.

expression for the sampling operation over the irradiance of the interferogram (Equation 11.1) as:

$$S[s(x,y)] = \left[s(x,y) * * \text{circ}\left(\frac{\rho}{d}\right) \right] \text{comb}\left(\frac{x}{x_s}, \frac{y}{y_s}\right),$$

$$\rho = \left(x^2 + y^2\right)^{1/2}$$

(11.20)

where the function $S[s(x,y)]$ is the sampling operator over the irradiance given by Equation 11.1, the symbol (**) indicates a two-dimensional convolution, and circ(ρ/d) is the circular size of the CCD detector. The comb function is an array of delta functions with the same spacing as the CCD pixels. The phase map of the sampled interferogram in Equation 11.20 can be obtained using, for example, three phase-shifted interferograms as follows:

$$s_1(x,y) = a(x,y) + b(x,y)\cos(\phi(x,y) + \alpha)$$

$$s_2(x,y) = a(x,y) + b(x,y)\cos(\phi(x,y))$$

(11.21)

$$s_3(x,y) = a(x,y) + b(x,y)\cos(\phi(x,y) - \alpha)$$

where α is the phase shift. Using well-known formulae, we can find the subsampled wrapped phase as:

$$\phi_w(x,y) = \tan^{-1}\left(\frac{1 - \cos(\alpha)}{\sin(\alpha)} \frac{S[s_1(x,y)] - S[s_3(x,y)]}{2S[s_1(x,y)] - S[s_2(x,y)] - S[s_3(x,y)]}\right) \times$$

$$\times \sigma(x,y)$$

(11.22)

where $\sigma(x,y)$ is an indicator function that equals one if we have valid phase data; zero, otherwise. As Equation 11.22 shows, the phase obtained is a modulo 2π of the true undersampled phase due to the arc tangent function involved in the phase-detection process. Figure 11.13 shows an example of a subsampled phase map of pure spherical aberration.

Figure 11.13 Subsampled phase map corresponding to pure spherical aberration.

11.4.2 Null Fringe Analysis of Subsampled Phase Maps Using a Computer-Stored Compensator

As mentioned earlier, one way to deal with deep aspherical wavefronts is to use an optical, diffractive, or software compensator. Optical or diffractive compensators reduce the number of aberration fringes so they can be analyzed without aliasing. To construct the compensator, we must have a good knowledge of the testing wavefront up to a few aberration fringes. The remaining aberration fringes constitute the error between the expected or ideal wavefront and the actual one from the testing optics. In this way, we can analyze the remaining uncompensated fringes using standard fringe analysis techniques. Fortunately, in optical shop testing, we typically have a good knowledge of the kind and amount of aberration expected at the testing plane (in the final stages of the manufacturing process). This knowledge allows us to construct the proper optical or diffractive compensator. In this section, we deal with another kind of compensator: the software compensator (Servín and Malacara, 1996). The software compensator does not have to be constructed (as an optical or diffractive compensator); instead, it is calculated by computer. This software compensator, however, does require a specially constructed CCD video array having small light detector size d with respect to the spatial separation, (x_s, y_s) (see Equation 11.20).

If we assume that the expected or ideal wavefront, $\phi_i(x,y)$, differs from the detected phase, $\phi_w(x,y)$, by only a few wavelengths, we can form an oversampled wrapped wavefront error, $\Delta\phi_w(x,y)$, as:

$$\Delta\phi_w(x,y) = \tan^{-1}\{\tan[\phi_w(x,y) - \phi_i(x,y)]\}\sigma(x,y) \quad (11.23)$$

We can then unwrap the wavefront error, $\Delta\phi_w(x,y)$, by using standard unwrapping techniques. To obtain the unwrapped testing wavefront, the unwrapped error and the ideal wavefront are added:

$$\phi(x,y) = [\phi_i(x,y) + \Delta\phi(x,y)]\sigma(x,y) \quad (11.24)$$

where $\Delta\phi(x,y)$ is the unwrapped phase error. As mentioned before, the limitation of the technique presented in this section resides in the fact that the error wavefront (Equation 11.19) must be oversampled. This requirement is the same as when an holographic or diffractive compensator is used. That is, the wavefront being tested must be close enough to the expected ideal wavefront to obtain a compensated interferogram having spatial frequencies below the Nyquist upper bound over the CCD array. In summary, the problem of building an optical or holographic compensator is replaced herein by the construction of a special-purpose CCD video array or construction of a mask of small holes in contact with the CCD array. The considerable benefit of this approach is that, when the CCD mask or the specially built CCD array is available, the need to build special-purpose diffractive or holographic compensators disappears. The use of this technique is illustrated in Figure 11.14. Figure 11.14a shows the analysis of a subsampled phase map. This phase map is then compared, using Equation 11.23, to the expected one shown in Figure 11.14b. Their phase difference (the phase error between them) is shown in Figure11.14c. As in the case of using an optical compensator, positioning of the CCD array used to collect the interference irradiance is very critical. A mispositioning of the compensator or, in this case, the CCD array can give erroneous measurements.

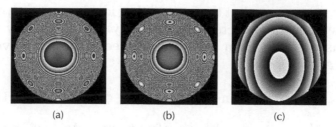

(a)   (b)   (c)

Figure 11.14 (a) Subsampled phase map obtained using Equation 11.19; (b) ideal or expected subsampled phase map; (c) phase error between the two phase maps according to Equation 11.21.

11.4.3 Unwrapping of Smooth Continuous Subsampled Phase Maps

In the last subsection, we have discussed the problem of unwrapping undersampled phase maps. The method is based on having a good enough prior knowledge of the kind and amount of aberrations to perform null testing on the detected phase map. This section generalizes the problem of unwrapping undersampled phase maps to smooth wavefronts; that is, the only prior knowledge about the wavefront being analyzed is the smoothness. This is far less restrictive than the null testing technique presented in the last section. Analysis of interferometric data beyond the Nyquist frequency was first proposed by Greivenkamp (1987), who assumed that the wavefront being tested is smooth up to the first or second derivative. Greivenkamp's approach to unwrapping subsampled phase maps consists of adding multiples of 2π each time a discontinuity in the phase map is found. The number of times a 2π is added is determined by the smoothness condition imposed on the wavefront in its first or second derivative along the unwrapping direction. Although Greivenkamp's approach is robust against noise, its weakness resides in the fact that it is a path-dependent phase unwrapper.

The method of Servín and Malacara (1996) overcomes the path dependency of the Greivenkamp approach but preserves its noise robustness. In this case, an estimation of the local wrapped curvature (or wrapped Laplacian) of the sub-sampled phase map, $\phi_w(x,y)$ (Equation 11.22), is used to unwrap the interesting deep aspherical wavefront. When we have obtained the local wrapped curvature along the x and y directions we can use least-squares integration to obtain the unwrapped continuous wavefront. The local wrapped curvature is obtained as:

$$L_x(x_i,y_j) = V\Big[\phi_w(x_{i-1},y_j) - 2\phi_w(x_i,y_j) + \phi_w(x_{i+1},y_j)\Big]$$
$$L_y(x_i,y_j) = V\Big[\phi_w(x_i,y_{j-1}) - 2\phi_w(x_i,y_j) + \phi_w(x_i,y_{j+1})\Big] \tag{11.25}$$

If the absolute value of the discrete wrapped Laplacian given by Equation11.25 is less than π, its value will be unwrapped. We can then obtain the unwrapped phase, $\phi(x,y)$, by means of the function that minimizes the following quadratic merit function (least squares):

$$U = \sum_{(x,y)\in\sigma(x,y)} U_x(x,y)^2 + U_y(x,y)^2 \tag{11.26}$$

where $\sigma(x,y)$ is an indicator or mask function that equals one if we have valid phase data; zero, otherwise. The functions $U_x(x,y)$ and $U_y(x,y)$ are given by:

$$U_x(x_i,y_j) = L_x(x_i,y_j) - \Big[\phi(x_{i-1},y_j) - 2\phi(x_i,y_j) + \phi(x_{i+1},y_j)\Big]$$
$$U_y(x_i,y_j) = L_y(x_i,y_j) - \Big[\phi(x_i,y_{j-1}) - 2\phi(x_i,y_j) + \phi(x_i,y_{j+1})\Big] \tag{11.27}$$

The minimum of the merit function given by Equation 11.26 is obtained when its partial with respect to $\phi(x,y)$ equals zero; therefore, the set of linear equations that must be solved is:

$$\frac{\partial U}{\partial\phi(x,y)} = U_x(x_{i-1},y_j) - 2U_x(x_i,y_j) + U_x(x_{i+1},y_j) + \tag{11.28}$$
$$+ U_y(x_i,y_{j-1}) - 2U_y(x_i,y_j) + U_y(x_i,y_{j+1})$$

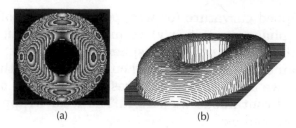

(a)   (b)

Figure 11.15 (a) Subsampled phase map of a wavefront with a central obstruction. (b) Wire mesh of the unwrapped phase map according to the least-squares integration of wrapped phase curvature presented in this section.

Several methods can be used to solve this system of linear equations; among others is the simple gradient descent shown below:

$$\phi^{k+1}(x,y) = \phi^k(x,y) - \eta\frac{\partial U}{\partial\phi(x,y)} \qquad (11.29)$$

where the parameter η is the rate of convergence of the gradient search. The simple gradient descent is quite slow for this application, so we have used a conjugate gradient to speed up the computing time. Figure 11.15a shows a subsampled phase map, and Figure 11.15b shows the unwrapped phase in wire mesh.

11.4.4 Unwrapping the Partial Derivative of the Wavefront

Another method for unwrapping an oversampled interferogram is to simulate a lateral shear interferogram, as shown by Muñoz et al. (2003, 2004). Essentially this method is equivalent to calculating a lateral shear interferogram, where the slopes are smaller than in the original wavefront. A lateral shear interferogram can be digitally obtained from a Twyman–Green-like interferogram with phase differences $\phi(x,y)$, which are written here as ϕ_{ij}, by creating a new phase map given by $\phi_{ij} - \phi_{ij+1}$. This phase map can be obtained with the following trigonometric expression:

$$\phi_i - \phi_{i+1} = \tan^{-1}\left(\frac{\sin\phi_i \cos\phi_{i+1} - \cos\phi_i \sin\phi_{i+1}}{\cos\phi_i \cos\phi_{i+1} - \sin\phi_i \sin\phi_{i+1}}\right) \quad (11.30)$$

where

$$\tan\phi_i = \left(\frac{N_i}{D_i}\right) \quad (11.31)$$

and

$$\tan\phi_{i+1} = \left(\frac{N_{i+1}}{D_{i+1}}\right) \quad (11.32)$$

Hence, the sin and cosine values of ϕ_i and ϕ_{i+1} can be obtained from:

$$\sin\phi_i = \left(\frac{N_i}{\sqrt{D_i^2 + N_i^2}}\right) \quad (11.33)$$

and

$$\cos\phi_i = \left(\frac{D_i}{\sqrt{D_i^2 + N_i^2}}\right) \quad (11.34)$$

and in an identical manner for the pixel $(i + 1)$. When these functions have been obtained, they are substituted in Equation 11.30 to obtain the desired phase map. This map can be interpreted as a lateral shear interferogram with a shear equal to one pixel.

11.5 CONCLUSIONS

In this chapter, we have analyzed some important techniques for unwrapping phase maps of continuous and smooth functions. We presented two algorithms to unwrap good-quality phase maps; the first one applies only to full-field phase maps while the second one can be applied to a phase map bounded by an arbitrary single connected shape. We have also presented the unwrapping technique utilizing least-squares integration of phase gradients to obtain the continuous phase being sought. The main limitation of this approach is estimation of the phase gradient as the wrapped difference of two consecutive pixels along the x and y directions. This gradient phase estimation works well only for relatively small phase noise because a very noisy phase map can have differences between two adjacent pixels that exceed π or $-\pi$ rad.

Next we discussed the two-dimensional regularized phase tracking (RPT) phase unwrapping system, which is capable of unwrapping severely degraded phase maps. This unwrapping system tracks the instantaneous phase and its gradient, adapting a plane to the estimated wrapped and unwrapped phases simultaneously. In other words, the system fits the best least-squares tangent plane at each pixel in the wrapped and unwrapped phase space within a small neighborhood $(N_{x,y})$. When the least-squares best plane is found at a given location, the constant term of this plane, $\phi(x,y)$, gives the estimated unwrapped phase at the (x,y) location, and the slope, (ω_x, ω_y), estimates the local frequency.

Finally we analyzed two techniques for dealing with subsampled interferograms. One of these is a null unwrapping technique in which we must have information about the wrapped wavefront up to a few wavelengths. The second technique is more general; the only prior assumption about the testing wavefront is smoothness up to its second derivative.

REFERENCES

Bone, D.J., Fourier fringe analysis: the two dimensional phase unwrapping problem, *Appl. Opt.*, 30, 3627–3632, 1991.

Bryanston-Cross, P.J. and Quan, C., Examples of automatic phase unwrapping applied to interferometric and photoelastic images, in *Proceedings of the 2nd International Workshop on Automatic Processing of Fringe Patterns*, Jüptner, W. and Osten, W., Eds., Akademie Verlag, Bremen, 1993.

Buckland, J.R., Huntley, J.M., and Turner, S.R.E., Unwrapping noisy phase maps by use of a minimum-cost-matching algorithm, *Appl. Opt.*, 5100–5108, 1995.

Fried, D.L., Least-squares fitting a wave-front distortion estimate to an array of phase difference measurements, *J. Opt. Soc. Am.*, 67, 370–375, 1977.

Ghiglia, D.C. and Romero, L.A., Robust two dimensional weighted and unweighted phase unwrapping that uses fast transforms and iterative methods, *J. Opt. Soc. Am. A*, 11, 107–117, 1994.

Ghiglia, D.C., Mastin, G.A., and Romero, L.A., Cellular automata method for phase unwrapping, *J. Opt. Soc. Am.*, 4, 267–280, 1987.

Greivenkamp, J.E., Sub-Nyquist interferometry, *Appl. Opt.*, 26, 5245–5258, 1987.

Hudgin, R.H., Wave-front reconstruction for compensated imaging, *J. Opt. Soc. Am.*, 67, 375–378, 1977.

Hunt, B.R., Matrix formulation of the reconstruction of phase values from phase differences, *J. Opt. Soc. Am.*, 69, 393–399, 1979.

Huntley, J.M., Noise-immune phase unwrapping algorithm, *Appl. Opt.*, 28, 3268–3270, 1989.

Huntley, J.M., Phase unwrapping: problems and approaches, in *Proc. FASIG, Fringe Analysis '94*, York University, U.K., 1994.

Huntley, J.M. and Saldner, H., Temporal phase-unwrapping algorithm for automated interferogram analysis, *Appl. Opt.* 21, 3047–3052, 1993.

Huntley, J.M., Cusack, R., and Saldner, H., New phase unwrapping algorithms, in *Proceedings of the 2nd International Workshop on Automatic Processing of Fringe Patterns*, Jüptner, W. and Osten, W., Eds., Akademie Verlag, Bremen, 1993.

Itoh, K., Analysis of the phase unwrapping algorithm, *Appl. Opt.* 21, 2470–2473, 1982.

Kreis, T., Digital holographic interference-phase measurement using the Fourier-transform method, *J. Opt. Soc. Am. A*, 3, 847–855, 1986.

Macy, W. Jr., Two-dimensional fringe pattern analysis, *Appl. Opt.*, 22, 3898–3901, 1983.

Marroquín, J.L. and Rivera, M., Quadratic regularization functionals for phase unwrapping, *J. Opt. Soc. Am. A*, 12, 2393–2400, 1995.

Muñoz, J., Stroknik, M., and Páez, G., Phase recovery from a single undersampled interferogram, *Appl. Opt.*, 42, 6846–6852, 2003.

Muñoz, J., Páez, G., and Stroknik, M., Two-dimensional phase unwrapping of subsampled phase-shifted interferograms, *J. Mod. Opt.*, 51, 49–63, 2004.

Noll, R.J., Phase estimates from slope-type wave-front sensors, *J. Opt. Soc. Am.*, 68, 139–140, 1978.

Servín, M. and Malacara, D., Sub-Nyquist interferometry using a computer stored reference, *J. Mod. Opt.*, 43, 1723–1729, 1996a.

Servín, M. and Malacara, D., Path-independent phase unwrapping of subsampled phase maps, *Appl. Opt.*, 35, 1643–1649, 1996b.

Ströbel, B., Processing of interferometric phase maps as complex-valued phasor images, *Appl. Opt.*, 35, 2192–2198, 1996.

Su, X. and Xue, L., Phase unwrapping algorithm based on fringe frequency analysis in Fourier-transform profilometry, *Opt. Eng.*, 40, 637–643, 2001.

Takajo, H. and Takahashi, K., Least squares phase estimation from phase differences, *J. Opt. Soc. Am. A*, 5, 416–425, 1988.

12

Wavefront Curvature Sensing

12.1 WAVEFRONT DETERMINATION BY SLOPE SENSING

Wavefront slopes can be measured by using testing methods that measure the transverse ray aberrations in the x and y directions, which are directly related to the partial derivatives of the wavefront under analysis. Many of these tests use screens; two typical examples are the Hartmann and the Ronchi tests described in Chapter 1. Another system that measures the wavefront slopes is the lateral shearing interferometer, also described in Chapter 1. The transverse aberrations are related to the wavefront slopes. To obtain the shape of the testing wavefront we must use an integration procedure as described before. In this chapter, we describe another method to obtain the wavefront by measuring local curvatures using diffraction images.

12.2 WAVEFRONT CURVATURE SENSING

The observation of defocused stellar images, known as the star test, has been used for many years as a sensitive method for detecting small wavefront deformations. The principle of this method is based on the fact that the illumination in a defocused image is not homogeneous if the wavefront has deformations. These deformations can be interpreted as variations in the local curvature of the wavefront. If the focus is shortened, the light energy will be concentrated at a shorter focus and vice versa. An obvious consequence is that the illuminations at the two planes being observed, located symmetrically with respect to the focus, have different illumination densities. For a long time, this test was used primarily as a qualitative visual test.

12.2.1 The Laplacian and Local Average Curvatures

Roddier (1988) and Roddier et al. (1988) proposed a quantitative wavefront evaluation method indirectly based on the star test principle which measures wavefront local curvatures. The local curvatures c_x and c_y of a nearly flat wavefront in the x and y directions are given by the second partial derivatives of this wavefront as follows:

$$c_x = \frac{\partial^2 W(x,y)}{\partial x^2} \quad \text{and} \quad c_y = \frac{\partial^2 W(x,y)}{\partial y^2} \tag{12.1}$$

Hence, the Laplacian defined by:

$$\nabla^2 W(x,y) = 2\rho(x,y) = \frac{\partial^2 W(x,y)}{\partial x^2} + \frac{\partial^2 W(x,y)}{\partial y^2} \tag{12.2}$$

is twice the value of the average local curvature $\rho(x,y)$. This expression is known as the Poisson equation. To solve the Poisson equation to obtain the wavefront deformations $W(x,y)$, the following must apply:

1. The average local curvature distribution, $\rho(x,y)$, is a scalar field and no direction is involved (as in the wavefront slopes).
2. The radial wavefront slopes at the edge of the circular pupil are used as Neumann boundary conditions.

As described by Roddier et al. (1988), the simplest method to solve the Poisson equation when the Laplacian has been determined is the Jacobi iteration algorithm. Noll (1978) showed that Jacobi's method is essentially the same as that derived by Hudgin (1977) to find the wavefront from slope measurements. Equivalent iterative Fourier methods to obtain the wavefront without having to solve the Poisson equation directly are described in Section 12.3.4.

12.2.2 Irradiance Transport Equation

Let us consider a light beam propagating with an average direction along the z-axis after passing through a diffracting aperture (pupil) on the x,y plane. The irradiance as well as the wavefront shape continuously change along the trajectory. As proved by Teague (1983), the wave disturbance $u(x,y,z)$ at a point (x,y,z) can be found with good accuracy, even with a diffracting aperture with sharp edges, using the Huygens–Fresnel diffraction theory if a paraxial approximation is taken. This approximation considers the Huygens wavelets to be emitted in a narrow cone and uses a parabolic approximation for the wavefront shape of each wavelet. This can be considered a geometrical optics approximation. Teague (1983) and Steibl (1984) showed that if we assume a wide diffracting aperture, much larger than the wavelength, the disturbance at any plane with any value of z can be found with the differential equation:

$$\nabla^2 u(x,y,z) + 2k^2 u(x,y,z) + 2ik\frac{\partial u(x,y,z)}{\partial z} = 0 \qquad (12.3)$$

where $k = 2\pi/\lambda$. We can consider a solution to this equation of the form:

$$u(x,y,z) = I^{1/2}(x,y,z)\exp(ikW(x,y,z)) \qquad (12.4)$$

where $I(x,y,z)$ is the irradiance. If we substitute this disturbance expression into the differential equation, after some algebraic steps we can obtain a complex function that should be made equal to zero. Then, equating real and imaginary parts to zero, we obtain:

$$\frac{\partial W}{\partial z} = 1 + \frac{1}{4k^2 I} \nabla^2 I - \frac{1}{2} \nabla W \bullet \nabla W - \frac{1}{8k^2 I^2} \nabla I \bullet \nabla I \quad (12.5)$$

and

$$\frac{\partial I}{\partial z} = -\nabla I \bullet \nabla W - I \nabla^2 W \quad (12.6)$$

where the (x,y,z) dependence has been omitted for notational simplicity and the Laplacian (∇^2) and gradient (∇) operators work only on the lateral coordinates x and y. The first expression is the phase transport equation, which can be used to find the wavefront shape at any point along the trajectory. The second expression is the irradiance transport equation. Ichikawa et al. (1988) demonstrated phase retrieval based on this equation. Following an interesting discussion by Ichikawa et al. (1988), we can note in the irradiance transport equation the following interpretation for each term:

1. The gradient $\nabla W(x,y,z)$ is the direction and magnitude of the local tilt of the wavefront, and $\nabla I(x,y,z)$ is the direction in which the irradiance value changes with maximum speed. Thus, their scalar product, $\nabla I(x,y,z) \bullet \nabla W(x,y,z)$, is the irradiance variation along the optical axis z due to the local wavefront tilt. Ichikawa et al. (1988) referred to this as a *prism* term.
2. The second term, $I(x,y,z) \nabla^2 W(x,y,z)$, can be interpreted as the irradiance along the z-axis caused by the local wavefront average curvature. Ichikawa et al. (1988) referred to this as a *lens* term.

In sum, these terms describe the variation of the beam irradiance caused by the wavefront deformations as it propagates along the z-axis. This means that the transport equation is a geometrical optics approximation, valid in the absence of sharp apertures and as long as the aperture is large enough compared to the wavelength. To gain even greater insight into the nature of this equation, we can rewrite it as:

$$-\frac{\partial I(x,y,z)}{\partial z} = \nabla \bullet \left[I(x,y,z) \nabla W(x,y,z) \right] \quad (12.7)$$

and, recalling that ∇W is a vector representing the wavefront local slope, we can easily see that the transport equation represents the law of light energy conservation, which is analogous to the law of mass or charge conservation, frequently expressed by:

$$-\frac{\partial \rho}{\partial t} = \nabla \bullet (\rho v) \tag{12.8}$$

where ρ and v are the mass or charge density and the flow velocity, respectively.

12.2.3 Laplacian Determination with Irradiance Transport Equation

Roddier et al. (1990) used the transport equation to measure the wavefront. Let $P(x,y)$ be the transmittance of the pupil which is equal to one inside the pupil and zero outside. Furthermore, we assume that the illumination at the plane of the pupil is uniform and equal to a constant I_0 inside the pupil. Hence, the irradiance gradient $\nabla I(x,y,0) = 0$ everywhere except at the edge of the pupil where:

$$\nabla I(x,y,0) = -I_0 \boldsymbol{n} \delta_c \tag{12.9}$$

where δ_c is a Dirac distribution around the edge of the pupil, and $\boldsymbol{n}$ is a unit vector perpendicular to the edge and pointing outward. Substituting this gradient into the irradiance transport equation we obtain:

$$\left(\frac{\partial I(x,y,z)}{\partial z}\right)_{z=0} = -I_0 \bullet \left(\frac{\partial W(x,y,z)}{\partial n}\right)_{z=0} \delta_c - \\ -I_0 P(x,y)\nabla^2 W(x,y,z). \tag{12.10}$$

where the derivative on the right-hand side of the expression is the wavefront derivative in the outward direction, perpendicular to the edge of the pupil. Curvature sensing consists of taking the difference between the illuminations observed in two planes located symmetrically with respect to the diffracting

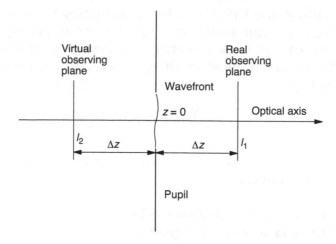

Figure 12.1 Irradiance measured in two planes placed symmetrically with respect to the pupil.

stop, as shown in Figure 12.1. Thus, the measured irradiances at these two planes are:

$$I_1(x, y, \Delta z) = I_0 + \left(\frac{\partial I(x, y, z)}{\partial z} \right)_{z=0} \Delta z$$

$$I_2(x, y, -\Delta z) = I_0 - \left(\frac{\partial I(x, y, z)}{\partial z} \right)_{z=0} \Delta z$$

(12.11)

When the wavefront is perfectly flat at the pupil, the Laplacian at all points inside the pupil and the radial slope at the edge of the pupil are both zero. Then, $I_2(x,y,-\Delta z)$ is equal to $I_1(x,y,\Delta z)$. Having obtained these data, we can form the so-called sensor signal as:

$$s(x, y\Delta z) = \frac{I_1 - I_2}{I_1 + I_2} = \frac{1}{I_0} \left(\frac{\partial I(x, y, z)}{\partial z} \right)_{z=0} \Delta z$$

(12.12)

Substituting Equation 12.27 into Equation 12.29 yields:

$$\frac{I_1 - I_2}{I_1 + I_2} = \left(\frac{\partial W(x, y)}{\partial n} \delta_c - P(x, y) \nabla^2 W(x, y) \right) \Delta z$$

(12.13)

Thus, with the irradiances I_1 and I_2 in two planes located symmetrically with respect to the pupil ($z = 0$), we obtain the left-hand term of this expression. This gives us the Laplacian of $W(x,y)$ (average local curvature) for all points inside the aperture and the wavefront slope, $\partial W/\partial n$, around the edge of the pupil, $P(x,y)$, as a Neumann boundary condition, to be used when solving Poisson's equation.

The two planes on which the irradiance has to be measured are symmetrically located with respect to the diffracting pupil. In other words, one plane is real because it is located after the pupil, but the other plane is virtual, because it is located before the pupil. In practice, this problem has an easy solution because the diffracting aperture is the pupil of a lens to be evaluated, typically a telescope objective.

As we see in Figure 12.2, a plane at a distance l inside the focus is conjugate to a plane at a distance Δz after the pupil. On the other hand, if a small lens with focal length $f/2$ is placed at the focus of the objective, a plane at a distance l outside the objective focus is conjugate to a plane at a distance Δz before the pupil. In both cases, the distance Δz and the distance l are related by:

$$\Delta z = \frac{f(f-l)}{l} \tag{12.14}$$

Roddier and Roddier (1991b) pointed out that a small lens with length $f/2$ is not necessary if l is small compared with f. We must take into account that one defocused image is rotated 180° with respect to the other, as well as any possible difference in the magnification of the two images. The important consideration is that the subtracted and added irradiances in the two measured images must correspond to the same point (x,y) on the pupil.

The measurements of the irradiance have to be made close enough to the pupil so the diffraction effects are negligible and the geometric approximation remains valid. Let us assume that the wavefront to be measured has some corrugations and deformations of scale r_0 (maximum spatial period). With the diffraction grating equation we see that these corrugations spread

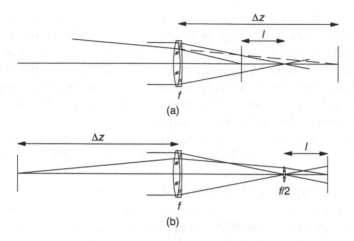

Figure 12.2 Two conjugate planes, one plane before refraction on the optical system, at a distance Δz from the pupil, and the second plane after refraction, at a distance l from the focus of the system: (a) with the first plane at the back of the pupil and the second plane inside of focus; and (b) with the first plane at the front of the pupil and the second plane outside of focus, using an auxiliary small lens with focal length $f/2$.

out the light over a narrow cone with an angular diameter α $= \lambda/r_0$. Thus, the illumination in the plane of observation can be considered a blurred pupil image. Let us now impose the condition that the maximum allowed blurring at a distance Δz is equal to $r_0/2$. With this condition it is possible to show that the geometrical optics approximation implied in the transport irradiance equation is valid only if Δz is sufficiently small, so that the following condition is satisfied:

$$\Delta z \ll \frac{r_0^2}{2\lambda} \qquad (12.15)$$

It is interesting to see that the distance Δz is one fourth the Rayleigh distance in Talbot autoimaging, as described in Chapter 1. This result is to be expected, as then the shadow of the grating is geometrical. If the light angular diameter spread (α) is known (for example, if this is equal to the atmospheric light seen in a telescope), then we can also write:

$$\Delta z << \frac{\lambda}{2\alpha^2} \qquad (12.16)$$

When measuring in the converging beam, this condition implies that the defocusing distance l should be large enough so we have:

$$l >> \frac{f}{1 + \dfrac{r_0^2}{2\lambda f}} \qquad (12.17)$$

In conclusion, the minimum defocusing distance depends on the maximum spatial frequency of the wavefront corrugation we want to measure. This frequency also determines the density of sampling points to be used to measure the irradiance in the defocused image.

12.2.4 Wavefront Determination with Iterative Fourier Transforms

Hardy et al. (1977) measured slope differences to obtain the curvatures from which the Poisson equation can be solved to obtain the wavefront. The curvature in the x direction is taken as the difference between two adjacent tilts in this direction, and in the same manner the curvature along the y-axis is obtained. The average of these curvatures can then be calculated. They used the Hudgin (1977) algorithm to obtain this solution.

Roddier and Roddier (1991a) and Roddier et al. (1990) reported a method for obtaining the wavefront deformations, $W(x,y)$, from a knowledge of the Laplacian operator by solving the Poisson equation using iterative Fourier transforms. To understand this method, let us take the Fourier transform of the Laplacian operator of the wavefront as follows:

$$F\{\nabla^2(x,y)\} = F\left\{\frac{\partial^2 W(x,y)}{\partial x^2}\right\} + F\left\{\frac{\partial^2 W(x,y)}{\partial y^2}\right\} \qquad (12.18)$$

On the other hand, from the derivative theorem in Section 2.3.4, we have:

$$\mathrm{F}\left\{\frac{\partial W(x,y)}{\partial x}\right\} = i2\pi f_x \mathrm{F}\{W(x,y)\} \qquad (12.19)$$

and similarly for the partial derivative with respect to y. In an identical manner we can also write:

$$\mathrm{F}\left\{\frac{\partial^2 W(x,y)}{\partial x^2}\right\} = i2\pi f_x \mathrm{F}\left\{\frac{\partial W(x,y)}{\partial x}\right\} = -4\pi^2 f_x^2 \mathrm{F}\{W(x,y)\} \quad (12.20)$$

Thus, it is easy to prove that

$$\mathrm{F}\{\nabla^2(x,y)\} = -4\pi^2 \mathrm{F}\{W(x,y)\}\left(f_x^2 + f_y^2\right) \qquad (12.21)$$

Hence, in the Fourier domain the Fourier transform of the Laplacian operator translates into a multiplication of the Fourier transform of the wavefront $W(x,y)$ by $f_x^2 + f_y^2$.

The wavefront can be calculated if measurements of the slopes along x and y are available, as in the case of the Hartmann and Ronchi tests:

$$W(x,y) = -\frac{i}{2\pi}\mathrm{F}^{-1}\left\{\frac{f_x \mathrm{F}\left\{\dfrac{\partial W(x,y)}{\partial x}\right\} + f_y \mathrm{F}\left\{\dfrac{\partial W(x,y)}{\partial y}\right\}}{f_x^2 + f_y^2}\right\} \qquad (12.22)$$

This simple approach works for a wavefront without any limiting pupil. In practice, however, the Laplacian operator is multiplied by the pupil function to take into account its finite size; thus, its Fourier transform is convolved with the Fourier transform of the pupil function. As a result, this procedure does not give correct results. To extrapolate the fringes outside of the pupil an apodization in the Fourier space (i.e., a filtering of the frequencies produced by the pupil boundaries) is necessary, as in the Gershberg algorithm described earlier in this book. Dividing by $f_x^2 + f_y^2$ produces this filtering. As a result of this filtering, just as in the Gershberg algorithm, and after taking the inverse Fourier transform, the wavefront extension is not restricted to the internal region of the pupil

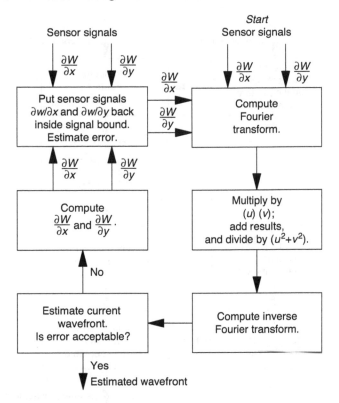

Figure 12.3 Iterative Fourier transform algorithm used to find the wavefront from the measured slopes. (Adapted from Roddier and Roddier, 1991b.)

but extends outside the initial boundary. The complete procedure to find the wavefront is thus an iterative process, as described in Figure 12.3.

We can also retrieve the wavefront by taking the Fourier transform of the wavefront Laplacian operator, dividing it by $f_x^2 + f_y^2$, and taking the inverse Fourier transform as follows:

$$W(x,y) = -\frac{1}{4\pi^2} \mathrm{F}^{-1}\left\{ \frac{\mathrm{F}\left\{ \nabla^2 W(x,y) \right\}}{f_x^2 + f_y^2} \right\} \qquad (12.23)$$

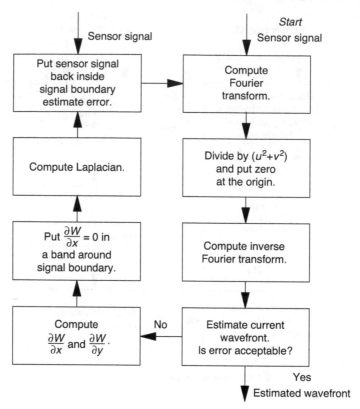

Figure 12.4   Iterative Fourier transform algorithm used to find the wavefront from measurement of the Laplacian operator. (Adapted from Roddier and Roddier, 1991b.)

An iterative algorithm quite similar to the one just described, based on this expression, has also been proposed by Roddier and Roddier (1991b), as shown in Figure 12.4. The Laplacian is measured by the method described earlier with two defocused images. The Neumann boundary conditions are taken by setting the radial slope equal to zero within a narrow band surrounding the pupil. To better understand this boundary condition we can consider the wavefront curvature on the edge of the pupil as the difference between the slopes on each side of the edge of the pupil. If the outer slope is set to zero, the curvature has to be equal to the inner slope. In other words, the edge radial slope is not arbitrarily separated from the inner curvature if this external slope is made equal to zero.

12.3 WAVEFRONT DETERMINATION WITH DEFOCUSED IMAGES

If the defocusing distance cannot be made large enough, the geometrical optics approximation assumed by the irradiance transport equation is not satisfied. In this case, diffraction effects are important, just as in the classical star test. The method described in the preceding section cannot be applied, so different iterative methods must be used. Gershberg and Saxton (1972) described an algorithm using a single defocused image:

1. An arbitrary guess of the wavefront deformations (phase and pupil transmission) is made. The pupil transmission is frequently equal to one and the phase can be anything.
2. The defocused image (amplitude and phase) in the observation plane is computed with a fast Fourier transform.
3. The calculated amplitude is replaced by the observed amplitude (square root of the observed intensity), keeping the calculated phase.
4. An inverse Fourier transform gives a new estimate of the incoming wavefront amplitude and phase (deformations).
5. The calculated input amplitude is replaced by the known input amplitude (pupil transmission), keeping the calculated phase.

These steps are iterated until a reasonable small difference between measured and calculated amplitudes is obtained. This algorithm quickly converges at the beginning but then tends to stagnate.

Based on the work by Fienup and Wackermann (1987) and Misell (1973a,b), an improved method that converges more easily using two defocused images was described by Roddier and Roddier (1991a). This method was used to test the defective primary mirror of the Hubble telescope.

12.4 CONCLUSIONS

In this chapter, we have presented the most important techniques for testing optical wavefronts by estimating the slope and curvature changes as the wavefront propagates along the experimental setup. We have seen that the main advantage of the screen and curvature methods (especially if one is using a low-resolution CCD camera to capture the desired data) is the wider measuring dynamic range. That is, these methods allow us to measure a greater number of aberrant waves than standard interferometric methods such as temporal phase shifting. This increase of measuring range comes at the price of a proportional sensitivity reduction. While commercial phase-shifting interferometers can have a sensitivity as high as $\lambda/100$, slope and curvature test typically can reach a $\lambda/10$ accuracy. An important advantage of curvature sensing over all other testing methods analyzed in this book is its capacity to measure large optics *in situ*, without the need for any special experimental arrangement other than the optics where the lenses or mirrors are used.

REFERENCES

Dörband, B. and Tiziani, H.J., Testing aspheric surfaces with computer generated holograms: analysis of adjustment and shape errors, *Appl. Opt.*, 24, 2604–2611, 1985.

Fienup, J.R. and Wackermann, C.C., Phase-retrieval stagnation problems and solutions, *J. Opt. Soc. Am. A*, 3, 1897–1907, 1986.

Freischlad, K., Wavefront integration from difference data, *Proc. SPIE*, 1755, 212–218, 1992.

Freischlad, K. and Koliopoulos, C.L., Wavefront reconstruction from noisy slope or difference data using the discrete Fourier transform, *Proc. SPIE*, 551, 74–80, 1985.

Fried, D.L., Least-squares fitting of a wave-front distortion estimate to an array of phase-difference measurements, *J. Opt. Soc. Am.*, 67, 370–375, 1977.

Gershberg, R.W. and Saxton, W.O., A practical algorithm for the determination of phase from image and diffraction plane pictures, *Optik*, 35, 237, 1972.

Ghiglia, D.C. and Romero, L.A., Robust two dimensional weighted and unweighted phase unwrapping that uses fast transforms and iterative methods, *J. Opt. Soc. Am. A*, 11, 107–117, 1994.

Hardy, J.W., Lefebvre, J.E., and Koliopoulos, C.L., Real-time atmospheric compensation, *J. Opt. Soc. Am.*, 67(3), 360–369, 1977.

Horman, M.H., An application of wavefront reconstruction to interferometry, *Appl. Opt.*, 4, 333–336, 1965.

Hudgin, R.H., Wave-front reconstruction for compensated imaging, *J. Opt. Soc. Am.*, 67, 375–378, 1977.

Hung, Y.Y., Shearography: a new optical method for strain measurement and nondestructive testing, *Opt. Eng.*, 21, 391–395, 1982.

Hunt, B.R., Matrix formulation of the reconstruction of phase values from phase differences, *J. Opt. Soc. Am.*, 69, 393–399, 1979.

Ichikawa, K., Lohmann, A.W., and Takeda, M., Phase retrieval based on the irradiance transport equation and the Fourier transport method: experiments, *Appl. Opt.*, 27, 3433–3436, 1988.

Misell, D. L., An examination of an iterative method for the solution of the phase problem in optics and electron optics. I. Test calculations, *J. Phys. D., Appl. Phys.*, 6, 2200, 1973a.

Misell, D.L., An examination of an iterative method for the solution of the phase problem in optics and electron optics. II. Sources of error, *J. Phys. D., Appl. Phys.*, 6, 2217, 1973b.

Noll, R.J., Phase estimates from slope-type wave-front sensors, *J. Opt. Soc. Am.*, 68, 139–140, 1978.

Roddier, C. and Roddier, F., Reconstruction of the Hubble space telescope mirror figure from out-of-focus stellar images, *Proc. SPIE*, 1494, 11–17, 1991a.

Roddier, C., Roddier, F., Stockton, A., and Pickles, A., Testing of telescope optics: a new approach, *Proc. SPIE*, 1236, 756–766, 1990.

Roddier, F., Curvature sensing and compensation: a new concept in adaptive optics, *Appl. Opt.*, 27, 1223–1225, 1988.

Roddier, F., Wavefront sensing and the irradiance transport equation, *Appl. Opt.*, 29, 1402–1403, 1990.

Roddier, F. and Roddier, C., Wavefront reconstruction using iterative Fourier transforms, *Appl. Opt.*, 30, 1325–1327, 1991b.

Roddier, F., Roddier, C., and Roddier, N., Curvature sensing: a new wavefront sensing method, *Proc. SPIE*, 976, 203–209, 1988.

Steibl, N., Phase imaging by the transport equation of intensity, *Opt. Commun.*, 49, 6–10, 1984.

Teague, M.R., Deterministic phase retrieval: a Green's function solution, *J. Opt. Soc. Am.*, 73, 1434–1441, 1983.

Index

Milton Keynes UK
Ingram Content Group UK Ltd.
UKHW020005071024
449327UK00031B/2666